T0360525

NEWTON AND THE GREAT WORLD SYSTEM

NEWTON AND THE GREAT WORLD SYSTEM

Peter Rowlands

University of Liverpool, UK

World Scientific

NEW JERSEY · LONDON · SINGAPORE · BEIJING · SHANGHAI · HONG KONG · TAIPEI · CHENNAI · TOKYO

Published by

World Scientific Publishing Europe Ltd.

57 Shelton Street, Covent Garden, London WC2H 9HE

Head office: 5 Toh Tuck Link, Singapore 596224

USA office: 27 Warren Street, Suite 401-402, Hackensack, NJ 07601

Library of Congress Cataloging-in-Publication Data

Names: Rowlands, Peter, author.

Title: Newton and the great world system / by Peter Rowlands (University of Liverpool, UK).

Description: Singapore ; Hackensack, NJ : World Scientific, [2017] |
   Includes bibliographical references and index.

Identifiers: LCCN 2017024280| ISBN 9781786343727 (hc ; alk. paper) |
   ISBN 178634372X (hc ; alk. paper) | ISBN 9781786343734 (pbk ; alk. paper) |
   ISBN 1786343738 (pbk ; alk. paper)

Subjects: LCSH: Newton, Isaac, 1642-1727--Knowledge--Mathematics. | Physics.

Classification: LCC QA29.N4 R69 2017 | DDC 531--dc23

LC record available at https://lccn.loc.gov/2017024280

British Library Cataloguing-in-Publication Data

A catalogue record for this book is available from the British Library.

Desk Editors: Suraj Kumar/Jennifer Brough/Koe Shi Ying

Typeset by Stallion Press
Email: enquiries@stallionpress.com

Printed in Singapore

Preface

The first book in this series, *Newton and Modern Physics*, explored how Newton created an extraordinarily powerful method of scientific thinking based on taking concepts to the ultimate level of abstraction and generality, and used it in a penetrating analysis of many unexplained physical phenomena. The problems, however, because of their complexity, remained largely inaccessible to the equally powerful mathematical structures he had developed simultaneously. The modernity of the work was forced upon him because he had to use creative analytical thinking to make progress where there was no obvious deductive mathematical procedure leading from generic ideas to particular applications, while avoiding facile hypotheses in the search for the generic. It has much more resonance for the present than it could have had for his own time.

When he was faced with the problems of dynamics and planetary motion, however, he was investigating the phenomena most amenable to the mathematical techniques he had been developing from the beginning of his studies. In fact, the mathematics was itself created largely in the pursuit of solving the problems of 'motion', or, in modern terms, dynamics, and was ideally fitted to the use he made of it. Mathematics is, in many ways, the most generic and abstract of all systems of human thought, and once Newton found he could describe dynamics and planetary motions using abstract generic mathematical laws, he was able to present the world with a system which didn't require his powers of analysis and could progress using more deductive processes under its own momentum.

The special significance of this work was that, because Newton succeeded for the first time in discovering universal mathematical laws that applied to the entire system of the world, this could be considered as the moment in history when humans first realised that the whole of Nature was accessible to them using reasoning based on mathematics and experimental observation. Given time and ingenuity every aspect of Nature would find its explanation.

Newton himself clearly recognised this because he repeatedly stated how aspects of chemistry, biology and even human thought could be accessed by his method, and, almost immediately after his time, the Newtonian method began to be applied to many subjects outside of physics, including chemistry, physiology and philosophy. Newton also knew how immense the task would be, involving many contributors over a time period of many centuries, but the system was in place and it could be extended indefinitely.

The breakthrough was neither an obvious development nor an inevitable one. For all its employment of a powerful mathematical structure applicable to a wide variety of problems, the Newtonian world picture still required a quite unprecedented approach to the philosophy of physics which continues to have significance today. Appreciation of this has largely been lost for two reasons. The first is that the methodology largely succeeded and so became adopted as the standard one without needing further philosophical justification, even though this took another fifty years. The second is that the spectacular success of general relativity in the twentieth century, seemingly using an entirely different physical theory, meant that aspects of the Newtonian theory which are essential for the development of modern physics were thought to have been superseded by less fundamental approaches.

However, even though a 'revolution' was proclaimed in 1919, the reality was somewhat different, for Newtonian theory has remained an essential component of general relativity, and the use of a mathematics of curvature has nothing to say about the intrinsic nature of physical space or of gravity. It even has a Newtonian precedent. In addition, historical research shows that significant aspects of general relativity were anticipated by pre-relativistic physicists using minimal extensions to Newtonian methods. The true fundamental relation between Newtonian theory and general relativity has never been extensively studied, even though the lack of such a relation has been a major barrier to understanding the true basis of either theory. It will be proposed here that it is closely connected with the relation between gravity and inertia, a subject of particular interest to Newton himself.

As with *Newton and Modern Physics*, this book has benefited from the cooperation and support of many people, in particular Niccoló Guicciardini for his profound comments and suggestions on the mathematical sections, and also Mike Houlden, Colin Pask, Mervyn Hobden, John Spencer and my wife Sydney.

Peter Rowlands
Oliver Lodge Laboratory
University of Liverpool

About the Author

 **Dr. Peter Rowlands** obtained his BSc and PhD degrees from the University of Manchester, UK. He then spent some time working in industry and further education. He became a Research Fellow at the Department of Physics, University of Liverpool in 1987, and still works there. He has also been elected as Honorary Governor of Harris Manchester College, University of Oxford, a post he has held since 1993. Dr. Rowlands has published around 200 research papers and 12 books. Some of his recent works, published by World Scientific, include *Zero to Infinity The Foundations of Physics* (2007), *The Foundations of Physical Law* (2014), and *How Schrödinger's Cat Escaped the Box* (2015). As a theoretical physicist, his research interests include, but are not limited to, foundations of physics, quantum mechanics, particle physics, and gravity. He has also done extensive work on the subjects of history and philosophy of science, and has published several books on these topics.

Contents

Chapter 1

Metaphysics and Methodology

1.1. Newton and Hypotheses

To work at the most fundamental concepts with the sustained success shown by Newton over a period of more than 50 years would be impossible without both a powerful methodology and a strong metaphysical basis. Great science at this level cannot be done without a powerful system of philosophy which goes beyond the usual scientific method. Newton certainly had such a system, and, without it, he would not have created the unique style of science with which he has always been credited, but, unlike, say, his contemporary Leibniz, he has never had independent recognition as a philosopher. Robert DiSalle, who credits Newton with a general system of philosophy, says that: 'Because Newton never drafted a treatise on, or even a digest of, this general system, his stature as one of the great philosophers of the seventeenth century, indeed, of all time, is no longer widely appreciated'.[1]

Newton's metaphysics, however, is one of his most remarkable achievements, and his desire to attain ultimate metaphysical truth can be seen as the real driver behind his science.[2] Consequently, his views on space, time and motion, and their interpretation in a metaphysical context, are the essential basis of everything he achieved in physics, as well as being an extremely profound approach to the most fundamental truth that a human intelligence can hope to attain. Earlier commentators, confident that the science had developed its own truth beyond any previous philosophical origins, looked on the metaphysical background as an accident of history, now superseded, and saw little significance for such philosophy in the autonomous discipline of modern physics. However, the seemingly continual reversion of modern

---

[1] DiSalle (2004).

[2] Newton, however, would not have used the term 'metaphysics' in this context, as to him it seemingly referred to attempts at a direct explanation of how God acted in the world, and so, in this sense, was always to be rejected (Levitin, 2013).

physics to prototypes with a manifestly Newtonian origin — what we may call a 'Newtonian attractor' — suggests otherwise, as the common basis for all this work seems to be the Newtonian metaphysics. In addition, the metaphysics was also the basis for Newton's scientific *methodology*, which was also profoundly innovative but which was so successful in its effects that subsequent physicists had no option but to adopt it if they wanted to achieve successes of a similar kind.

The *Principia* and *Opticks* undoubtedly reveal that Newton was a most remarkable scientific thinker; but the manuscripts reveal one who was even more remarkable, a daring speculator whose extraordinary intuitions require an idea of science far beyond a procedure dominated by the rigid application of some established order and method. Philosophy of science has not yet developed a language adequate to deal with them. According to a philosophy of science much in vogue today, science advances by a 'hypothetico-deductive' process: hypotheses are suggested to explain observed facts and are then tested by their experimental predictions; hypotheses which fail in their predictions are discarded, hypotheses which succeed in their predictions are retained. Newton would not have recognised this as a valid process for fundamental physics, and, though he did put forward hypotheses, these were not the main sources of his creative thought.

Much of our knowledge of Newton's methodology comes from a revealing episode early in his career. This came after Newton made the ground-breaking discovery that a prism *dispersed* white light into the rainbow colours of the spectrum because the rays producing each colour had a different refractive index, and so a different velocity in a dispersive medium. This idea that white light is *intrinsically* composite is so familiar today that it is hard to believe that it was ever controversial, but at the time it caused a storm of criticism of such ferocity that Newton threatened to give up science altogether. Newton himself felt that his discovery of the different refrangibilities of the components of white light, and his proof that they were intrinsic to the light rays creating the sensations of different colours, was an outstanding contribution; he knew that he had made a discovery of an entirely new kind, 'the oddest if not the most considerable detection wch hath hitherto beene made in the operations of Nature'.[3] Everything that we know about Newton's early career suggests that at this period he was content to lead the life of a private scholar, and that he had no desire to seek publicity for his discoveries, but by 1671 the Royal Society had news of his reflecting

---

[3]Newton to Oldenburg, 18 January 1672; *Corr.* I, 82.

telescope, and he resolved on presenting them with a much more extensive account of his optical discoveries, written in a bold and forthright style. He sent to the Society a paper on 'A New Theory of Light and Colours',[4] which was read on 8 February 1672, 'with a singular attention and an uncommon applause'.[5]

The paper was a brilliant account of his discovery of the analysis of white light, presented as a chain of deductive inferences from an ordered sequence of experiments. Though we may feel that this is how science is supposed to happen, we know, in fact, that presentation of results in this style is merely part of the didactic process which is necessary to the acceptance of a scientific discovery. The engagingly autobiographical style gives the impression that the paper was a straightforward account of work undertaken with no theoretical preconceptions, but with logical inferences derived directly from experimental results. The autobiographical' sections are quite famous and often quoted verbatim as an historical account of his work: 'in ye beginning of ye year 1666. . . . I procured me a triangular glass Prisme, to try there wth ye celebrated phaenomena of colours', and so forth.[6]

While the individual autobiographical facts may be true, the account connecting them is a fiction, a classic example of the reconstruction of a scientific discovery in the form most likely to convince a new audience — a procedure which is now fundamental in the writing of all scientific papers. Newton wrote that he was 'surprised' to find that the spectrum of light from his prism and circular slit was oblong rather than circular, though it was exactly what he had expected and had carefully positioned his apparatus to produce. He then outlined a series of tests made in a logical and systematic order leading to an 'experimentum crucis' in which specifically coloured rays from the first prism were separately passed through a second prism and shown to refract at different angles. 'Colours', he concluded, 'are not *Qualifications of Light* derived from Refractions or Reflections of natural Bodies (as *'tis* generally believed), but *Original and connate properties, which in divers Rays are divers'.*[7] The conclusion, however, was shocking to its targeted audience. For 2000 years, philosophers had followed Aristotle in proclaiming whiteness as a symbol of purity and simplicity; colouration was held to be due to a modification by the medium, in this case the glass of

---

[4] *Corr.* I, 92–102.
[5] Oldenburg to Newton, 8 February 1672; *Corr.* I, 107.
[6] *Corr.* I, 92–102, text as quoted from CUL Add. 3970, ff. 460^r–466^v in *OP*, I, 10.
[7] 'A New Theory', 1672, *Corr.* I, 92–102.

the prism. Newton was now claiming that white light was heterogeneous and coloured light pure. The roles of colour and whiteness had been reversed.

Newton knew that his conclusion was boldly original, but, as an author entirely new to scientific publishing, he was completely unprepared for what followed. The paper was rightly praised for its ingenious experiments, but Robert Hooke, the Royal Society's curator of experiments, immediately attacked the 'hypothesis' on which Newton's theoretical treatment was based, supposedly the corpuscular theory which he had mentioned in passing as part of a mechanism which he immediately discarded.[8] In a pattern which has repeated itself all too often in the history of science, there was more than a hint in Hooke's remarks of the experienced metropolitan scientist putting the provincial amateur in his place, and, again, in a pattern which would repeat itself all too often with Hooke himself, the Royal Society curator tried to make out that Newton's novel experiments were merely variations on his own. For all his importance to the Royal Society, Hooke was a rather insecure man, keenly aware of his physical deformities and his status as a Society 'employee', and Newton's work had, in effect, undermined the views on light and colour that he had recently put forward in his *Micrographia*, which was that white light was 'nothing but a pulse or motion propagated through an homogeneous, uniform and transparent medium', and that 'colour was nothing but the disturbance of yt light by the communication of that pulse to other transparent mediums, that is by the refraction thereof".[9] So, a ray incident obliquely on a refracting surface created an obliquity in the light pulse. A red sensation was produced when the strongest part of the pulse preceded, and the weakest followed; a blue sensation when the position was reversed; the other colours were produced by combinations lying between these extremes.

Newton was shocked by this damaging 'peer review' from such an influential person. Like many later authors in the same position, he resolved to write a devastating reply, and spent three months carefully composing a letter addressed to the Society's Secretary, Henry Oldenburg. 'I was a little troubled', he wrote, 'to find a person so much concerned for an *Hypothesis*, from whom in particular I most expected an unconcerned & indifferent examination of what I propounded'.[10] Newton would not be the last scientist to fail to receive an 'unconcerned & indifferent examination' of his work, but

---

[8]Hooke to Oldenburg, 15 February 1672; *Corr.* I, 110–114; first published, Birch (1756–1757), 3: 10–15.

[9]Hooke (1665, 54–67), esp. 62–64; *Corr.* I, 110.

[10]11 June 1672; *Corr.* I, 171–188, 171.

he seems to have been generally taken aback when it happened. Hooke, he said, had completely misunderstood his methodology. He had not intended to put forward a hypothesis but had simply stated obvious truths based on his experiments. He stated, in fact, and with remarkable clarity, the difference between his own methodology and that of the mechanistic philosophers who dominated the Royal Society, and who had originally influenced him in his early work. What was at stake here was more than one particular theory. It was a whole new approach to doing science. Newton had, in fact, laid out the methodology which has been responsible for all the most important advances in fundamental physics down to the present day but it had been explicitly rejected, and confusingly presented as an example of the very thing that he had confidently set out to replace. If his work was rejected, it would be a catastrophic loss of a major opportunity for putting scientific thinking on a much more fundamental basis.

Of course, at this point, also, unknown to Hooke or anyone else, Newton was undergoing a personal and possibly spiritual crisis with his impending compulsory ordination as a Cambridge fellow. Whatever direction his spiritual studies had taken by this time, they led him ultimately to a rejection of the doctrine of the Trinity (in the College of the Holy and Undivided Trinity of all places!) in favour of a system of beliefs that might be called Arian or unitarian. Acutely aware of the controversies generated in the early Church and the persecution that Arius had suffered as a result of his beliefs, he clearly came to identify in some way with the fourth century theologian. He was deeply concerned by the choice he would have to make between accepting ordination within three years and signing the dreaded 39 Articles of the Anglican Church, in contradiction to his now strongly held beliefs, or abandoning any hope of a prolonged university career. Even without any doctrinal complication, ordination would have a damaging effect on a career dedicated to scholarly pursuits.[11] Hooke had now also succeeded in damaging his scientific credibility. Newton's reply was coloured by the pain of this rejection. The devastating onslaught on what Hooke had probably considered (for him) only a mild and casual criticism took the latter completely by surprise and led to a bitter quarrel between the two men which was never completely resolved. And this was far from the end of the matter. After Hooke, everyone else weighed in — Huygens, Pardies and a whole host of English Jesuits, many of them criticising his results as well as his conclusions — and no one came out in his support.

---

[11] Iliffe (2016).

The controversy lasted for six years and only ended when Newton decided he had had enough of it.

What had really gone wrong was that, though presenting his work as the inevitable conclusion drawn from a planned series of experiments, Newton had actually introduced a new principle of scientific reasoning while trying to give the impression that he was working within the established scientific tradition. Established systems are usually quick to detect threats to their structures, even if they don't fully understand them. The principle that appeared in Newton's work for the first time was that there were a few certain types of information which were more fundamental than others and that these were abstract and could be defined precisely in an abstract way without regard to any model of nature based on concrete terms. This had already had results in laying the foundation of Newton's whole development of dynamics, and the fact that, at this vital moment, he alone recognised it is a measure of his exceptional status in scientific history, for it is not a well-known fact, even today.

Newton had made his position clear in his first letter in a paragraph which Oldenburg unfortunately edited out:

A naturalist would scarce expect to see the science of those [colours] become mathematical, yet I dare affirm that there is as much certainty in it as in any other part of optics. For what I shall tell concerning them is not an hypothesis but most rigid consequence, not conjectured by barely inferring 'tis thus because not otherwise or because it satisfies all phenomena (the philosophers' universal topic) but evinced by the mediation of experiments concluding directly and without any suspicion of doubt.[12]

The theorist was to make theoretical conclusions by analytical thinking, which must be true by fundamental reasoning, not to make new hypotheses based on mechanical models, which included extra suppositions which would very likely be in themselves untrue.

Thus he wrote in reply to criticisms by the Jesuit Ignaz Pardies (with a notable degree of politeness):

I am content that the Reverend Father calls my theory an hypothesis if it has not yet been proved to his satisfaction. But my design was quite different, and it seems to contain nothing else than certain properties of light which, now discovered, I think are not difficult to prove, and which if I did not know to be true, I should prefer to reject as vain and empty speculation, than acknowledge as my hypothesis.[13]

---

[12]Passage edited out, *Corr.* I, 96–97 Cohen (1958a, 371).

[13]Newton to Oldenburg, 3 April 1672; *Corr.* I, 144.

In a second reply to the same author, he asserted that:

> Only after a thorough and experimental exploration of the properties of light, involving the examination both of the collateral and of the successive parts of the rays, is it time for hypotheses to be discussed, and to be rejected if they disagree with the facts. But it is a very easy matter to accommodate hypotheses to this doctrine... And nothing can be harsher in these hypotheses than the contrary supposition.[14]

For Newton, we should not accept an explanation simply because it is a plausible one, but because we cannot accept any other without multiplying assumptions. We can reject specific hypotheses without having to find a replacement. Many people who have an ingenious idea have often only come up with another speculative model, which makes more (hidden) assumptions than it solves problems. What Newton saw, but his contemporaries didn't, was that you could build virtually an infinite number of hypothetical models to explain a phenomenon; but the only thing that was really useful was to find an abstract, general way of thinking which did away with models altogether.

When he later suggested a particular optical hypothesis for the benefit of Hooke, he wrote: 'But whatever be the advantages or disadvantages of this Hypothesis, I hope Mr Hook will excuse me from taking it up, since I do not think it needfull to explicate my Doctrine by any Hypothesis at all'.[15] Of course you *could* say, he argued, that particles of light produced waves of different lengths in the aether. He had done exactly this in his own notebooks. And, of course, this assumption could produce the whole theory, if it was supposed true, but the theory was stronger than any of the particular hypotheses which could generate it. It existed at an abstract level independent of *all* mechanical conceptions, whether or not they were true.

He was equally explicit in a letter intended for Huygens: 'But to examin how colours may be thus explained Hypothetically is besides my purpose. I never intended to show wherein consists the nature and difference of colours, but onely to show that *de facto* they are originall & immutable qualities of the rays wch exhibit them, & to leave it to others to explicate by Mechanicall Hypotheses the nature & difference of those qualities...'.[16] Newton seems to have fully understood the abstract nature of his own

---

[14]Newton to Oldenburg, 10 June 1672; *Corr.* I, 170–171.

[15]Newton to Oldenburg, 11 June 1672; *Corr.* I, 177.

[16]Newton to Oldenburg, 3 April 1673; *Corr.* I, 264; '*An Extract of Mr. Isaac Newton's Letter, written to the Publisher from Cambridge April 3. 1673*', *Phil. Trans.*, no. 9, 6108–6111, 6 October 1673.

thought, for a few years later, he complained, in a letter to Oldenburg, that: 'I have observed the heads of some great virtuoso's to run much upon Hypotheses, as if my discourses wanted an Hypothesis to explain them by, & found, that some when I could not make them take my meaning, when I spake of the nature of light & colours abstractedly, have readily apprehended it when I illustrated my Discourse by an Hypothesis...'.[17] Newton was uncompromising in rejecting the easy but fallacious method of inventing new hypotheses when problems arose: 'To make an exception upon a mere Hypothesis is to feign an exception. It is to reject the argument from Induction, & turn Philosophy into a heap of Hypotheses, which are no other than a chimerical Romance'.[18]

In a series of 'Rules' for philosophical reasoning, placed at the beginning of the third book of the *Principia*, and successively refined and extended over the course of the three editions, Newton codified his belief that induction direct from experiment, and not hypothetical reasoning, was the most successful way to tackle the problems of nature. The correct way of reasoning, he said, was to admit the minimal number of causes that were 'true and sufficient to explain' phenomena (Rule I), and to assign these cases, as far as possible, to 'the same natural effects' (Rule II). Only those properties of bodies that were manifestly valid from experimental evidence should be accepted as universally true (Rule III). The argument of induction should not be 'evaded by hypotheses'. We should prioritise the results from induction, however many contrary hypotheses could be imagined, until new phenomena force us to make changes (Rule IV).[19]

A draft Rule V (never published) defines what is meant by a hypothesis as whatever is not taken directly from the senses or by the internal thoughts, which are as much 'phenomena' as the products of the senses:

> Whatever is not derived from things themselves, whether by the external senses or by the sensations of internal thoughts, is to be taken for a hypothesis. Thus I perceive that I am thinking, which could not happen unless at the same time I were to perceive that I exist. But I do not perceive that any idea whatever may be innate. And I do not take for a phenomenon only that which is made known to us by the five external senses, but also that which we contemplate in our minds when thinking: such as, I exist, I believe, I understand, I remember, I think, I wish, I am

---

[17] *Hyp.*, 7 December 1675, corrected in Newton to Oldenburg, 26 January 1676; *Corr.* I, 363 and 414; first published in Birch (1756–1757), 3: 247–305; *Corr.* I, 363.

[18] Draft of letter to Conti of 25 November 1715, Koyré and Cohen (1962), 113–115.

[19] Rules I, II (1687), III (1713), IV (1726). In the first edition, Rules I and II were called Hypotheses I and II.

unwilling, I am thirsty, I am hungry, I rejoice, I suffer, etc. And those things which neither can be demonstrated from the phenomenon nor follow from it by the arguments of induction, I hold as hypotheses.[20]

Newton famously wrote 'Hypotheses non fingo' ('I feign no hypotheses') in the General Scholium to the second edition of the *Principia* of 1713, but this does not mean that he *never* used hypotheses. In the same edition of the *Principia*, there is a clearly labelled 'Hypothesis' in Book II, and it is a significant one, being known today as Newton's law of viscosity.[21] What he means is that he didn't use hypotheses when he had an analytical theory which didn't need them, and, in the particular case of the General Scholium, he was referring to gravity, which he had managed to explain on an abstract basis that was beyond all hypotheses. He also believed that the only hypotheses that should even be considered are those that can be tested by experiment. So, again in the *Principia* of 1713, he says '... since the qualities of bodies are only known to us by experiments, we are to hold for universal all such as universally agree with experiments ...'[22] In his *Account of the Commercium Epistolicum*, an anonymously published attack on Leibniz written in 1714, he described his own position in relation to that of Leibniz in the guise of an objective and unprejudiced witness.

> It must be allowed that these two Gentlemen differ very much in Philosophy. The one proceeds upon the Evidence arising from Experiments and Phenomena, and stops where such Evidence is wanting; the other is taken up with Hypotheses, and propounds them, not to be examined by Experiments, but to be believed without Examination. The one for want of Experiments to decide the Question, doth not affirm whether the Cause of Gravity be Mechanical or not Mechanical; the other that it is a perpetual Miracle if it be not Mechanical.[23]

Things took a turn for the better in 1675, when Newton, in London in connection with his dispensation from taking orders, called in at the Royal Society to take up the fellowship to which he had been elected in 1672. He was surprised to receive a favourable reception, due to the apparent partial conversion of Hooke, who had tried out some of Newton's experiments, finding that they worked out after all, and was even prepared to look again

---

[20] A. Koyré (1965, 272), translation slightly modified, Rogers (1982, 223). The 'real' or phenomenological nature of human thought, an idea created in conscious opposition to the Cartesian dualism, is also discussed in a proposed Preface for the *Principia* (see 12.4).

[21] II, Section IX, Hypothesis.

[22] III, Rule III.

[23] 'Account', *Phil. Trans.*, 1715, 224.

at some of Newton's theories. To show that he was not against hypotheses as such, Newton sent to Oldenburg in the following December an extraordinary account of a 'theory of everything', entitled 'An Hypothesis explaining the Properties of Light', though not intending it for publication.[24] Here, a general system of nature is put forward, with the first hint of a universal gravitation based on the pressure of the all-pervading aether. But it also contained several statements which the hypersensitive Hooke interpreted as personal attacks. Another brief flare-up ended quickly with the blame being conveniently attached to Oldenburg, a long-time enemy of Hooke. A brief correspondence, including Newton's 'standing on the Shoulders of Giants', established an ostensible (but possibly bogus) reconciliation between the combatants.[25] It had certainly been a dispute between personalities, but something more profound was also at stake.

1.2. The Mediaeval and Theological Background

'God does not play dice', according to Einstein;[26] if we could discover the fundamental theory of physics, 'we would know the mind of God', according to Stephen Hawking.[27] Physicists are famous for invoking God as the ultimate arbiter of physical theory, and one (Paul Davies) has even claimed a substantial prize from a foundation devoted to achieving a better understanding between science and religion by incorporating God as a metaphor for ultimate knowledge in his books and essays. But modern physics is noted, in any case, for its quasi-theological tone — its making of *ex cathedra* statements that are inviolable (for example, in the case of various conservation laws), its aura of utter certainty while all around it is held in doubt, and its drive towards an ultimate goal based on a belief that such is to be found in the systems of physical laws and fundamental particles and will be recognizable when it is. A well-known contemporary physicist, Steven Weinberg, states that 'Christianity and Judaism teach that history is moving toward a climax — the day of judgement; similarly, many elementary particle physicists think that our work in finding deeper explanations of the nature of the universe will come to an end in a final theory toward which we

---

[24]*Hyp.*, 1675, *Corr.* I, 361–386. William R. Newman ascribes the title 'theory of everything' to the related but earlier and more explicitly alchemical *Of Natures Obvious Laws & Processes in Vegetation*, NPA.

[25]Newton to Hooke, 5 February 1676; *Corr.* 1: 416. See Westfall (1980, 272–274).

[26]Einstein to Max Born, 4 December 1926, *The Born-Einstein Letters*, translated by Irene Born, Walker and Company, New York, 1971.

[27]Hawking (1993).

are working'. Such physics 'is moving toward a fixed point. But this fixed point is unlike any other in science. That final theory toward which we are moving will be a theory of unrestricted validity, a theory applicable to all phenomena throughout the universe — a theory that, when finally reached, will be a permanent part of our knowledge of the world'.[28]

Such notions run utterly counter to the fashionable idea among sociologists that science is a mere social construction like other cultural phenomena, but they are held, either explicitly or implicitly, by all who work in the field of fundamental physics; the subject would be virtually impossible without them. And it is mainly the opponents of the concept of a fundamental theory who attack them as quasi-theological, with the implication that they are deliberately copying the procedures of theology in order to gain an authoritative tone which they would not otherwise deserve. But they are not *quasi*-theological at all; they are *actually theological*. Their origin is in the work of theologians of a particular period, which has been subsequently imported into physics for fundamentally theological reasons, and it is the necessary extremism of that theology which is responsible for their success in claiming *universal* application. Without this in-built extremism, physics would never have had its success in overtaking all other attempts at constructing a fundamental system of thought.

The theology that we are talking about is mediaeval, specifically late mediaeval, though it also has a tinge of late classical Neoplatonist philosophy. Physics, in the sense of the creation of a set of powerful and simple abstract concepts, applicable to a wide variety of cases, does not begin in Greek times, with the solution of specialised problems in hydrodynamics and optics, nor in the Renaissance, as is often supposed, but back in the fourteenth century, at the height of mediaeval scholasticism, at least a century and a half before the Renaissance is said to have begun. Mediaeval Christianity had added something during its revival of Greek thought which had not been present in the original — a feeling for the infinite derived ultimately from the all-powerful creator God of the Hebraic tradition, though to this was added an emphasis on an active all-pervading universal spirit which derived partly from the Neoplatonist philosophy of the third century writer Plotinus.

In Greek times, Plato and Pythagoras had believed that abstract mathematics was the key to the material universe, but their successor, Aristotle, had neglected mathematics and put forward a view of nature which was antithetical to that of the physicist; his form of 'natural theology',

---

[28]Weinberg (1998).

based on the existence of an effectively infinite number of verbal categories, was a virtual denial of the existence of general laws of nature.[29] In the thirteenth century, the rediscovery of Aristotle's works led to an attempt at an accommodation between Aristotelianism and the Christian theological tradition, most notably by Thomas Aquinas; and many people, following its deliberate promotion by Pope Leo XIII in the nineteenth century, consider the Aquinas synthesis to be the pinnacle of mediaeval thought. But Aquinas was not particularly well thought of in the Middle Ages, and his doctrines were formally condemned in 1277 by the Archbishops of Paris and Canterbury. The problem was that Aquinas, in accommodating Aristotle's categorizing to Christianity, had effectively limited the power of God; so theologians were encouraged thereafter to develop a new kind of theology which was more in keeping with the all-powerful and unlimited creator God of the Hebraic tradition. Thus the early fourteenth century saw the beginning of a rejection of the thirteenth century compromise and the inauguration of the most extreme and unorthodox period in the whole history of mediaeval thought. The key issue was the respective spheres of faith and reason. Aquinas had used them in support of each other, but the more extreme thinkers who came after him rejected the lessening in his theology of the role of faith and the infinite unknowability of God, and effected a more or less permanent separation of faith from the Aristotelian tradition of reason.

Working mainly at the great centres of Oxford and Paris, the fourteenth century schoolmen developed a terminology for dealing with abstractions in a way that was ultimately theological, but also, as a by-product, produced a powerful new mathematics and mathematical physics. This was because they based their most fundamental thinking about God and the universe on an investigation of the two great metaphysical concepts of space and time, which had long been recognised as the two most simple abstract concepts available to the human mind, and which, even today, are the ultimate and only source of all human knowledge of variation in nature. What we would call the increase and decrease of variable quantities, they called the 'intension and remission of forms'. The relationship between space and time variation became the discussion of motion, variously uniformly difform, nonuniformly difform, or difformly difform, leading to the modern concepts of velocity, acceleration, momentum, force, and also to the new mathematical techniques needed to describe such variations.

---

[29]Spencer (2012) has provided a strong case for the continuance and significance of the Platonic tradition in modern physics.

In this way, Thomas Bradwardine described the exponential function, Richard Swineshead (whose *Liber Calculator* later became the favourite mathematical authority of Leibniz) the use of infinite series, William Heytesbury the Merton mean speed theorem, $s = (u + v)t/2$ (later so important to Galileo's work on free fall);[30] Nicholas Oresme developed the use of graphs, and Jean Buridan the theory of impetus (a forerunner of the idea of momentum). Oresme, and later Nicholas of Cusa, presented startling arguments from first principles about possible world systems which could have influenced Copernicus. But more important than any of these technical contributions was the method which was encapsulated in William of Ockham's famous 'razor' (so-called by the nineteenth century mathematician William Rowan Hamilton, one of its greatest exponents): the idea that 'entities must not be multiplied unnecessary' — the idea that 'forms' must be as few as possible, that explanations must be refined to the greatest pitch of simplicity, using ideas as abstract as could be found. Only in this way could we begin to approach to a knowledge of the work of the all-powerful Creator and universal spirit.

In his drive towards the idea of an absolute and noncontingent God opposed to a contingent nature, Ockham had hacked away at the multifarious Aristotelian categories used, by such early mediaevals as Thomas Aquinas, to describe natural things, and reducing the 'entities' that had been 'multiplied'. Because he believed that theological knowledge was based on faith, not reason, so he separated faith and reason and excluded natural knowledge from faith; his God was not to be discovered through the laws of nature. Having removed theology from nature, he became an empiricist in natural things, with knowledge grounded in experience. At the end of the Middle Ages, Ockham helped in the creation of a thoroughly scientific attitude to nature precisely because he was such an uncompromising theologian. Of course, he could not describe nature without using some sort of categories or abstract conceptions but these would be reduced to a minimum, and would show the uniformity in nature, rather than an Aristotelian diversity.

The work of these late mediaeval philosophers is essential to the understanding of why modern science developed only in the West and only during the Renaissance which began in the late fifteenth century. While the Greeks

---

[30] A curious survival of a Mertonian idea in Newton's work occurs in Lemma 1 of *Principia*, Book II, where he states the well-known relation that 'Quantities proportional to their differences are continually proportional' because he needs it in his work on resisted motion in Proposition 2.

had a mathematics which they applied to astronomy and music, it was the mediaeval philosophers, with their precise discussions of the mathematical relations between space and time, who first mathematised physics as we know it today. Many societies, at various times, have developed science-like activities and have appreciated particular aspects of the phenomenon which we call science, but only in the West in the last 500 years has it had the capacity for being a self-sustaining enterprise. The lack of perfection in human reasoning, perceived by theologians who put all ordinary physical explanation second to the infinite power of God, created a situation in which scientific thought became an enterprise which required continual renewal, and was never allowed to fossilise into a doctrinaire pseudo-intellectual discipline shoring up the foundations of society, as it was with the followers of Aristotle, and with many cultures in the East.

The work of these cloistered theologians may have had no impact on secular society but the invention of printing made their mathematical techniques available to men with more practical objects in mind. At the same time, during the early Renaissance, the universities began to be the training places of more secular scholars; and the physical ideas of the late mediaeval philosophers, which were still being discussed and interpreted by university professors in the mid-sixteenth century, began also to influence the secular masters who had had their training under them. Such, at any rate, were Galileo Galilei, trained at the University of Pisa, and heir, through Domingo de Soto, and the Ockhamist John Major, to the fourteenth century tradition of the mathematical treatment of uniformly accelerated motion; and Thomas Harriot, educated at Oxford, and conscious inheritor of the mathematico-physical traditions of the great fourteenth century Mertonians, Bradwardine, Heytesbury and Swineshead. Galileo, of course, through a series of experiments and related calculations, showed that a whole series of simple laws could be derived which would explain the motions of falling bodies, on the assumption that gravity involved uniform acceleration. The final synthesis was Galileo's, but the terms had been set by his mediaeval predecessors, and he made specific reference to the work of Domingo de Soto. The study of motion through abstract concepts related by mathematical laws was entirely the creation of the fourteenth century scholars, and the concepts with which Galileo and his successors worked were taken over directly from their writings.

Perhaps surprisingly, the tradition which had led to the work of Galileo was rejected by his immediate successors in favour of a new 'mechanistic' style of thinking, whose ultimate inspiration was probably the classical poem

De Natura Rerum, by Lucretius, which had been printed in the fifteenth century from a single surviving copy, and which presented a persuasive, and utterly materialist, view of a universe, seen in terms of atoms, brought together and separated by random motions, accounting for all physical processes. The mechanists thought that the mediaeval schoolmen were hopelessly out of date, alongside Aristotle. Their chief propagandist was René Descartes, who took the opportunity to develop a kind of 'constructivist' view of physics, in which there would be little need for the direct action of the all-powerful creator or active universal spirit. In his purposely titled *Principia Philosophiae* (or *Principles of Philosophy*) of 1644, Descartes developed a 'dualistic' separation of matter and spirit, insisting that scientific philosophy was concerned only with the former, and not with the latter. The Cartesian dualism ensured that mind or spirit was not considered an active principle in nature. Natural philosophy was concerned with bodies (*res extensae*), not with minds (*res cogitantes*). Matter was wholly inert and made up simply of particles in motion. Bodies could only act on each other by some kind of direct impact — a 'push–pull' arrangement — making space itself full of dense matter.

The mechanistic style became dominant in two powerful organizations, the Royal Society of London and the Académie in Paris, achieving significant results in the hands of such masters as Huygens, Hooke and Boyle. So, when Newton began his scientific career, the consensus view was that scientific methodology should adopt the procedure of Descartes or, at least, some combination of Descartes and Gassendi, the main reviver of atomism. In this tradition, individual problems were solved by devising a particular hypothesis relating to the particles of matter or their properties and following it through to the desired conclusion. There was no suggestion of a universal method; each problem was tackled on its own merits. What changed all this was the startling success of Newton in applying a universal method to the problem of planetary motion and all other large-scale physical phenomena in 1687. Newton's work is often presented as though it were a synthesis of Galileo's terrestial and Kepler's celestial physics, but it was not; it contradicted both of these earlier systems in fundamental respects and was conceived in an entirely different spirit. It was not conceived by synthesis at all, but by *analysis*, and was opposed *in principle* by nearly all leading scientists of the time. It succeeded ultimately, not on account of its fundamental analytical validity, but because Newton was equally gifted at showing its *synthetic* application to a wide range of physical phenomena.

1.3. Newton's Method

Though a version of the mechanistic philosophy, based on hypotheses, was the dominant approach to investigating nature in his day, Newton's methodology came from an entirely different tradition, one based on theology rather than mechanism. At first Newton was a mechanistic philosopher, in the sense that he studied the best available models. He started as a Cartesian, but soon diverged. There was, however, no Damascene conversion, and there is no evidence that he was ever wholly committed to the mechanistic cause, even in his period as a scientific apprentice. Despite the prevalence of Cartesian views in physics at the time, and his citing of Descartes' works in the *Quaestiones quaedam philosophicae* inserted into his student notebook, Newton's thought developed from an early stage in an abstract analytical way in which he gradually worked his way from the Cartesian influence.

Abstraction was the natural mode of thinking for such a theologian as it never could be to a mechanistic philosopher. A memorandum by David Gregory implies that Newton believed that abstract and general ideas are simpler than concrete and particular. For Newton, though not for his mechanistic contemporaries, the ultimate causes of things were abstract rather than mechanical.

> In Mr Newtons opinion a good design of a publick Speech (and which may serve well at ane Act) may be to shew that the most simple laws of nature are observed in the structure of a great part of the Universe, that the Philosophy ought ther to begin, and that Cosmical Qualities are as much easier as they are more Universall than particular ones, and the general contrivance simpler than that of Animals plants &c[31]:

A famous passage in Query 31 of the *Opticks* suggested that the method of obtaining such general propositions could be derived by analogy with the processes of mathematics:

> As in Mathematicks, so in Natural Philosophy, the Investigation of difficult Things by the Method of Analysis, ought ever to precede the Method of Composition. This Analysis consists in making Experiments and Observations, and in drawing general Conclusions from them by Induction, and admitting of no Objections against the Conclusions, but such as are taken from Experiments, or other certain Truths. For Hypotheses are not to be regarded in experimental Philosophy. And although the arguing from Experiments and Observations by Induction be no Demonstration of general Conclusions; yet it is the best way of arguing which the Nature

---

[31]28 December 1691; *Corr.* III, 191.

of Things admits of, and may be looked upon as so much the stronger, by how much the Induction is more general. And if no Exception occur from Phænomena, the Conclusion may be pronounced generally. But if at any time afterwards any Exception shall occur from Experiments, it may be begin to be pronounced with such Exceptions as occur. By this way of Analysis we may proceed from Compounds to Ingredients, and from Motions to the Forces producing them; and in general, from Effects to their Causes, and from particular Causes to more general ones, till the Argument end in the most general.

As Newton himself wrote in the General Scholium: 'In this philosophy particular propositions are inferred from the phænomena, and afterwards rendered general by induction. Thus it was that the impenetrability, the mobility, and the impulsive force of bodies, and the laws of motion and of gravitation, were discovered'.

Newton's method separated the abstract *system* from physical *measurement*. The system had a perfection that could never be physically realised. The principle that appeared in his work for the first time was that there were a few certain types of information which were more fundamental than others and that these were abstract and could be defined precisely in an abstract way without regard to any model of nature based on concrete terms. For Newton, though not for his mechanistic contemporaries, the ultimate causes of things were abstract rather than mechanical. The laws describing the system did not depend on any physical hypotheses. As he said in Query 28:

> the main Business of natural Philosophy is to argue from Phaenomena without feigning Hypotheses, and to deduce Causes from Effects, till we come to the very first Cause, which certainly is not mechanical.... And though every true Step made in this Philosophy brings us not immediately to the Knowledge of the first Cause, yet it brings us nearer to it, and on that account is to be highly valued.

This was the origin of Newton's important distinction between theories, based on abstract general principles, and hypotheses, mere suppositions based on mechanistic ideas. History forces us to concede that Newton was right in making this distinction, but does not give us any criterion for deciding why and how he came upon the correct abstract principles. As outlined in *Newton and Modern Physics*, I believe that it comes from a particular method, and it is one that can be described. It stems from a totally different brain activity, which is nothing to do with following step by step in the process of synthesis from fundamental axioms. What we are unable to explain is why some people have this kind of abstract imagination, while many do not. The distinction between theory and hypothesis must have been obvious to Newton, but it was not obvious to his contemporaries, many of whom like Huygens, Leibniz and Hooke were men of real brilliance, who, for all their scientific achievements, lacked Newton's abstract vision.

Though Newton made a great deal of public show about basing his ideas on experiment, experiment, therefore, was not the fundamental basis of his natural philosophy. Clearly, experimental evidence was a major source of physical information, as Newton himself showed, and crucial experiments often decided Newton's theoretical positions. So, the lack of resistance to pendulums would prove there was no dense fluid aether, thin film effects would show that light was periodic, Hauksbee's experiments, in the early eighteenth century, would show that capillary action was not due to the repulsion of air molecules and that the emission of light was connected with electrical action. Nevertheless, though experiment was necessarily the initial source of all information about the natural world, it could not decide how that knowledge must be organised, and, for Newton, this involved the use of principles of a semi-abstract nature (his 'other certain truths') derived more by some kind of intuitive process which he couldn't describe other than as obvious consequences from experiment.

So, although Newton derived major aspects of his theories directly from experiments, it was also true that this was only possible because he had introduced into science a new and non-mechanistic style of thinking which had more in common with late mediaeval modes of thought than anything fashionable in his day. His contemporaries found his ideas difficult to accept because they thought that he was not being 'modern'. The mechanistic style — the invention of mechanistic or model-dependent hypotheses specific to particular circumstances — was modern; abstract categories were 'occult', and had been banished by Descartes, once and for all, from the study of nature. Ultimately Newton's ideas were based on a faith not shared by his contemporaries, one that has only slowly filtered through into science and has always been resisted, that nature is based on simple and abstract principles. The quasi-theological nature of this proposition has often been noticed with respect to modern science, along with its propensity to making *ex cathedra* statements about laws of physics which apply to the whole of Nature. It is not always realised that it does actually stem from theology and its application to metaphysics, and that it had a particular effectiveness in an individual whose thought-processes had a strongly recursive (as well as theological) component.

Newton, of course, was not always able to adduce the correct principles but his method gave him a kind of automatic confirmation when he had succeeded. He had the capacity to select the unifying abstract concept which represented the simplest conceivable reduction of a mass of experimental data without needing to know how this was derived — and, in fact, he couldn't possibly have done so. Francis Bacon called this 'induction' but Newton's use of the process was more precise and more profound than anything that Bacon had anticipated with his almost mechanical conception of the idea. It was certainly much more powerful than the kind of 'induction' that philosophers like Karl Popper rejected as the basis of scientific method.

It is, in fact, extraordinary how Newton so often managed to grasp the key features of anything he looked at, virtually first time. As soon as he looked at mechanics, he saw the need for conservation of momentum, laws of motion and the concept of mass, and found the formula for centrifugal force and a curvature method for orbits. He even did the same for rotational motion. As soon as he looked at optics, he developed a theory of colour. As soon as he looked at mathematics, he was on to calculus, and finding out significant things about ellipses for future reference. He had hardly begun to consider the planetary motions when he came up with an inverse square law for gravity. Often these flashes of inspiration were not followed up in detail for many years, as one subject after another grabbed his attention. Occasionally, he forgot his early breakthroughs once he had put them aside. On one or two occasions he even contradicted them. Overall, however, they surely point to a characteristic mode of thinking which was not defined by a laborious analysis of the possible hypotheses.

Newton's real objection to hypotheses was that he knew that the theories defined by his method, which we may identify as the recursive one (though not anyone else's, if defined by hypothetical thinking), represented a higher kind of truth. As de Morgan expressed it in the nineteenth century, Newton was 'so happy in his conjectures as to know more than he could possible have hope of proving'.[32] He had at once the analytic, and we could also say recursive, capacity to see further than his contemporaries into the laws of nature, and the theological cast of mind to recognise the true abstract nature of scientific thought. The recognition of this fundamental abstractness made it possible for him to employ for the first time in physics the powerful mathematical techniques, with their direct incorporation of infinities, which he had developed during the same period. He represented it as coming direct from experiments, telling Oldenburg in 1672 that: 'You know the proper Method for inquiring after the properties of things is to deduce them from Experiments. And I told you that the Theory wch I propounded was evinced

---

[32]De Morgan (1914, 49).

to me, *not by inferring tis thus because not otherwise*, that is not by deducing it onely from a confutation of contrary suppositions, but *by deriving it from Experiments concluding positively & directly*.[33] But this is not what he did, and his opponents knew it.

Though he used empirical evidence in his analytical reasoning, he was fully aware that it could not be done on the basis of simple inference, but on some process within the brain that established a pattern that was automatically satisfied. According to Cohen, the 'sharp distinction' Newton made between experimental philosophy and philosophy beyond empirical evidence 'may possibly rival any other contribution he made to science'.[34] Here he is in conscious opposition to Descartes. 'To dispute about the objects of ideas except in so far as they are phenomena is dreaming'.[35] It would certainly be wrong to imagine that Newton was a believer in the hypothetico-deductive method because he laid stress on experimental testing as a criterion for hypotheses. Testing and prediction, of course, are very important processes in the development of scientific theories, but it is a mistake to give them the primacy over the more fundamental principle of simplicity, decided automatically by the analytical thinker, especially when the predictions are based on extensive mathematical analyses.

For Newton, scientific knowledge was not, therefore, organised around hypotheses but around fundamental abstract laws which were not derived directly from experiment, but which were the basic parameters in terms of which experimental information was organised. Universal laws were abstract definitions and did not primarily describe nature. To be able to carry out experimental investigations, we had to assume that these laws were valid, and changes in these laws would require really major experimental evidence. Predictions were not normally direct tests of these laws but of models which used them in some particular way. Since they only concerned details and since assumptions of any kind could be made in hypotheses, experimental tests were not tests of the validity of fundamental laws. The history of physics has shown that, where successful, physics has largely kept to this programme. Anomalous facts are not automatically explained in terms of new hypotheses or changes in fundamental laws. The form of such laws as the conservation of mass and the conservation of energy, for example, has been retained in the supposedly revolutionary 4-vector physics, introduced by relativity,

---

[33]6 July 1672; *Corr.* I, 209, published as 'A Serie's of Quere's propounded by Mr. Isaac Newton...', *Phil. Trans.*, no. 85, 5004–5007, 15 July 1672.

[34]Cohen (1999, 59).

[35]Cohen (1999, 54).

by developing a new understanding of the fundamental concepts they contain.

Though the qualitative analytical method of thinking could be described as inductive, the certainty that the results were true was automatically checked by using a criterion of minimalism close to Ockham's razor. It would be interesting to know if Newton's Ockhamism stems in any way from his Cambridge education; the only direct statement I have so far found is in *Of Natures Obvious Laws & Processes in Vegetation*, a manuscript of the early 1670s, where Newton says 'Tis suitable with infinite wisdom not to multiply causes without necessity', though the early correspondence with Hooke, of the same period, has each man accusing the other of multiplying assumptions.[36] But there is no denying the effectiveness of Newton's Ockhamist reasoning. As a qualitative thinker, Newton was an Ockhamist who out-Ockhamed Ockham. Many results which he derived by purely qualitative Ockhamist thinking are startlingly like ideas associated only with much later periods, but the line of reasoning where it can be followed shows that they are not speculative guesses.

For Newton, as for Ockham, multiplication of categories denied the uniformity of nature (and the ultimate Oneness of God). In a letter to William Briggs, he wrote that 'Nature is after all simple, and is normally self-consistent throughout an immense variety of effects, by maintaining the same mode of operation',[37] and he repeated the phrase 'if Nature be simple and conformable to herself' endlessly in his writings.[38] This was a great metaphysical principle, such as all fundamental thought requires. In modern language, we call it self-similarity, and, in modern physics, it leads to such ideas as the renormalization group. It is a key aspect of his third law of motion. Newton sought self-similarity of ideas because nature was self-similar. He also had an instinctive belief in the importance of symmetry, another important unifying principle in modern physics, though his comments near the end of Query 31 in the *Opticks* suggest that he didn't know where it was leading. He absorbed metaphysical ideas from Neoplatonism, alchemy and Arianism into physics, making them coherent by subjecting them to his own abstracting process.

---

[36] *Of Natures Obvious Laws*, NPA. Hooke to Oldenburg, 15 February 1672; *Corr.* I, 110–114, 113; Newton to Oldenburg, 11 June 1672; *Corr.* I, 171–188, 176.

[37] 25 April 1685; *Corr.* II, 418.

[38] Q 31; draft Praefatio for the *Principia*, *Unp.*, 307; draft Conclusio for *Opticks*, Book IV, c 1690, in Westfall (1971, 379) and Westfall (1980, 393).

The significance of the success of Newton's qualitative analytical reasoning is not in whatever anticipations of later ideas he may have made, but in the confidence with which we can claim that his Ockhamist analyses were responsible for his success in creating the fundamental principles of his dynamical system. Newton's abstract laws, based on new mathematics, were the origin of the runaway success of modern physics, and its primacy over all previous systems of thought. If physics is written in the language of mathematics, then Newton saw that we must be interested in something that is fundamentally abstract. He realised that the physicist had to be an extremist. There was no reason to stop at any point and decide that was good enough, until the final answer was achieved. In this sense, Newton could be seen as the originator of the quest for a unified theory. His analysis led to a universal method, which was justified by synthesis, and created his own universal system of the world. The extremely abstract nature of his reasoning seems to point directly to the similarly abstract development of twentieth and twenty-first century physics. However, in his own time, and for a long time afterwards, Newton's mathematical theorems were accepted but his qualitative analyses never caught on. His startling and unexpected claim that the basic truths of even the *mechanical theory of physics* were fundamentally abstract was so extraordinary a conception that his contemporaries could not accept it and it still remains unaccepted in popular philosophy today.

Chapter 2

Mathematics

2.1. Calculus

Ultimately, physics assumes meaning when we can understand it in qualitative terms. Mathematical expressions become physical when they are processed by the human 'transducer', as we may call it. Significant discoveries may be expressed in mathematical form, but we only know that they are significant when they can be seen in terms of some qualitatively observable effect. However, to approach the most fundamental or analytical level, the qualitative concepts need to be taken to the utmost point of abstraction, and mathematics plays an extremely significant role in this process. Consequently, the qualitative and mathematical aspects can never be completely separated, and the most profound qualitative thinkers have nearly always had a deep understanding of mathematics. Even Faraday, though devoid of all mathematical training, made his most creative theoretical breakthroughs using a kind of quasi-mathematical visual geometry which later workers, such as Maxwell, were relatively easily able to translate into more regular mathematical language. In the case of Newton, we have a profound mathematical innovator, who forged his mathematical instruments in parallel with the physics he wanted to create, leaving his successors unable to account for their undeniable effectiveness with the rigour they would have expected — an outstanding example of the 'unreasonable effectiveness of physics in mathematics'. It was no exaggeration when, in reply to an enquiry from the Queen of Prussia in 1701, his great rival 'Leibnitz said that taking Mathematicks from the beginning of the world to the time of Sir I. What he had done was much the better half — & added that he had consulted all the learned in Europe upon some difficult point without having any satisfaction & that when he wrote to Sir I. He sent him answer by the first post to do so & so & then he would find it out'.[1]

When Newton started on his mathematical career, there were three main branches of mathematics available to him: Euclidean geometry, codified in the third century BC; a version of algebra and the theory of equations which had reached maturity only with the Renaissance; and the even more recently introduced analytic geometry of René Descartes, which, as subsequently interpreted by Frans van Schooten and other commentators, was a kind of combination of the other two, in which functions between algebraic variables could be mapped as figures on orthogonal coordinate axes.[2] Newton made significant contributions to all three, besides effectively creating and massively developing (along with his contemporary and rival, Leibniz) yet another branch: the differential and integral calculus. All these branches of mathematics played a major part in his physical theorising, and Descartes' coordinate geometry was a strong early influence. Initially Newton paid little attention to Euclidean geometry, but later on, with his rejection of the philosophy and physics of Descartes, he began to think that the ancient 'synthetic' geometry of Euclid was fundamentally superior to the modern 'analytic' geometry of Descartes.[3]

Newton's mathematics is largely directly related to his being a creative thinker at the frontier of physics. He creates new mathematics to try to understand the nature of space, time and motion. This is especially true of his creation of the new mathematics of the calculus. Calculus reflects his physical requirements: his first main tract on the subject was entitled 'To resolve problems by motion'.[4] He always thought of the variation that calculus requires in terms of the only two truly variable fundamental quantities in

---

[1]Westfall (1980, 721), according to the eyewitness testimony of Sir A. Fontaine. The historian A. R. Hall is among a number of authorities who have given support to this view: 'Well before 1690... [Newton] had reached roughly the point in the development of the calculus that Leibniz, the two Bernoullis, L'Hospital, Hermann and others had by joint efforts reached in print by the early 1700s' (1980, 136).

[2]Newton read Descartes' *La Géométrie* (1637) in the annotated Latin translation published by van Schooten in 1659–1661, and perhaps also in the earlier version of 1649. Descartes did not use the now well-known 'Cartesian coordinates' or draw curves derived from algebraic equations. His coordinates were oblique ones derived from 'certain directions embedded in the given figure' and his curves were created from specific 'geometrical definitions' or by specifically defined motions (Guicciardini, 2009, 32).

[3]This development began relatively early with the *Observations on Kinckhuysen* that he was persuaded to do by his correspondent, John Collins (1670, *MP* II, 295–447).

[4]'To Resolve Problems by Motion', October 1666, MP I, 400–448. It was preceded by tracts dated 20 May 1665, 13 November 1665, 14 May 1666 and 16 May 1666; MP I, 272–280, 382–390, 390–392, 392–394.

physics, space and time, even when he was concerned with pure mathematics rather than kinematics or dynamics. His efforts at making mathematics respond more closely to physical meaning are paralleled by a lifelong effort at reducing physics, even in its most qualitative aspects, to a quasi-mathematical pattern, by avoiding mechanistic hypotheses and introducing an ever-increasing degree of abstraction. Galileo said that the 'Book of Nature' was written in the language of mathematics, while physicists today talk about the 'unreasonable effectiveness of mathematics in physics' and the 'unreasonable effectiveness of physics in mathematics'. In Newton's work the two come together more closely than in the work of any contemporary and, indeed, of that of many of his successors.[5]

Calculus has two main branches, differential and integral, and two main procedures associated with them. Differentiation means finding the rate of change of one quantity with another, say y with respect to x, or the gradient at any point of the tangent to the graph of y against x. Integration of y with respect to x means the finding of the area under the graph of y against x. The realisation that these are inverse procedures is the fundamental theorem of the calculus, known to Newton as early as 1665, and proved by him in geometrical form at that date.[6] Newton developed a standard algorithm for the differentiation of an algebraic function $y = f(x)$, or even $f(y, x) = 0$, symbolised by a cursive $\mathscr{X}$, which could be related to the finding of normals, tangents and curvatures. He was able to find maxima, minima and points of inflection — the points of zero gradient. He found differentials for the trigonometric functions and constructed tangents to such mechanical (or transcendental) curves as the quadratrix of Dinostratus, the ellipse, the cycloid, and the spiral traced out by a point in uniform motion along a line which is itself in uniform rotation about one of its ends. His demonstration of the isochronous or regularly periodic property of the cycloid (the curve generated by 'a point on the circumference of a circle rolling along a fixed straight line') was an 'easier' and 'more natural' demonstration than that of Huygens.[7] It was also fundamentally dynamic, 'equating instantaneous acceleration in the direction of motion to the component' of the gravitational force which was generating it', in contrast to Huygens' kinematic treatment, and totally original in this respect.[8]

---

[5] Galileo, *The Assayer*, 1623.

[6] *MP* I, 298 ff, mid-1665.

[7] Hall and Hall, *Unp.*, 170.

[8] Whiteside, *MP* III, 391, 420–423, quoting 422–423, VI, 403. Whiteside describes it as 'radically novel, indeed unprecedented', and says it was to lead to the theory of central force orbits in *Principia*, I, Proposition 41, which is the most advanced in the book.

His early work, which was both algebraic and geometric, includes, according to Whiteside's interpretation, a full symbolism for partial derivatives using a dot notation, along with 'the five first- and second-order partial derivatives of a two-valued function',[9] though Newton, like his contemporary Leibniz, lacked any real concept of a function, whether single-valued or multivariate.[10] He also wrote down an expression for the third derivative. He produced rules for finding the differentials of products and quotients, and also roots and powers, with a clear statement of the product rule. His rectification methods could find arc lengths, such as that of the quadratrix, and the general arcs of the epicycloid and hypocycloid. He gave the first account of elementary differential geometry. He was also a pioneer in vector analysis, both in recognising the application of the parallelogram law of addition for a number of physical quantities, and in understanding that such quantities could be resolved into components in arbitrary directions, though the true algebraic definition of a vector quantity would only emerge after another two centuries. He showed that results derived by analysis (algebra) could be converted into ones derived by synthesis (geometry) and vice versa.

Newton produced an extensive list of integrals, of functions such as $y = ax^{m/n}$, of polynomials, of infinite series, and curves such as the circle, parabola, hyperbola, cycloid, quadratrix, cissoid, Gutschoven quartic, and conchoid. Polynomials could be integrated by series expansion, a method that he was ready to divulge to Leibniz. The more advanced methods of integration, over which he was much more secretive, included 'antidifferentiation', or using the fundamental theorem of calculus to reverse the differentiation process, and using a substitution of variables to convert the area of a given curve to that of a conic section with a new ordinate; one example of this was the cissoid whose area could be calculated, by

---

Kulviecas (1993) argues that Newton's paper began 'the systematization' in mechanics of the concept of instantaneous acceleration at a point, in advance of the nineteenth century formulation of Poncelet, and was the first work to use the term 'in the sense of [a] concrete physical quantity', the "effectiveness of gravity' condition[ing] a certain change in the body speed (dv) in reference to a fixed time (dt)', 'as in contemporary kinematics'.

[9]Whiteside, *Mathematical Works*, I, x. These are in a general 'curvature algorithm', which was first 'developed implicitly' in print in Hayes (1704, 190) (Whiteside, 1967/1970, 81).

[10]That is, in the sense later defined by Euler. The fluxions could be considered as introducing functions of time, so providing a closer approach to the later concept than Leibniz's more geometrically-inspired conceptions. Guicciardini (personal communication, quoting Engelsman, 1984), sees Newton's 'lateral dots notation as abbreviations for 'subroutines'' of the well-known algorithm that Johannes Hudde had derived for finding the maxima and minima (and in an unpublished letter) the tangent for a polynomial function.

substitution, as that of the circle. Integration by parts also occurs as a component of these methods.

Newton wrote down the first differential equations and found their solutions. They are used extensively in the *Principia*, notably in Book I, Propositions 39–41 and Proposition 97, Corollary 1; and in Book II, Propositions 2–4 and in the Scholium to Proposition 34. He made extensive use of infinite series, using an iterative procedure to reduce affected equations, that is, finite polynomials in products of powers of both x and y, equated to 0, to an infinite series for y in powers of x. He devised a general method for transforming trinomials into readily-integrable binomials. *Principia*, Book II, Propositions 2–4 'appeal to logarithmic integrations whose justification is achieved by variations on a canonical limit-increment argument' from James Gregory.[11] *Principia*, Book II, Propositions 25–30 'shed unique illumination on his mature power to formulate viable exact solutions to 'infinitesimal' equations not in immediately quadrable form'.[12]

Principia, Book I, Proposition 41 includes a significant use of double integration; multiple integrals are also used in Book I, Section XII, though expressed in geometrical form, for example the triple integrals in Propositions 79–81. In Lemma 29 an integral over two variables is reduced to an integral over a single variable by reversing the order in which the integrations are carried out and then finding the integral over the other variable; the triple integral is reduced to a process of evaluating just a simple single integral. While Newton's integration procedures generally follow the system made rigorous by Bernhard Riemann in the nineteenth century, of dividing the area under a curved graph into narrow vertical strips and then taking the limit as the strip widths or abscissae are reduced, something closer to a Lebesgue integral, dividing the ordinates rather than abscissae, is used in Proposition 76.[13]

He had three different versions of the calculus, all of which were used in the *Principia*. (The method of finite differences used in interpolation theory for obtaining results by approximation might be considered a fourth.) Infinitesimals, which appeared, early, in *De analysi*,[14] are based on increments of algebraic variables or segments of geometrical lines, which can be made indefinitely small and ultimately reduced to zero. An alternative

---

[11] Whiteside (1970, 130).
[12] Whiteside, *MP* VI, 439.
[13] Chandrasekhar (1995, 281).
[14] *De analysi*, 1669, published 1711, *MP* II, 206–247.

method of moments, described in the *Methodus fluxionum*,[15] used products of *fluxions*, or instantaneous speed, and time, with integrals being described as *fluents*. The free-flowing time-like parameter providing the continuous variation was not necessarily to be equated with physical time. It was also not an observable. It was what we would now call an 'independent variable', one which acts independently to regulate the variation of the other, which might be a physically observable quantity.

Mathematically, Newton's fluxions and Leibniz's differentials are indistinguishable, but they appear to have different physical meanings, in the sense that the fluxions are generated by a kinematical process. Intrinsically, Newton believed that the fluxions represented something physically real while Leibniz thought they were mathematical constructs which did not exist in reality. The fluxions represent a way of making infinitesimals become finite. The third method, the geometrical theory of 'first and last' or prime and ultimate ratios of lines and curves, which he described in *De quadratura*,[16] extended into a theory of limits, with a concept of limit values which can be approached more closely than any defined difference. This involved an idea of absolute continuity.

In the Scholium to *Principia*, Book I, Section I, the curvature at any point on a curve is compared with that of the osculating circle to be applied to the curved lines in Lemmas 1–11, which make up the section. A 'particularly original' concept; this provides the basic foundation of the idea of the limit, and the method of taking limits, and the existence of a limiting value is illustrated kinematically in the Scholium. Contrary to a good deal of subsequent opinion, Newton had a clear definition of limit in this Scholium: 'Those ultimate ratios ... are not actually ratios of ultimate quantities, but limits ... which they can approach so closely that their difference is less than any given quantity...',[17] while, Lemma 11, according to Bruce Pourciau, may contain 'the first algebraic epsilon-argument ever given'.[18]

---

[15] *De methodis fluxionum*, 1670–1671, published 1736, *MP* III, 32–353.

[16] *Tractatus de quadratura curvarum*, 1693, published 1704, *MP* VII, 588–645 and VIII, 92–159. There were earlier versions presented in the Addendum to *De methodis* (1671); in *Geometria curvilinea* (c 1680); and in *Principia*, Book I, Section I (1687).

[17] Cohen and Whitman (1999, 442), Pourciau (2001b, 19).

[18] Cohen and Whitman (1999, 442), Pourciau (2001b, 28). Pourciau, who showed that the Cajori translation of Lemma 11 of 1934 had introduced an error not in the Motte version of 1729, says that the correct Lemma 11 demonstrates 'that by 1687 Newton had acquired a surprisingly clear conception of the limit process. We remain impressed, here in this proof and in the early sections of the *Principia* generally, by Newton's mastery of the basic idea'. (28) Kinematic notions, in particular, appear only in Newton's descriptions, not in

Lemma 1, with its *ab absurdam* proof, is the foundation of the method: 'Quantities, and the ratios of quantities, which in any finite time converge continually to equality, and before the end of that time approach nearer the one to the other than by any given difference, become ultimately equal'. Lemma 2 uses this idea of limit to improve on Barrow's demonstration that the area under a curved line can be found by dividing into inscribed and circumscribed rectangles and allowing the number of these rectangles to tend to infinity.[19] Lemma 3 says that the ultimate ratios are still equal even when the rectangles are of different widths, while Lemma 4 and its Corollary show that two curves must be identical if the ultimate ratios of their inscribed parallelograms are equal.

Versions of some of the lemmas had been produced by other authors such as Barrow and Gregory, but here they are subsumed to become integral parts of Newton's overall limit construction. Lemma 5 simply states that: 'In similar figures, all sorts of homologous sides, whether curvilinear or rectilinear, are proportional; and the areas are in the duplicate [squared] ratio of the homologous sides'. Lemmas 6–8 share the same diagram, which has a curved arc ACB, with a radius AR drawn from apex A and a tangent AD perpendicular to this, with the line joining these, RD, going through point B. Lemma 6 says that if the points A and B at the edges of the chord and arc approach each other and meet, then the angle between the chord and tangent $(B\hat{A}D)$ will ultimately vanish, meaning that the chord and arc will coincide with the tangent. Lemma 7, using the method called 'Newton's microscope', shows that the ultimate ratios of the arc, the chord, and the tangent to each other are ones of equality, which, according to Corollary 3, allows any of these lines to be used in place of one another in any limit argument. Lemma 8 says that the triangles RAB, $RACB$, RAD formed, respectively, with the chord, arc and tangent will also ultimately become equal as the points A and B approach and ultimately coincide, meaning, according to the Corolllary, that, in finding the limit, any of the three triangles can be used in place of any other.

Section I of the *Principia* is, in effect, a small treatise on the method of first and last ratios or the calculus of limits presented in a geometric

---

the proofs of his lemmas, which freed the concept from physics well before d'Alembert (1789) and Lacroix (1802, 29).

[19] See Guicciardini (2009, 219–223). Guicciardini says that the proof of Lemma 2 is 'magisterial in is simplicity', and that 'Its structure can still be found in present-day calculus textbooks in the more general and abstract definition of the Riemann integral' (221).

form. According to Pourciau's analysis of the eleven lemmas (and nineteen corollaries), 1 defines the limit, and 2–3 the integral, 4 and 5 then give the vertical and horizontal expansion properties of the integral, 6 (and also 9) define the first derivative, 7 and 8 the derivative of sine, 9 the fundamental theorem of the calculus, and 10 and 11 the second derivative. Not included are the limit of a difference, the demonstration of the existence of the integral, and its translation invariance property.[20]

The method, in fact, was not purely geometrical. An algebraic form of a limit argument is presented in *De quadratura*, where Newton finds the limit of the ratio of $(x + o)^n - x^n$ to the finite (not infinitesimal) quantity as both tend simultaneously to zero, to be nx^{n-1}.[21] The Scholium to Lemma 10 of Section I in discussing ratios of 'indetermined quantities of different sorts' uses an algebraic form to express direct and inverse relations, so making the connection between geometrical and algebraic arguments. The method also played a major role in Newton's physics. Two of the Lemmas from Section I, 9 and 10, have been described as 'central to Newton's mathematization of force'. Lemma 9 says that if a curve and a vertical straight line intersect each other at A, and horizontal lines BD and CE are drawn from two points B and C on the curve to the vertical line, then, as B and C together approach towards A, the areas of the triangles ABD and ACE will be in the ratio of the squares of the sides. In Lemma 10, Newton supposes that the abscissae AD and AE represent time and the ordinates DB and EC velocity. Then, at the beginning of the motion, under *any* force, uniform or nonuniform, the distances traversed will be proportional to the squares of the times, as Galileo had found out for uniform acceleration; and so, at any particular point in the motion, the velocity will effectively increase or decrease linearly with time. This became crucial to Newton's demonstration by a limit argument in Proposition 1 that motion of a body under the action of a centripetal force would require the radius sweeping out equal areas in the same plane in equal times, as Kepler's empirical second law of planetary motion required.

Newton also had a number of different notations, which he introduced at different times, including the dot notation for fluxions, and Q (for *quadratura*) as a substitute for Leibniz's $\int$ (for *summa*). According to Rouse Ball: 'If a flowing quantity or fluent were represented by x, Newton denoted its fluxion by $\dot{x}$, the fluxion of $\dot{x}$ or second fluxion of x by $\ddot{x}$, and so on.

---

[20]Pourciau (1998).

[21]*MP* VIII, 126–129. Guicciardini (2009, 227) calls it 'strikingly modern'.

Similarly the fluent of x was denoted by $\boxed{x}$, or sometimes by x' or $[x]$'.[22] In the version he used most often, the higher order fluents were expressed in the form $x'', x''', \ldots$, corresponding to the higher order fluxions $\ddot{x}, \dddot{x}, \ldots$. Newton's equivalent to Leibniz's dx was $\dot{x}o$, with o being equivalent to dt, and what we now call dy/dx, in the Leibniz notation, would be represented by $\dot{y}/\dot{x}$.[23] Newton, however, was a good deal less concerned with symbolism than Leibniz, who always considered it an essential part of his algorithm. Newton's symbolism was only systematised in his later works, beginning with *De quadratura* in 1691, though it was extended retrospectively to the published versions of the earlier ones.

Modern theory, by comparison with Newton, has non-standard analysis corresponding to fluxions or differentials, and standard analysis based on limits.[24] It is now established that neither method is intrinsically superior to the other. Any theorem can be proved by either method but some have turned out to be easier in one formalism and some in the other. Both are physically real, but mean different things, respectively determined by differentiation with respect to space and time. Newton is the only early mathematician to create a full algorithm for calculus employing methods related to both.[25]

The Scholium to *Principia*, Book I, Section I incidentally hints at a Zeno-type argument against Newton's use of an 'ultimate proportion of evanescent quantities', 'because the proportion before the quantities have vanished, is not the ultimate, and when they have vanished is none'. In the same way, 'it may be alledged that a body arriving at a certain place, and there stopping, has no ultimate velocity: because the velocity, before the body comes to the place, is not the ultimate velocity; when it has arrived, is none'. However, he considers that 'the answer is easy; for by the ultimate velocity is meant that with which the body is moved, neither before it arrives at its last place and the motion ceases, nor after, but at the very instant it

---

[22]Rouse Ball (1908). The box notation for the fluent began early and was used for one example in *De analysi*. It was also used in front of the integrand in the same way as Leibniz's $\int$ in 'To Resolve Problems by Motion'.

[23]This, of course, implies a version of the chain rule, though this is more visible in the Leibniz notation. Newton's original notation for dx/dt, dy/dt and dz/dt was p, q and r, so his dy/dx was q/p.

[24]Robinson (1974) is the major text on non-standard analysis. Cauchy (1821) is considered to have made the theory of limits rigorous using an epsilon-delta definition.

[25]That is, in published form. Leibniz (1993, 2002) produced a manuscript analysis of quadrature methods in terms of limits with some similarities to Newton's method of first and ultimate ratios. Knobloch (1989), cited by Guicciardini (2009, 332).

arrives; that is, that velocity with which the body arrives at its last place, and with which the motion ceases'. It is significant here that the paradoxes of Zeno, concerned with such things as the race between Achilles and the tortoise where Achilles can never catch his slower rival once he has given him a start, and the flying arrow whose motion can never even begin, are considered to be resolved only by an appeal to the calculus of limits, that is, by standard analysis. They can be resolved only by a realisation that time has fundamentally different properties to those of space, and it is interesting that Newton instinctively realises that differentiation with respect to time must be the basis for a calculus which has a more fundamental connection with continuity and problems of physical motion than a calculus based on infinitesimals.[26]

From an early stage in his work, Newton developed a 'calculus of curvature'; it became a significant component in his dynamics, and remained so, even after he developed alternative methods. He devised a general method for finding axes for any curved line, or for showing, as with cubics, that they did not exist. He also carried out investigations into vertices and asymptotes, and discovered that reducing equations and algebraic curves to standard Cartesian coordinates helped to simplify the problems. Other advances include the determination of curvature in the form of the reciprocal of the radius of curvature and a formula for this radius. He devised a structured and completely general procedure for finding the radius of curvature to each of the conic sections; the various stages in the procedure were explained in geometric terms and characterised by the circles that could be drawn to cut the parabola or other conic section and tangent to it. Newton's method of computing crookedness allowed him to rectify the evolutes, which were the loci of the centres of curvature of some specified curves. A particular example was the semicubical parabola $(y^2 = ax^3)$, which was the evolute of the parabola. He also produced a general equation for defining the circle of curvature, from which one could determine the points of both greatest and least curvature. He found that the radius of curvature became zero at the end of a cycloid.

2.2.   Application to Optics and the Calculus of Variations

Newton's mathematical methods were applied early to optics and theoretical physics. A neat construction appended to his predecessor, Isaac Barrow's

---

[26] This aspect of Zeno's paradoxes is discussed in Rowlands (2007, 46–47).

optical Lecture XIII, could be used to determine the tangential image-point for the refraction of an oblique pencil of rays at a spherical surface, that is, rays incident upon a lens at a distance from its axis. This addition to Barrow's work provided a simple way of designating by geometry the position of the image formed by a lens in any given case, and a construction of the lens to project an image at a given point. Since 'the refracted rays are tangent to the caustic curve' at the point 'where infinitely close rays intersect', this becomes equivalent to determining 'the general diacaustic condition'. Another 'elegant and convenient' construction by Newton, of the caustic point, was included at the end of Barrow's Lecture XIV.[27]

In both his Lucasian *Lectures on Optics* (1670–1671) and in *Principia*, Book I, Proposition 97, Newton sets out to determine what shape of refracting surface through a given point C would focus at A the rays issuing from point B.[28] Describing an oval of Descartes, Corollary 1 states that 'By causing the point A or B to go off sometimes *in infinitum*, and sometimes to move towards other parts of the point C, will be obtained all those figures which *Cartesius* has exhibited in his Optics and Geometry relating to refractions. The invention of which *Cartesius* having thought fit to conceal, is here laid open in this Proposition'. In the Corollary, 'the problem is not reduced to quadratures but to an algorithm by which the solution curve can be extended, given the derivative of the curve at any point, that is to a *first-order differential equation*',[29] though it is not entirely certain that this was the solution that Newton intended. A second Corollary was aimed at finding the shape of refracting surface through C which would refract all the rays which issued from some given point A and make them into a parallel stream of rays in a given direction.

Another result in geometrical optics occurs both as a single '*Problema*' in Newton's 'Waste Book' and as Proposition 98 in *Principia*, Book I: 'Given any refracting surface CD which is to refract rays diverging from the point A in any manner whatever, to find a second surface EF which will make all refracted rays DF converge on some point B'.[30] The version in *Principia*

---

[27] Appended to Barrow (1669), Lecture XIII, §XXVI, 26, *Works*, 117. *Optica*, I, Lecture 14, *OP*, I, 412–413 (image point). The derivation was 'more direct' than that of Barrow, though longer. (Shapiro, *OP*, I, 38) Shapiro, *OP*, I, 24, 412 (diacaustic). *Optica*, I, Lecture 14, *Works*, 127–128 (caustic point).

[28] *Optica*, I, Lecture 14, *OP*, I, 417.

[29] Chandrasekhar (1995, 331).

[30] *MP* III, 496–497 and plate facing 530; *Principia*, 232–233. An earlier version is in *Optica*, I, Lecture 14, Proposition 34, *OP*, I, 416–417.

is followed by a Scholium which extends the discussion to the problems of lenses for optical telescopes.

> In the same manner one may go on to three or more superficies [surfaces]. But of all figures the sphaerical is the most proper for optical uses. If the object glasses of telescopes were made of two glasses of a sphaerical figure, containing water between them, it is not unlikely that the errors of the refractions made in the extreme parts of the superficies of the glasses may be accurately enough corrected by the refractions of the water. Such object glasses are to be preferred before elliptic and hyperbolic glasses, not only because they may be formed with more ease and accuracy, but because the pencils of rays situate without the axis of the glass would be more accurately refracted by them. But the different refrangibility of different rays is the real obstacle that hinders optics from being made perfect by sphaerical or any other figures. Unless the errors thence arising can be corrected, all the labour spent in correcting the others is quite thrown away.

Here he is saying that *spherical* aberration, or distortion due to the lens's shape, can be corrected using a compound lens, but, unless we can find a way of correcting the more serious problem of *chromatic* aberration, due to the spreading out of light of different colours, then all such efforts are useless.

Newton's fluxional algorithms and infinite series expansions were also applied to optical examples, the first derivative in elementary problems of refraction and the second in diacaustics. He discovered that two image points were formed when oblique pencils were refracted; through this discovery, he set the study of astigmatism on a firm mathematical foundation stimulating the later work of Thomas Young.[31] Newton found an 'equivalent limit-increment proof of the diacaustic'.[32] He extended the use of oblique pencils to finding a general solution which gave the radius of a rainbow of any order. For the rainbow, Newton calculated geometrically the exact radii of the primary and secondary bows (as well as the colours and their reversal in the secondary bow), at 40° 17′/42° 2′ and 50° 57′/54° 7′. His formula was the first that could compute radii for bows of any order n, and for a medium with any index of refraction.[33] He further extended the concept of pencil to cases involving three dimensions, and found an equivalent to the modern idea of the circle of least confusion.[34] A solution to the anaclastic problem,

---

[31] *Lectiones opticae*, Lecture 12, 129, *OP*, I, 214–217.

[32] Shapiro (1990, 148).

[33] *Optica*, I, Lectures 14–15, *OP*, I, 418–429, II, Lecture 16, *OP*, I, 592–603; *Opticks*, Book I, Part 2, Proposition 9.

[34] *Optica*, I, Lecture 14, 144–151, *OP*, I, 418–425 (rainbow); *Optica*, I, Lecture 13, Corollary 6, *OP*, I, 408–409 (least confusion).

to 'find the point of refraction' by tracing the path of a ray from a point in one medium to a point in another, was given in the first version of his optical lectures, but abandoned in the revised version.[35] Making use of his anaclastic solution, an 'elegant' demonstration in both the *Lectiones opticae* and in *Optica* showed that, because of the 'varying index of refraction', a point source 'viewed across a plane surface' gave an 'extended image' of the point in the shape of a 'Dioclean cissoid', with equation $x^3 = y^2(2a - x)$ and a cusp at the point of refraction.[36]

In both versions of Newton's optical lectures, there are sections which are largely concerned with the refraction of monochromatic light at spherical surfaces, and which include a wealth of important results in geometrical optics, frequently using advanced mathematical techniques, such as infinite series and the identification of maxima and minima for algebraic functions.[37] In the second version, in *Optica*,[38] Newton first finds, in Proposition 29, the image point for rays parallel to the principal axis arriving at the surface of a spherically convex lens, in a mathematical construction which is 'equivalent to the Gaussian formula', $n/s + n'/s' = (n - n')/\rho$.[39] Then, after showing how 'to determine the intersection of the axis and the nearest normal' for 'any given curve' (Lemma 9),[40] he generalises his result, in Proposition 30, to any curved surface by using the centre of curvature close to the incident rays rather than the centre of the refracting spherical surface. Then, in the following Proposition, he finds both the spherical aberration and the circle of least confusion for a plano-convex lens.[41]

In the next two propositions, Newton, creates an 'elegant' derivation of 'the location of the primary image point, or caustic locus', when rays fall obliquely on spherical, and then on any curved surface, while also locating the secondary image point which he had discovered. Following this, as in *Principia*, I, Proposition 97, he finds the Cartesian oval or 'aplanatic surface' 'that will refract homogeneous rays, whether they are parallel or terminate

---

[35]Shapiro, *OP*, I, 38–39, *Lectiones opticae*, Lecture 13, 141, *OP*, I, 226–229.

[36]Shapiro, *OP*, I, 38. *Lectiones opticae*, Lecture 13, 142–144, *OP*, I, 228–233, *Optica*, I, Lecture 9, 73–75, *OP*, I, 360–365.

[37]Shapiro, *OP*, I, 40. Shapiro describes the section as the 'highpoint' of the entire work, 'an intimate blend of mathematics and physics consistently yielding novel, significant results'.

[38]Lecture I, 13–15, *OP*, I, 401–429.

[39]Shapiro, *OP*, I, 40.

[40]*OP*, I, 404–405.

[41]Shapiro says that, 'Because of its algebraic formulation, this proposition is particularly accessible to the modern reader and provides a fine example of Newton's application of mathematics to physics'. (*OP*, I, 40–41). Spherical aberration is also calculated in 'Of Refractions', *MP*, I, 572–574.

at some common point', so that the refracted rays will all meet in a single point, a problem solved by Descartes and investigated by Newton as early as September 1664.[42] The section is concluded with Newton's highly original calculations relating to the rainbow and chromatic aberration.[43] Newton's work on geometric optics, additionally, includes an accurate account of the anatomy of the eye and a qualitative account of the main eye defects.[44]

The calculus of variations, another of Newton's major innovations, seeks to find the path, curve, surface, etc., for which a given function has a stationary value, which, in physical problems, is usually a minimum or maximum. He used this calculus in the proof of the first of the three propositions which are stated in the Scholium to *Principia*, Book II, Proposition 34, dealing with the solid of revolution of least resistance, where he gave a demonstration that the shape which suffered least resistance was a truncated cone. Anticipating a special case of the Euler-Lagrange equations, Newton's is the unique solution for 'curves with continuously varying tangents',[45] while the physical assumptions that Newton makes are valid for bodies travelling at supersonic speeds.[46] The Scholium 'gives, in the form of a geometrical relation', the first-order 'differential equation defining the meridian curve of this solid, thus reducing, for the first time in the history of mathematics, a variational problem to its Eulerian relation'.[47] Though Newton's fluxional equation in the *Principia* was given without proof, there is a derivation in manuscript, which 'combines a partial differentiation — in preferred limit-increment form — followed by an integration to produce a first-order fluxional equation, which is then solved parametrically'.[48]

---

[42]Shapiro, *OP*, I, 41. *Optica*, I, Lecture 14, 141–142, *OP*, I, 416–419. *Principia*, I, Proposition 97.

[43]The *Lectiones opticae* and *Optica* have calculations for the chromatic aberration of the image of a point observed through a prism (*Lectiones opticae*, Lecture 18, Proposition 8, 190, *OP*, I, 276–279, *Optica*, I, Lecture 12, 115, *OP*, I, 398–401) a plano-convex lens (*Optica*, I, Lecture 15, Proposition 37, 153–154, *OP*, I, 424–429) and, a thin lens (*Optica*, II, Lecture 14, 127–128, *OP*, I, 574–579, dated October 1672).

[44]*Opticks*, Book I, Part 1, Axiom 7.

[45]Chandrasekhar (1995, 570).

[46]Newton's solution requires defining an integration constant by using a result which he had obtained in the first part of the Scholium. This employs the more familiar mathematics of maxima and minima to find the shape of a frustum of known height and base which offers the least resistance. Modern reconstructions of the Newtonian argument include those of Chandrasekhar (1995, 555–570), and of Edwards (1997), who applies modern computer algebra to retrace Newton's steps.

[47]Néményi (1963).

[48]MS version, Add 3965.10, 107^v/134^v; Whiteside (1970, 130).

Because it produces so many different solutions based on the starting assumptions, the problem has aroused a great deal of interest in the twentieth and twenty-first centuries. One author, James Ferguson, says that, 'Apart from giving birth to an entire field, a further point of interest about Newton's study of motion in a resisting medium is that it is actually one of the most difficult problems ever tackled by variational methods until the twentieth century'. Apart from the restricting physical assumptions that need to be made, 'the problem can possess solutions that have a corner (i.e. a discontinuous slope) which, when expressed parametrically, may not have a solution in the ordinary sense'. Referring to the earlier work of Goldstine (1980), Ferguson says: 'This foreshadows twentieth century developments in optimal control theory',[49] as applied to real engineering problems, where such conditions often occur. Authors such as Torres and Plakhov have actually proclaimed Newton's problem as 'the birth of optimal control'.[50]

Newton's analysis may at first have seemed baffling, but it did not remain hidden for long. Huygens came close to reconstructing Newton's argument in April 1691.[51] Leibniz was unable to make progress when he examined the *Principia* in detail in 1689, but Fellmann believes that his annotations, including the apparently invented word 'isolabis' instead of 'isoperimetris', may indicate that he had sensed a 'historical break-through'.[52] Though Newton generally preferred not to publish his analytical results, he gave a revised version of his solution to David Gregory in July 1694, and it was subsequently publicised that autumn in Gregory's Oxford lectures.[53] Fatio de Duillier published a reconstruction of Newton's fluxional derivation in an appendix to his *Lineae brevissimi* of 1699, and Charles Hayes compared a

---

[49]Ferguson (2004), Goldstine (1980). Newton seems to have been aware that he could not define an absolute minimum but had to specify some limiting conditions for a solution (*MP* VI, 461).

[50]Torres and Plakhov (2006); see also Scriba and Torres (2006). A textbook on *Variational Analysis and Aerospace Engineering* (Buttazzo and Fredianao, 2009) has a chapter (by Buzzatto) titled 'A Survey on the Newton Problem of Optimal Profiles' (33–49). A philosophical work by Markku Roinila with the title *Leibniz on Rational Decision-Making* states that Newton was right in thinking that 'his solution might be useful in the building of ships', as 'optimisation problems' have become 'an essential feature not only in ship-building but also in the building of submarines and aeroplanes' (2007, 37).

[51]*MP* VI, 466, note 25.

[52]Fellmann (1988, 24).

[53]Gregory, 'Isaaci Newtoni Methodus Fluxionum ubi Calculus Differentialis Leibnitij, et Methodus Tangentium Barrovij explicantur, et exemplis plurimis omnis generic illustrantur', autumn 1694, MS St Andrews QA33G8D12; autograph fair copy, Christ Church, Oxford; transcripts by John Keill and William Jones, CUL, Lucasian Papers, Packet No. 13 and MS in private possession.

number of by-then published methods, including Newton's, in his *Treatise of Fluxions* of 1704.[54]

The calculus of variations was also used in Newton's answer to a challenge problem set by Johann Bernoulli which he solved in one night in January 1697: the curve of swiftest descent, or 'brachistochrone', is a cycloid.[55] The problem has a direct connection with *Principia*, I, Proposition 50, on how 'To cause a pendulous body to oscillate in a given cycloid'.[56] Whereas the least resistance problem involved the minimising of the integral $\int F ds$, over the arc length S, where $F = y\dot{y}^3/(\dot{x}^2 + \dot{y}^2)$, the brachistochrone calculation minimised $\int_0^t F(t)dt$, over the time t, where $F = \sqrt{(\dot{x}^2 + \dot{y}^2)/2gy}$.

The problems are connected in a particularly intriguing way. The least resistance calculation finds that, for the integral to be minimised, $y\dot{y}^3\dot{x}/(\dot{x}^2 + \dot{y}^2)^2$ must be a constant, say K, so $F = K(\dot{x}^2 + \dot{y}^2)/\dot{x}$. So, for the integral to be a minimum for an object with constant mass, the integrand $F ds$ must have the dimensions of action, the product of mass, velocity and distance. In some sense a hidden concept of least action is at work. Another way of looking at the problem is to say that Newton's own model of resistance effectively requires the total resistance to be minimised to be the integral of $\rho v^2 \cos^2 \phi 2\pi x dx$, where the density ρ and the velocity v are constant and ϕ is the angle between the axis and the tangent to the meridian curve. Since $\cos^2 \phi = (dx/ds)^2 = 1/(1 + (dy/dx)^2) = 1/(1 + u^2)$, the integral to be minimised is of the form $(x/(1 + u^2))dx$ or, since v is a constant, $(t/(1 + u^2))dt$, meaning that $t/(1 + u^2)$ can be taken as proportional to the effective 'Lagrangian' L in the minimisation of the integral of $L dt$.[57] Newton's 'kernel-function' $F = K(\dot{x}^2 + \dot{y}^2)/\dot{x}$ also 'satisfies the identity' $F = \dot{x}(\partial F/\partial \dot{x}) + \dot{y}(\partial F/\partial \dot{y})$, a single equation which can be taken as an integral form of each of the two Euler–Lagrange equations.[58]

In the case of the brachistochrone, where the time integral was derived by integrating $ds/\sqrt{2gy}$ between the start and end of the trajectory, F is the square root of a ratio of kinetic and potential energies, so the integrand

---

[54]Guicciardini (1999, 90), Fatio (1699b), Hayes (1704, 146–150). Gregory's version was included as an appendix in Motte's English translation of the *Principia* (1729, v–vii). The manuscript version of Newton's solution was published in *Corr.* II, 375–377, 380–382, and *MP* VI, 456–480.

[55]*De ratione temporis*, *Phil. Trans.*, January 1697, *MP* VIII, 72–79.

[56]Chandrasekhar (1995, 573–574).

[57]Newton actually considered that other possible resistance conditions might apply, leading, for example, to the integration of $(t/(1 + u^n))dt$, with $n \geq 1$ (Comte and Diaz, 2004, 2).

[58]Chandrasekhar (1995, 567).

Fdt has the dimensions of time, and the problem clearly involves Fermat's principle of least time, which had been first defined for the path of a light ray in the process of refraction. While Bernoulli claimed that his own solution to the problem (which used the principle of least time but not the calculus of variations) had created a link between optics and mechanics,[59] Newton's generic use of the calculus of variations for two quite different mechanical problems may be seen as the first (if unrecognised) indication that there was a fundamental connection between the principles of least action and least time.

The solution to the less significant, second part of Bernoulli's challenge, which Newton solved simultaneously, required the two segments of a straight line drawn from a given point through a curve being raised to any given power, and taken together to make everywhere the same sum. The original challenge led to important consequences. The brachistochrone problem was solved in different ways by a number of prominent mathematicians, and the *Acta Eruditorum* for May 1697 contained the solutions of Leibniz (205), Johann Bernoulli (206–211), Jakob Bernoulli (211–214) and Newton (in a Latin translation) (223). Another solution, by the Marquis de l'Hôpital, was, for some reason, left in manuscript. The *Acta Eruditorum* publication, though seemingly presented as an example of the success of international cooperation in the search for ultimate truth, led to a bitter public row between the brothers Johann and Jakob Bernoulli, with Jakob issuing his own challenge to Johann to solve the problem under more stringent conditions, and with Johann then offering counter-challenges. Out of this quarrel emerged the version of the calculus of variations which Euler was to systematise in the mid-eighteenth century, but which Newton had already pioneered in two different problems.

2.3. Algebra and Algebraic Geometry

Newton's first major mathematical discovery, the general binomial theorem or expansion of $(a+b)^n$, where n is any rational number, integral or fractional, is one of the most powerful and significant in the whole of mathematics. It led to many further applications and to a development of many aspects of infinite series. His general exposition of series gave the power series for e^x, $\sin x$, $\cos x$, arc $\sin x$; $z = r\sin^{-1}[x/r]$ and its inverse $x = r\sin[z/r]$; the versed sine $r(1-\cos[z/r])$; and $x = e^{z/b} - 1$, the inverse of $z = b\log(1+x)$, also known as the Mercator series. Some individual series were already known, sometimes

---

[59]Bernoulli (1697).

only in computational or iterative numerical, rather than algebraic, form, but Newton's methods could generate virtually any series at will. His deduction of the series for sin x from that for $\sin^{-1} x$ was 'the earliest known instance of a reversion of series'. His *Epistola Prior*, sent to Leibniz on 13 June 1676 includes this result as well as an 'expression for the rectification of an elliptic arc in an infinite series'.[60] It also contains a novel series for π:

$$1 + \frac{1}{3} - \frac{1}{5} - \frac{1}{7} \cdots = \frac{\pi}{2\sqrt{2}}.$$

Newton made the significant observation that all operations on infinite series were carried out 'in the symbols just as they are commonly carried out in decimal numbers'.[61] He used series to determine areas, volumes, and so forth, and showed the reduction of mechanical (or transcendental) curves, such as the quadratrix, to series. This was a major extension to the Cartesian system which had limited algebraic representations to geometrical curves which could be expressed by a finite polynomial.

Newton was aware of the importance of convergence as a necessary condition for expansion in terms of an infinite series. He set down the first Taylor expansion of a function, seeing that, in expanding the function to include an increment in the variable, the fluxions of increasing order, divided by their respective factorials, would act as coefficients of the powers of the fluent in the potentially infinite series generated.[62] Taylor subsequently generalised the process in *Methodus incrementorum directa et inversa* (1715) into what is now called 'Taylor's theorem'. *Principia*, Book II, Proposition 10 openly introduces Newton's method of solving problems via the terms of a converging series and also employs his method of second differences.[63] This work was completely analytic and not, in any sense, synthetic or geometric. Newton further demonstrated that he knew how to set up problems using differential equations and how to solve problems using integration. The Scholium to *Principia*, Book I, Proposition 93 implicitly requires the second

---

[60]Rouse Ball (1908); *Epistola prior*, 13 June 1676, *Corr.* II, 20–47, translation 32–41.

[61]*Epistola prior*, 13 June 1676, *Corr.* II, 20–47, translation 32–41, 32.

[62]According to Whiteside, the first 'explicit enunciation of the Taylor expansion of a general function' is in Newton's *De quadratura*. The 'Maclaurin form', where the base is zero, is in Corollary 3, the general theorem in Corollary 4. James Gregory had used the Maclaurin case in trigonometric expansions in the winter of 1670–1671 without explicit development (*MP* VII, 19–20). A version of a Taylor expansion is used in *Principia*, II, Proposition 10 (263–269 in the 1687 edition).

[63]The proposition (a very extensive one incorporating 2 Corollaries, 4 Examples, a Scholium and 8 Rules) requires the 'equivalent of a third order differential equation' (Whiteside, *MP* VI, 91).

derivative of a function; the general motion of a body attracted to a plane and moving in a curved line can be obtained by 'compounding' this 'motion with an uniform motion performed in the direction of lines parallel to that plane'. It uses a convergent (Laurent) series, with independent even and odd solutions.[64]

His introduction of finite-difference techniques into the study of the factorization of polynomials created the basis for interpolation theory as we know it today. Interpolation is equivalent to finding, at the desired point, the ordinate of a curve that passes through the given values between which one is interpolating, and Newton produced a 'systematic exposition of interpolation by means of central differences'.[65] He derived a number of interpolation formulae, including the Gregory-Newton formula (with sketched proof), and also the Newton–Gauss, Newton–Stirling and Newton–Bessel formulae.[66] While numerical work on interpolation had a long history and a number of special cases were known involving particular techniques, Newton once again showed how a systematic approach could revolutionise a subject. *Principia*, Book III, Lemma 5, which examines the locus through a set of specified points, requires a method of interpolating a function which has been widely used, and states a theorem of 'notable mathematical interest'.[67] Lemma 6 applies the algorithm generated in Lemma 5 to the *longe difficillium* (most difficult) problem of using three given observations to find the orbit of a comet moving in a given parabola. Interpolation techniques could also be used to find areas. A manuscript of the mid-1690s shows how to calculate the approximate area under a curve of any kind, given values for a number of the ordinates, by interpolation using a polynomial function.[68]

Newton found a solution of the problem of summation of harmonic series or musical progressions; he invented methods of summing up a finite number of terms in any such series and a method by which the answer could be approximated. He found a series for calculating the zone of a circle, and a method for 'aggregating' the terms of musical progressions by logarithms,

---

[64] I, Proposition 93, Scholium, 225; Chandrasekhar (1995, 320).

[65] *Methodus differentialis*, 1676, published 1711, *MP* IV, 52–69 and VIII, 244–255. Newton to John Smith, 8 May, 24 July, 27 August 1675, *Corr.* I, 342–344, 348–349, 350–351, and *Regula differentiarium*, 1676, first published 1927, *MP* IV, 36–50. Newton was using interpolation as early as his discovery of the binomial theorem.

[66] Also known as the forward-difference, backward-difference and central-difference formulae.

[67] Cohen (1975–1980, 68).

[68] 'Of Quadratures by Ordinates', *MP* VII, 690–699, 700–702. A Corollary to Lemma 5 explains that this is possible. Guicciardini (2009, 210), says: 'The Newton-Cotes formula originates from this research'.

'using the mean value theorem in a very unusual form'. In a reversal of the usual procedure, he worked backwards from what could be taken as the limit of the initial approximation. He used the smooth curve to find the discrete rectangular ordinates; and then moved the ordinates laterally through half their width c, to improve the rapidity of the approximation; to a large extent the errors were therefore compensatory and cancelled each other out. Both of Newton's procedures were subsequently used by Maclaurin.[69]

A particularly remarkable result is contained in *Principia*, Book I, Lemma 28: there is no oval, that is closed convex curve, whose area, cut off by straight lines drawn 'at pleasure', can be described by an equation with a finite number of terms.[70] Now known as Newton's theorem about ovals, it implies that the area cut off a smooth convex oval (meaning an 'infinitely differentiable convex curve') by a secant, with an equation such as $ax + by = c$, will not be an algebraic function of the secant, that is, of a, b, and c.[71]

This has been described as a 'striking' proof, 'centuries ahead of its time', and 'an astonishingly modern topological proof of the transcendence of Abelian integrals'.[72] The 'first impossibility proof concerning the algebraic nonintegrability of curves', it uses 'the topology of Riemann surfaces', and implies the result now known as Bézout's theorem that algebraic curves of respective degrees m and n can intersect in no more than mn points.[73] Local algebraicity and local algebraic integrability are concepts related to the reducibility of curves, and imply even more related theorems: 'No algebraic oval is algebraically integrable even locally. Any local algebraically integrable oval is algebraic'. And 'The total area bounded by a self-intersecting closed locally algebraically integrable curve (taking into account signs) is zero'.[74] 'Far ahead of its time, this lemma is a glimpse into the existence of what later were called transcendental numbers'.[75]

---

[69] Turnbull (1947). See Newton to Collins, January 1670, *Corr.* I, 16–17, 'Newton's Mathematical Correspondence, 1670–1673', *MP* III, 561 ff. For the Euler–Maclaurin theorem, see Colin Maclaurin, *A Treatise of Fluxions*, 1742, ii, p. 672.

[70] See *MP* VI, 302–310.

[71] Arnol'd (1989), Arnol'd and Vasil'ev (1989). The implication of a counter-result would have been that such an oval would have been a more natural curve for the planetary orbits than an ellipse.

[72] Chandrasekhar (1995, 133), Arnol'd and Vasil'ev (1989).

[73] Arnol'd (1990), (2013), Chandrasekhar (1995, 136). Arnol'd (2013) considers Newton as 'Probably the first person to consider' the 'topology of abelian integrals with the help of their Riemann surfaces' (72).

[74] Chandrasekhar (1995, 137).

[75] Pesic (2001, 215).

For V.I. Arnol'd and V.A. Vasil'ev, the proof 'was incomprehensible both for Newton's contemporaries and for 20th century mathematicians who were bred on set theory and the theory of functions of a real variable, and who were *afraid of multivalued* functions'.[76] Bruce Pourciau commented:

> Two hundred years ahead of its time, Newton's argument for the noninte-grability of ovals in Lemma 28 embodies the spirit of Poincaré: a concern for existence or nonexistence over calculation, for global properties over local, for topological and geometric insights over formulaic manipulation, for proof by contradiction over direct computation. How remarkable to find mathematics of such modern character in a work of the seventeenth century![77]

Newton used it, in a Corollary, to separate curves into the geometrically rational and the geometrically irrational, in which he included spirals, quadratrixes and cycloids.

Another major area of Newton's mathematics was the classification of cubic and other algebraic curves. According to Rouse Ball's account:

> In analytical geometry, he introduced the modern classification of curves into algebraical and transcendental; and established many of the fundamental properties of asymptotes, multiple points, and isolated loops, illustrated by a discussion of cubic curves.... He begins with some general theorems, and classifies curves according as their equations are algebraical or transcendental; the former being cut by a straight line in a number of points (real or imaginary) equal to the degree of the curve, the latter being cut by a straight line in an infinite number of points. Newton then shews that many of the most important properties of conics have their analogues in the theory of cubics, and he discusses the theory of asymptotes and curvilinear diameters.[78]

Newton classified many algebraic curves, relying on the concept of degree as the basis for his subclassification. He created a systematic procedure for reducing the linear transformations of coordinates to canonical forms. The algorithm he developed for constructing subnormals at points on Cartesian curves was particularly significant in developing his original ideas on differentiation and in his realisation that it was the reverse process to

---

[76] Arnol'd and Vasil'ev (1989).

[77] Pourciau (2001a, 498).

[78] Rouse Ball (1908). Newton really maintained Descartes' own distinction between geometrical and mechanical curves, rather than using Leibniz's later terminology of algebraic and transcendental, but he extended the application of analytical geometry by showing that mechanical curves could be expressed algebraically by infinite series.

integration.[79] There are 78 species of cubic curve and it seems that Newton knew all of them in one form or another — only five had been identified previously.[80] His knowledge also extended to a considerable number of curves of even higher degree and, as we have seen, to curves of transcendental order. He reduced the general cubic equation to just five standard forms, or five divergent cubical parabolas, of equation $y^2 = ax^3 + bx^2 + cx + d$, from which all non-trivial cubics could be generated by optical projection or 'casting of the shadow', just as all three types of conic section could be generated by the projection of a circle onto a plane. The general theorem in which cubic curves are generated by the optical projection of the divergent parabola was first proved in 1731 by François Nicole and A. C. Clairaut; a more rigorous proof was supplied by Patrick Murdoch in 1746. Rouse Ball states that: 'In the course of the work Newton states the remarkable theorem that, just as the shadow of a circle (cast by a luminous point on a plane) gives rise to all the conics, so the shadows of the curves represented by the equation $y^2 = ax^3 + bx^2 + cx + d$ give rise to all the cubics'. Also that: 'In this tract Newton also discusses double points in the plane and at infinity, the description of curves satisfying given conditions, and the graphical solution of problems by the use of curves'.[81]

According to Ball: 'In algebra and the theory of equations he introduced the system of literal indices, established the binomial theorem, and created no inconsiderable part of the theory of equations'.[82] *De quadratura curvarum* contains 'the earliest use in print of literal indices'.[83] Algebra was regarded by Newton as a *language* in which equations became the method of translation. He described geometry as synthetic, algebra as analytic, though he also thought that geometry could be analytic. In his view, modern analysis, for example, that of Descartes, had failed to distinguish between geometry and arithmetic. Newton succeeded in reducing equations by one degree by using exponents as multipliers. He found that, for an x axis constituting the diameter of a curve bisecting all the ordinates, odd powers of the

[79]The subnormal to a point on a curve is the distance on the horizontal axis between the perpendicular from the point and the 'radius' or line drawn perpendicular to the tangent.

[80]*MP* VII, 194, 565. Newton's works on cubics include *Enumeratio curvarum trium dimensionum* (1667–1668 or 1670); a series of annotations of the late 1670s, *MP* IV, 366–401; and *Enumeratio linearum tertii ordinis* (1695, published 1704). Gjertsen (1986, 186–188) provides a convenient summary of the results. Guicciardini (2009, 109–136), analyses the methods by which Newton may have achieved them.

[81]Rouse Ball (1908).

[82]Rouse Ball (1908).

[83]*Tractatus de quadratura curvarum*, 1693, published 1704, *MP* VII, 588–645 and VIII, 92–159; Rouse Ball (1908).

corresponding y will not appear in the equation because a negative root of 'equal absolute value' will exist for every positive root. He wrote two works on the geometrical resolution of equations up to the fourth and higher orders.[84] His theory of equations and the identification of their roots, both real and complex, includes the statement of 'Newton's equations' or 'Newton's identities' for determining the ith powers of the roots of a polynomial equation, for any integer i, together with a rule that gave an upper bound on the positive roots. He also devised a method of 'approximating ye roots of affected Æquations by Gunter's line';[85] solved an equation using three movable rulers; and proposed also that a 'cross line' 'may be carried over them with parralel motion' — a 'runner for the slide rule'.[86] This was 'a very powerful step relating any conic to a straight line'.[87]

Among the many series that Newton discovered or investigated is a generalisation of the ordinary power series to one with negative and fractional exponents, which is known as the Puiseux series, and which was first discussed in Newton's *Epistola posterior* of 24 October 1676.[88] For integer values of k, positive integer values of n, and indeterminate T, a Puiseux series is the summation from $i = k$ to $i = \infty$ of $a_i T^{i/n}$. Newton had already defined what is now known as the 'Newton polygon' or 'Newton's parallelogram', a method for observing the behaviour of the roots of a power series equated to zero, that is, a polynomial equation of the form $P(x, y) = 0$. In Newton's algorithm, y is represented by an infinite series in fractional powers of x. A succession of linear approximations can then be constructed on two-dimensional (2-D) axes with the aid of a ruler and 'small parallelograms'.[89] A theorem variously known as Puiseux's theorem or the Newton–Puiseux theorem, which is implicit in Newton's work on the Newton polygon, states that, a polynomial equation of this form will provide solutions in y as functions of x, which can be expanded as a Puiseux series converging close to, but not coinciding with, the origin.

The Newton–Raphson method, which relates to the Newtonian procedure for resolving affected equations in two unknowns, gives the standard iterative solution of equations such as $x^p - N = 0$, where N is a given

---

[84] *Problems for Construing Equations*, sent to Collins on 20 August 1672, MP II, 450–516; *De constructione problematum geometricorum*, c 1706, MP VIII, 200–218.

[85] Newton to Collins, August 1672, *Corr.* I, 229–232, *MP* II, 156 ff.

[86] Whiteside, 'Newton's Mathematical Correspondence, 1670–1673', *MP* III, 561.

[87] *Corr.* I, 229–232.

[88] *Epistola posterior*, 24 October 1676, *Corr.* II, 110–161.

[89] *De methodis fluxionum et serierum*, 1670-1, published 1736, *MP* III, 32–353; *Epistola posterior*, 24 October 1676, *Corr.* II, 110–161; *Geometria Analytica*.

number, or to any polynomial equated to zero, that is a method of extracting the roots (expressed as infinite series) of affected equations. The root (zero) of a function is found iteratively using the derivative of the function and an assumed starting solution. Where there are more roots than one, the answer that emerges will depend on the original starting point. If $f(x)$ is the function, $f'(x)$ its derivative, and $x = a$ is the first attempted solution for $f(x) = 0$, then the next stage in the iteration is $a' = a - f(a)/f'(a)$, and so on, until the successive a values converge.[90] A fractal structure (Newton's fractal), which connects the initial and final solutions, has now been identified in this procedure. Starting with a good approximation for a produces a rapid convergence, but, if the starting a is far from the correct value, a self-similar fractal structure emerges, especially when complex roots are involved.[91]

According to Ball again: 'He uses the principle of continuity to explain how two real and unequal roots may become imaginary in passing through equality, and illustrates this by geometrical considerations; thence he shews that imaginary roots must occur in pairs', a fact that is of tremendous importance in fundamental physics, but a development far in the future which he could not possibly have foreseen. 'Newton also here gives rules to find a superior limit to the positive roots of a numerical equation, and to determine the approximate values of the numerical roots'.[92] Newton's rule for discovering the imaginary roots of such a polynomial, that is, his extension of Descartes' rule of signs to imaginary roots, was not proved until 1853.

Newton considered how to combine means and averages to obtain better approximations. He devised a 'rule for evaluating S_p, the sum of the pth powers of the roots of an equation rationally in terms of the coefficients of the equations. On raising each root to the same power p, summing these powers and then extracting the pth root of the sum, he obtained a kind of mean of the roots. As p tends to infinity this mean tends to the greatest root as its limit'. The rate of approximation improved 'by taking mixed means derived from these expressions'. In the field of higher degree equations, Newton used intersecting conics to solve polynomial equations of the third and fourth degree. He devised a general method for locating 'the rational factors, if any, of a polynomial in one unknown and with integral coefficients'.[93]

---

[90] *De analysi, MP* II, 218–232.

[91] Drexler *et al.* (1996) is one of many on-line discussions of this subject.

[92] Rouse Ball (1908).

[93] Turnbull (1945, 49–50), Cohen (1975–1980, 51).

In coordinate geometry, he constructed a quartic curve from two Cartesian ovals. He produced organic constructions for conics, generalising individual methods of generating conics into a device which could describe the general conic with a continuous, uninterrupted motion. A number of 'beautiful theorems' included Lemma 15 on the tangent to a central conic and the focal *radii vectores*; and the construction of a central conic of known 'species' or eccentricity — a hyperbolic, elliptic or parabolic curve — and given focus and major axis, so that it passed through specific points and touched specific lines: 'About a given conic that shall pass through given points and touch right lines given by position'.[94] He succeeded in resolving Apollonius's three-circles problem, with the construction of a circle tangent to three given circles, his method being an 'elegant variant' on Viète's solution.

Newton produced a completely general theory of coordinate transformations, with 'ordered series of equations' generating the lengths of the subnormals to any curve. He made a significant contribution to analytic geometry in drawing tangents 'according to the various relationships of curves to straight lines', that is, according to the 'modes' or coordinate system used for specifying the curve. These included standard bipolar coordinates, applicable to Cartesian ovals, and polar coordinates, which could be used to construct spirals. He generated equations which could transform between different coordinate systems, for example, from polar to Cartesian or rectangular, and developed formulas for the curvature of conics, spirals, and other curves, in both these systems. The *Methodus Fluxionum* of 1671 includes transformations between ten different coordinate systems. The Corollary to *Principia*, Book I, Proposition 68 recommends the use of what have subsequently been called 'Jacobi coordinates'.[95]

As early as 1665, he gave the equation of a plane section of a hyperboloid of revolution in an application to optics, anticipating the work of de Sluse and Towneley. He also devised an 'elegant construction of the cissoid as the locus of a point on a moving 'angle'', with one side constrained 'to slide through a fixed pole, while the end-point of the other slides along a fixed line'.[96] He described the tractrix with the method of finding its equation. He showed that the sections of a parabolic conoid, parallel to the

[94] Whiteside (1982); *MP* IV, 301.

[95] Chandrasekhar (1995, 265), citing D. Brouwer and G. Clemence, *Celestial Mechanics*, Academic Press, New York, p. 588.

[96] Algebraic Lectures, Problem 19; *MP* V, 214.

axis, are equal to the generating parabola,[97] and found (with proof) that the conjugates of the quadratrix are infinite in number. He stated that, if the generation of a catenary through motion were known, an infinite number of conjugates would be recognised. His 'demonstration that an upright catenary makes the strongest arch' showed that a thin incompressible material arc of uniform density supported at points A and B under its own weight was equivalent to the simpler problem of a uniform flexible string suspended from the same points acting under the tension produced by its own weight, the curves being 'equal and opposite catenaries'.[98] He devised an unpublished proof for constructing a circle of curvature for a given point on a conic section.[99]

In his investigation of higher plane curves, Newton made a considerable effort at finding the most general equation, in Cartesian coordinates, for the quartic trefoil, assuming that the curve would remain invariant when rotated through 120° about its centre; the transformed equation required the reduction of 106 terms.[100] In the problem of orthogonal trajectories, posed as a challenge by Leibniz in 1715, he found the family of curves which cut a given family of curves at a constant angle or at an angle which varied, according to a specified law, with each curve of the given family. The problem is connected with optics in which the paths of rays cut the wavefronts orthogonally, and the equations of the wavefonts will differ in different media. The method is fairly similar to the one now normally used, though Newton's version required ordinary second-order differential equations.[101] Newton, however, did not solve the general problem for any orthogonal trajectory, as Leibniz and Johann Bernoulli had really required but failed to state clearly in the wording of the challenge. This was accomplished by a number of Continental analysts, including Nicolaus Bernoulli, nephew of Johann and Jakob, but no British mathematician came near to a solution.[102]

2.4. Geometry and Numerical Analysis

In geometry, Newton devised a geometrical model to represent angle sections, and derived the algebraic equations which resulted from this. A major

[97] D. Gregory, 'Random jottings on Newton's thoughts', ? May 1694, *Corr.* III, 346.

[98] *ibid.*

[99] *Epistola posterior*, 24 October 1676, *Corr.* II, 110–161, 148, 160, 329.

[100] 'Miscellaneous Analytical Investigations of Cubic and Quartic Curves', c 1695?, *MP* VII, 656–671, 666.

[101] *Phil. Trans.*, 29, 399–400, February 1716, *MP* VIII, 424–434.

[102] Engelsman (1984).

investigation concerned the classical problems of finding loci, 'the pinnacles of Greek geometry'.[103] Depending on the conditions, the locus to be investigated is either a solid locus, that is, a straight line, circle or conic, or a linear locus, that is a curve of higher dimension than the conic. He worked out how to show when a locus problem became a solid locus, and generated a number of theorems of considerable generality on plane loci, or loci constituted of straight lines or circles. The beginning of Section V in the *Principia*, where this work is featured, was taken directly from his unpublished work on the *Solution of the Ancients' Problem of the Solid Locus*.[104] The first seven propositions of the treatise appear in the *Principia* as Lemmas 17–21 and Proposition 22. Newton reduced the four-line or solid locus problem to the organic construction he had previously generated for conics, so attacking the problem of demonstrating when a locus problem is a solid locus. His approach required such modern concepts as taking limiting values as finite increments of variables tended to zero when two points approached each other, or, in other cases, to infinity. Here, he employed and further developed the branch of mathematics we now know as projective geometry.[105]

The first proposition in the *Solutio*, which became Lemma 17 of the *Principia*, uses a result found in Apollonius's *Conics* to prove that, for a quadrilateral $ABDC$ inscribed in a conic, lines PQ, PR, PS and PT from a point P on the conic at given angles to the respective sides AB, CD, AC and BD, of the quadrilateral will be such that $PQ \times PR$ is in a fixed ratio to $PS \times PT$. This is the condition needed to solve the so-called Pappus three- or four-line problem geometrically, as Newton required, rather than using Descartes' algebraic solution in his *Géometrie* of 1637. The second proposition of the *Solutio*, which became Lemma 18 of the *Principia*, then proved the converse.[106]

Newton then solved the Pappus four-line-locus problem, in Lemma 19, by constructing a conic through five given points. In Rouse Ball's words:

> The problem of Pappus, here alluded to, is to find the locus of a point such that the rectangle under its distances from two given straight lines shall be

---

[103] Westfall (1980, 378).

[104] *Solutio Problematis Veterum de Loco Solido*, *MP* IV, 282–320, translation, *Solution of the Ancients' Problem of the Solid Locus*, *MP* IV, 283–321.

[105] Recognising the 'pioneering modernity' of this work, Whiteside declared that it was 'far in advance of contemporary thought' (*MP* IV, 226).

[106] See Guicciardini (2009a, 53–56, 91–92). A scholium following this lemma in the *Principia* defines conics as including rectilinear sections though the vertex as well as ones parallel to the base.

in a given ratio to the rectangle under its distances from two other given straight lines. Many geometricians from the time of Apollonius had tried to find a geometrical solution and had failed, but what had proved insuperable to his predecessors seems to have presented little difficulty to Newton who gave an elegant demonstration that the locus was a conic.[107]

His solution did not use 'an analytical calculus but a geometrical composition, such as the ancients required'.[108] He created an organic generation of conic sections via the intersections of straight lines in relative motion (Lemmas 20 and 21), which he then extended to construct a conic which passed through five points (Proposition 22, with two Corollaries and Scholium), another which passed through four points and was tangent to a single line (Proposition 23), and another which passed through three points and was tangent to two lines (Proposition 24). Lemmas 20 and 21 have been seen by Whiteside and other commentators as constituting a proof of Steiner's theorem, an important result of nineteenth century projective geometry,[109] of which he had 'an an intuitive if not explicit grasp'.[110] Lemma 22, showing how 'To change [plane] geometric curves into others of the same class', 'by compounding an affine translation and a perspectivity', then comes before Proposition 25, which constructs conic sections through two points to be tangent to three lines, and Proposition 26, which constructs the sections through a single point to be tangent to four lines.

Newton also demonstrates that parallel tangents to a conic are anharmonic (Lemma 24), using a demonstration close to that of Apollonius, *Conics*, III, 42, but generalised (in Lemma 25) for the case in which the tangents are no longer parallel, effectively achieving the same property (Corollary 1). He connects the properties of a quadrilateral to those of the conics, using the fact that the centres of conics which are tangent to four lines

---

[107]Rouse Ball (1908). This may not be strictly correct. Descartes had apparently mistranslated Pappus in the *Géometrie* and Apollonius may actually have had a solution to the 4-line locus. Newton certainly thought so, but, if it had existed in Apollonius's time, it was now lost. (Niccolò Guicciardini, personal communication).

[108]I, Lemma 19, Corollary 2.

[109]Whiteside, *MP* IV, 275–276; Guicciardini (2009, 99, 101), who cites Di Steno and Galuzzi (1989) and Shkolenok (1972), and says that Whiteside and the other authors see Newton as showing an understanding of the theorem as 'the foundation of the properties and organic descriptions presented' in section V and in the precursor *Solutio Problematis Veterum de Loco Solido.*

[110]Bloye and Huggett (2011). These authors also cite Shkolenok (1972) as stating that 'the transformations effected by the organic construction' in Lemma 21 'are in fact birational maps from the projective plane to itself, now known as Cremona transformations'. They allowed Newton 'to resolve singularities of plane algebraic curves'.

lie on a straight line (now called Newton's line) drawn through the midpoint of the three diagonals, and finds the conic which touches five lines — a 'beautiful theorem' (Proposition 27).[111] He fits a given triangle with one corner on each of three given straight lines (Lemma 26), and, similarly, a given 'trajectory', with a point on each of the lines (Proposition 28). He also sets a quadrilateral 'given in species' with each corner on one of four given straight lines (Lemma 27); the construction is still valid even when the quadrilateral becomes a single straight line, which is then divided by the four lines into sections with given ratios to each other (Corollary). In Proposition 29, he shows that a quadrilateral given in species can be similarly fitted to the given lines, with an added Scholium 'providing an ingenious variant technique of constructing the problem', 'using the meets of auxiliary rectilinear loci'.[112] It seems that Newton included these propositions and lemmas into the *Principia*, because he originally intended to use Wren's construction of a comet's orbit as a straight line from four observations, though, in the event, he finally constructed the orbit as a conic section.

Another purely mathematical section preceding this one in the *Principia* (IV) contained lemmas and propositions for the 'finding of elliptic, parabolic, and hyperbolic orbits, from the focus given', and was intended to be of direct use in applying to the cases of the orbital motion of astronomical objects under gravity. Lemma 15, a late addition, which states that the tangent to a point on a central conic 'bisects the angle contained there by the focal *radii vectores*',[113] follows from a version used by Seth Ward in the 1650s for the elliptical case. Proposition 18 shows that, from the focus and the principal (major) axis of either an elliptic or hyperbolic curve, the conic itself can be constructed to pass through specific points and to touch specified lines. (Book III, Proposition 16 referred the reader back to I, Proposition 18 for finding the eccentricities and aphelions of the planets.) Proposition 19 makes the same construction for a parabola. In Proposition 20, a conic of any given type or eccentricity, is constructed to pass through specific points and to touch specified lines; four possible cases are identified and discussed.[114] Lemma 16 gives his solution of Apollonius's problem.[115] Proposition 21 aims to describe the trajectory about a given conic 'that shall pass through given

---

[111]Whiteside (1982).

[112]Whiteside (1982).

[113]Whiteside (1982). A Scholium following this Proposition describes how to find axes and foci given the trajectory.

[114]69–72. Whiteside describes Case 1 as 'mathematically interesting' but without 'practical purpose' (*MP* VI, 234).

[115]See 2.3.

points and touch right lines given by position'. A Scholium, which follows, shows a conic being drawn with its focus at S and made to pass through the points B, C and D by a method which Newton described as being similar to one by that 'excellent geometer M. de la Hire'. Though sections IV–V were intended to be used in finding the orbits of comets, Newton, in the event found a method for solving this problem which turned out to be simpler, and only this Scholium found a use, in Book III, Proposition 42, for showing how, 'by arithmetical operations', a parabola could be drawn in the plane of an orbit with the Sun at its focus.

It is in connection with a theorem in conics (*Principia*, Book I, Lemma 12) that William Whiston made his famous remark that

> Sir Isaac, in Mathematicks, could sometimes see almost by Intuition, even without Demonstration; as was the Case in that famous Proposition in his Principia, that All parallelograms circumscribed about the Conjugate Diameters of an Ellipsis are equal; which he told Mr Cotes he used before it had ever been demonstrated by any one, as it was afterward. And when he did but propose Conjectures in Natural Philosophy, he almost always knew them to be true at the same Time.[116]

Lemma 12 was apparently added, along with two preliminary lemmas on parabolic geometry, Lemmas 13–14, at the persuasion of Halley, to help readers understand the propositions that were to follow. The first two are ascribed to the writers on conic sections, but, although Lemma 12 can be equated to Theorem 31 of Book VII of Apollonius's *Conics*, it was probably not derived from that source, which Newton had probably never studied in any depth. Whiteside has conjectured that Newton's 'mathematical intuition let him see that this Lemma' was, for an ellipse, 'all but self-evident by orthogonal projection at a fixed angle from a circle inscribed in a square'.[117]

One question that Newton considered was how legitimate conic sections were in solid geometry, that is as constructible in three-dimensional space, as they could not be produced geometrically in two dimensions, and so were, in this sense, 'illegitimate' as constructs of plane geometry. He was concerned to separate such synthetic geometric constructions from the equivalent versions generated by analytic algebra. The simplest conic section in synthetic geometry (other than the circle) would be the ellipse, but, in analysis, it would be the parabola. 'Simplicity in figures', he wrote, 'is dependent on the simplicity of their genesis and conception, and it is not its equation

---

[116]Whiston (1749, 39).
[117]Whiteside (1982).

but its description (whether geometrical or mechanical) by which a figure is generated and rendered easy to conceive'.[118]

In his *Inventio Porismatum, De compositione et resolutio veterum geometrarum*, and several other works, Newton attempted to restore the lost books of Euclid's *Porismata*.[119] Porisms were an ancient heuristic method of analysis, known mainly from a synopsis in the seventh book of Pappus of Alexandria's fourth century *Collectio*, but no one has conclusively established how the method worked. Newton thought that porisms were related to the construction of loci and that they might be connected to projections of the conic sections. He obtained results which anticipate some of those included in the later 'reconstructions' of Euclid by Robert Simson and Michel Chasles.[120] His concentration on such problems led to a belief that the ancients had a more advanced form of analysis than the available evidence suggests they did, enhancing the possibility that a geometrical form of advanced analysis could be achieved.

Just as Newton thought of calculus as being intended to resolve problems by motion, so also he thought of geometry as a branch of mechanics. In the Preface to the *Principia*, he wrote:

> for the description of right lines and circles, upon which geometry is founded, belongs to mechanics. Geometry does not teach us to draw these lines, but requires them to be drawn; for it requires that the learner should first be taught to describe these accurately, before he enters upon geometry; then it shows how by these operations problems may be solved. To describe right lines and circles are problems, but not geometrical problems. The solution of these problems is required from mechanics; and by geometry the use of them, when so solved, is shown; and it is the glory of geometry that from those few principles, brought from without, it is able to produce so many things. Therefore geometry is founded in mechanical practice, and is nothing but that part of universal mechanics which accurately proposes and demonstrates the art of measuring.

The idea that mathematics was as much physical as physics was mathematical was a key component in his success in combining them into virtually a single enterprise.

---

[118] *Lectures in Algebra*, 1673, I, 2, §1, *MP* V, 477.

[119] *MP* VII, 230–246; *De compositione et resolutio veterum geometrarum*, *MP* VII, 304–309. Other works of the period which attempted to restore the supposedly 'lost' analysis of the ancients included *Geometriae Libri Tres* (*MP* VII, 248–400, *Proemium*, 230–247) and *Geometriae Libri Duo* (*MP* VII, 402–561).

[120] Guicciardini (2009, 81–89), citing Simson (1776) and Chasles (1860, 82).

In the codification of plane and spherical trigonometry, a 1683 paper includes a 'group of Napierian half-angle theorems which he proved in a wholly novel way by projecting the spherical triangle stereographically into a corresponding plane (non-rectilinear) triangle, making good use of both the circle- and angle-preserving properties of the projection'.[121] Classical in origin, spherical trigonometry had a long history owing to its importance in navigation; it is in fact an aspect of a non-Euclidean geometry, in the same sense as elliptical and hyperbolic geometries, but no one at the time would have thought of it in those terms.

In numerical analysis, Newton explained how to work out compound interest to David Gregory; he devised a formula for the compound interest on an annuity continuing for a given number of years, possibly by an iterative method.[122] He disputed the kissing number problem for spheres with Gregory at Cambridge in 1694, in connection with an investigation of stellar statistics, correctly arguing that one sphere could touch just 12 identical surrounding spheres as opposed to Gregory's 13.[123] Kepler had previously shown how a sphere could touch 12 surrounding spheres in two different ways. The proof of the theorem was not established until 1953.

In an early manuscript, probably influenced by Huygens' *De Ratiociniis in ludo aleae* (1657), Newton was the first to consider geometrical probability, in order to accommodate irrational proportions, and applied it to the case of an irregular die.[124] Newton 'opts for a frequency theory of probability; that is, the absolute probabilities are not given a priori but are determined as the asymptotic limit of the numerical probability observed over a succession of occurrences of a state'.[125] A letter to Pepys explained why the chance of one six in a roll of six dice is higher than two sixes with twelve, or three sixes with eighteen, etc.[126] He also discussed such things as the movement of the planets one way in concentric orbs in probability terms — the probability of

---

[121] 1683, *MP* IV, 11.

[122] David Gregory, 'Random jottings on Newton's thoughts', ? May 1694, *Corr.* III, 346.

[123] David Gregory, Note 13, May 4, 1694; *Corr.* III, 317, also quoting from Gregory's Note-book, Christ Church, Oxford, p. 81.

[124] *MP* I, 58–61.

[125] 'Loose Annotations on Huygens' *De Ratiociniis in Ludo Aleæ*', summer 1665?, Whiteside, *MP* I, 61.

[126] Pepys to Newton, 22 November, 9 December, 21 December 1693; Newton to Pepys, 26 November, 16 December, 23 December, 1693, *Corr.* III, 293–303. Stigler (2007) says that Newton's numerical solution is 'elegant and flawless', but the extended logical argument he also invokes is invalid 'because the truth of the proposition depends upon the probability measure assigned to the sequences and the argument did not', However, 'modern probabilists should admire the spirit of the attempt'.

both these things happening independently would be $(1/2)^6 \times (1/2)^6$, which is less than 1 in 4,000, though he doesn't give any numerical figure, as Daniel Bernoulli and Laplace did subsequently, almost certainly with Newton's argument in mind.[127] John Arbuthnot was the first to make a specific calculation of this kind, referring to the relative probabilities of the births of boys and girls, in a paper of 1710.[128] Karl Pearson subsequently considered that Newton's argument for a Deity 'formed the foundation of statistical development'.[129] However, de Moivre and the statisticians who followed Newton were 'more influenced by his theology than by his mathematics'.[130]

Newton had knowledge of both systematic and random errors in experiments, though he made no specific distinction between them. His early studies, particularly in astronomy and optics, had suggested to him that the human senses were not a reliable guide in experimental work. He came to the apparently novel conclusion that the result of a series of observations, which may be considered individually unreliable, must be taken from the average value of a large number of measurements, rather than from the seemingly most accurate one. In this way a result could be achieved that was untainted by prior theoretical prejudice. A classic example of his procedure is seen in the experiments on diffraction published in the *Opticks*. Average values had been taken before, but as Buchwald and Feingold stress, in quite different contexts. Newton's procedure was unique and 'unprecedented in natural philosophy' before his 'work in optics in the late 1660s and early 1670s'.[131]

---

[127] Sheynin (1972, 222). Sheynin quotes the 'most beautiful system of the sun, planets, and comets, could only proceed from the counsel and dominion of an intelligent and powerful Being'. 'Whence is it that planets move all one and the same way in orbs concentrick, while comets move all manner of ways in orbs very excentric...' (*Opticks*) 'Now by the help of these principles, all material things have seem to have been composed of the hard and solid particles... variously associated in the first creation, by the counsel of an intelligent Agent. For it became Him who created them, to set them in order'. (Q 31) And 'blind Fate could never make all the planets move one and the same way in orbs concentrick, some inconsiderable irregularities excepted, which may have risen from the mutual actions of comets and planets upon one another, and which will be apt to increase, till this system wants a reformation. Such a wonderful uniformity in the planetary system must be allowed the effect of choice. And so must the uniformity in the bodies of animals' (Q 31).

[128] Arbuthnot (1710).

[129] Pearson (1926).

[130] *ibid*. A practical form of statistics also entered into Newton's work at the Mint.

[131] Buchwald and Feingold (2012), chapters 1 and 2, quoting p. 12. For these authors, Newton 'alone had developed a method that not only permitted, but actually urged, the experimental production and subsequent amalgam of discordant numbers' (12). They also describe how, in examining ancient astronomical data in 1700, 'Newton's extraordinary

Writing to Anthony Lucas in 1676 about discrepancies in outcomes between experiments by different observers or in repetitions by the same observer, Newton says:

> there may be many various circumstances which conduce to it; such as are not only the different figures of prisms, but also the different refractive power of glasses, the different diameters of the sun at divers times of the year, and the little errors that may happen in measuring lines and angles, or in placing the prism at the window; though...I took care to do these things as exactly as I could.[132]

The Sun's spectrum could also be expected to change with the varying brightness of the sky.[133] In his work on atmospheric refraction, he said that the altitudes measured for celestial objects should be adjusted for changes in the refraction due to the air temperature. He noticed a number of cases in experiments where errors were small enough to be neglected.[134] He also recognised the value of using an assistant to divide the colours of the spectrum in order to prevent the bias introduced by his preconceived idea of their harmonic ratios.[135] The method that he devised for measuring indices of refraction by finding the minimum deviation, which occurred when the refractions were equally on each side of the prism, is still the most accurate one available because it successfully minimises the errors.[136] His table of refractive indices in the *Opticks*, determined by this method, gives values as good as any achieved before the nineteenth century.[137] Newton, of course, was not only his century's outstanding mathematician; he was also its outstanding experimenter.

---

procedure took his error-slaying method into new territory for he had in effect used the original data to correct itself...' (103).

[132] Newton to Anthony Lucas, 16 August 1676, *Corr.* II, 76, originally published as '*Mr. Newton's Answer to the precedent Letter, sent to the Publisher*', *Philosophical Transactions of the Royal Society*, no. 128, 698–705, 25 September 1676.

[133] Sheynin (1972, 222).

[134] Whiteside (1980); Lehn (2008).

[135] *Hyp.*, *Corr.* I, 376; *Opticks*, Book I, Part II, Proposition 3, Experiment 7, 126.

[136] *Lectiones opticae*, Lecture 9, §102–106, *OP*, I, 177–185, *Optica*, I, Lecture 5, 29–35, 316–327.

[137] *Opticks*, Book II, Part III, Proposition 10. Shapiro, *OP*, I, 182–183.

Chapter 3

Space, Time and Motion

3.1. Absolute Space and Time

The issues involved in Newton's definitions of space, time and motion, once thought to be a seventeenth-century metaphysical backwater, are in fact of direct relevance to our understanding of physics at the deepest level today. If we want truly to understand physics, we need to understand these quantities at a deeper level than operationalist definitions will allow. Contrary to Descartes, who defines matter as extension, Newton separates space, time and body (matter) as independent entities, a point disputed by Leibniz, Huygens and Berkeley among his contemporaries. The extreme case of reducing all observation to extension on the Descartes model would seem to be Einstein's General Theory of Relativity of 1915, and his subsequent attempt at formulating a unified field theory, but Newton, though he privileged geometrical methods as much as Einstein ever did, chose not to go down this route. Though only length can be measured, the natural world seems to require that we use other parameters too. This leads to a crucial breakthrough in defining mass, the route to universality in physics. Many people today find it difficult to accept that space and time have as much 'reality' or are as 'ontological' in nature as material objects, but it is one of the keys to Newton's success in mathematising physics.

Newton's definitions of absolute space and time, introduced in a famous Scholium in the *Principia*, undoubtedly have a theological tone, and have been much criticised for this, as well as for their supposed inapplicability to post-relativity physics, especially in the case of the time parameter. However, these criticisms are based upon the misconception that relativity uses only relative space and time, and on ignorance of the crystal clear and careful distinction that Newton makes between the absolute and relative concepts. Never at any time does he suggest that the absolute quantities are the same as the ones we *measure*, and he even actually uses a well-known relativistic transformation in the case of relative space. The fundamentally absolute space and time could be said to apply to the abstract system, while the relative space and time apply to measurement. For Robert DiSalle, Newton's definitions are 'exemplary of the way in which science gives empirical meaning to theoretical notions'.[1]

Space and time were the fundamental parameters through which the mediaeval philosophers had sought to know 'the mind of God' and it is inconceivable that Newton would have regarded them in any but an absolute sense, but their definition as absolute concepts also has a physical reason. The indeterminacy of physical measurement is the real reason why Newton introduces the absolute concepts, and distinguishes them from the relative quantities commonly used in measurement. Famously, he writes:

> Absolute, true, and mathematical time, of itself, and from its own nature, flows equably without relation to anything external, and by another name is called duration: relative, apparent, and common time, is some sensible and external (whether accurate or unequable) measure of duration by the means of motion, which is commonly used instead of true time; such as an hour, a day, a month, a year.[2]

But, as he also writes:

> It may be, that there is no such thing as an equable motion, whereby time may be accurately measured. All motions may be accelerated and retarded, but the true, or equable, progress of absolute time is liable to no change. The duration of perseverance of the existence of things remains the same, whether the motions are swift or slow, or none at all: and therefore it ought to be distinguished from what are only sensible measures thereof. . . .[3]

Of course, a complete causal sequence can never be established, because we never have complete information about a physical situation through measurement, but such would certainly exist on a cosmic scale. Relative time becomes the order of measured events made without taking into account an infinite number of interactions.

Absolute time, to Newton, is an order of events, not a 'sensible' measure. A similar thing applies to space, but in a more subtle way: 'As the order of the parts of time is immutable, so also is the order of the parts of space. . . . All things are placed in time as to order of succession; and in place as

---

[1] DiSalle (2002, 51).

[2] *Principia*, Definitions, Scholium.

[3] *ibid.*

to order of situation. ....' Absolute space is also to be distinguished from relative space.

> Absolute space, in its own nature, without relation to anything external, remains always similar and immovable. Relative space is some movable dimension or measure of the absolute spaces; which our senses determine by its position to bodies; and which is vulgarly taken for immovable space; such is the dimension of a subterraneous, an æreal, or celestial space, determined by its position in respect of the earth. Absolute and relative space, are the same in figure and magnitude; but they do not remain always numerically the same. For if the earth, for instance, moves, a space of our air, which relatively and in respect of the earth remains always the same, will at one time be one part of the absolute space into which the air passes; at another time it will be another part of the same, and so, absolutely understood, it will be perpetually mutable.[4]

Place can be equally absolute or relative:

> Place is a part of space which a body takes up, and is according to the space, either absolute or relative. I say, a part of space; not the situation, nor the external surface of the body. For the places of equal solids are always equal; but their superfices, by reason of their dissimilar figures, are often unequal. Positions properly have no quantity, nor are they so much the places themselves, as the properties of places. The motion of the whole is the same thing with the sum of the motions of the parts; that is, the translation of the whole, out of its place, is the same thing with the sum of the translations of the parts out of their places; and therefore the place of the whole is the same thing with the sum of the places of the parts, and for that reason, it is internal, and in the whole body.[5]

Absolute motion is also to be distinguished from relative motion. Newton uses here the example of the motion of a ship, previously put forward by Galileo:

> Absolute motion is the translation of a body from one absolute place into another; and relative motion, the translation from one relative place into another. Thus in a ship under sail, the relative place of a body is that part of the ship which the body possesses; or that part of its cavity which the body fills, and which therefore moves together with the ship: and relative rest is the continuance of the body in the same part of the ship, or of its cavity. But real, absolute rest, is the continuance of the body in the same part of that immovable space, in which the ship itself, its cavity, and all that it contains, is moved. Wherefore, if the earth is really at rest, the

---

[4] *ibid.*
[5] *ibid.*

body, which relatively rests in the ship, will really and absolutely move with the same velocity which the ship has on the earth. But if the earth also moves, the true and absolute motion of the body will arise, partly from the true motion of the earth, in immovable space; partly from the relative motion of the ship on the earth; and if the body moves also relatively in the ship; its true motion will arise, partly from the true motion of the earth, in immovable space, and partly from the relative motions as well of the ship on the earth, as of the body in the ship; and from these relative motions will arise the relative motion of the body on the earth. As if that part of the earth, where the ship is, was truly moved toward the east, with a velocity of 10,010 parts; while the ship itself, with fresh gale, and full sails, is carried towards the west, with a velocity expressed by 10 of those parts; but a sailor walks in the ship towards the east, with 1 part of the said velocity; then the sailor will be moved truly in immovable space towards the east, with a velocity of 10,001 parts, and relatively on the earth towards the west, with a velocity of nine of those parts,[6]

Newton's conceptions also implicitly require the fundamental continuous symmetries of space and time translation and space rotation, which still retain their absolute validity in contemporary physics. The translation symmetry of space is quite explicitly invoked in a document of 1692, as well as the self-similarity which makes it possible for investigators to relate observations on one scale to those on another:

space itself has no parts which can be separated from one another, or be moved amongst themselves, or be distinguished from one another by any inherent marks. Space is not compounded of aggregated parts since there is no least in it, no small or great or greatest, nor are there more parts in the totality of space than in any place which the very least body occupies. In each of its points it is like itself and uniform nor does it truly have parts other than mathematical points, that is everywhere in number and nothing in magnitude.[7]

The rotation symmetry of space is, likewise, specifically invoked in Corollary 22 to *Principia*, Book I, Proposition 66 (see 4.4). The translation symmetry of time is also defined in the document of 1692: 'The duration of each thing flows, but its enduring substance does not flow, and is not changed with respect to before and after, but always remains the same'.[8] It is an immediate consequence of a quantity that 'flows equably without relation to anything external'. The moments of Newtonian time have no

---

[6] *Principia*, Definitions, Scholium.
[7] Hoskin (1977, 90).
[8] Hoskin (1977, 90).

'identity' outside their order, just as the elements of space have no 'identity' outside their position.[9] Such lack of identity is characteristic of quantities which are translation (and rotation) symmetric. The discrete symmetries of space reflection and time reversal also feature in some of Newton's dynamical theorems. These symmetries are valid for all of the physics known to Newton, being violated only by the weak interaction, discovered long after his time.

It is precisely because Newton felt that he needed to make space, time and matter into separate physical quantities that he also needed the concepts of absolute space and absolute time. The absolute concepts then had to be separate from the way we measure them, which introduced relative space and relative time. He believed that absolute space and absolute time, separated from matter, were required by an omnipresent and eternal God, though he would have argued that our theological knowledge was derived from our observation of nature rather than our natural knowledge being derived from the attributes of God. So, although theological truth may have been Newton's ultimate purpose, his absolute space was not a *consequence* of this theology, as is often thought, but actually *informed* it.[10] The view of Descartes and others before him (particularly Aristotle), that there is no void or empty space and that the material world is a plenum, was in direct contradiction to Newton's position. When we talk about space, in Cartesian philosophy, we are talking about the positioning of bodies in the plenum, while time is observed only when we see changes in the world. Newton's views on space and time have been called 'substantivalism' as opposed to the 'relationism' of his opponents.[11] The separate identities of space, time and matter, however, proved to be important because they enabled physics to be established on abstract fundamentals rather than on arbitrary models.

3.2. Relative Space and Time

Having established the ideas of absolute space and time, Newton uses relative space and time for most of the *Principia*. In the Scholium to the Definitions, he says: 'relative quantities are not the quantities themselves, whose names they bear, but those sensible measures of them (either accurate or inaccurate), which are commonly used instead of the measured

---

[9]The lack of identity is referred to by DiSalle (2002, 39).
[10]Stein (2002, 256–307, 261–262, 268–272).
[11]Rynasiewicz (2011).

quantities themselves'. Again: 'instead of absolute places and motions, we use relative ones; and that without any inconvenience in common affairs; but in philosophical disquisitions, we ought to abstract from our senses, and consider things themselves, distinct from what are only sensible measures of them'. Newton never claimed that we could *measure* absolute space and time; the absolute quantities had to be distinguished from the 'sensible measures' used in 'common affairs'. Newton's definitions of absolute space and time have been seen, in fact, as essentially philosophical concepts, rather than components of his scientific theories, though they did have a fundamental scientific significance, and challenges to them were usually made on philosophical, rather than experimental, grounds.

The classical dynamical theory associated with Newton was not based on absolute space or absolute time, and, like that of Galileo, was fully 'relativistic', requiring relativistic transformations between the coordinates of systems in uniform relative motion. In classical mechanics, a distance measured in a stationary system is not normally identical to one measured in a system moving with uniform velocity with respect to the observer. The Galilean-Newtonian system requires the application of a transformation equation to convert distances measured in one system to those measured in the other, which enables physicists to preserve the form of Newton's laws — which recognise the special nature of accelerated, but not of unaccelerated motion — whether the system being observed is at rest or in uniform motion with respect to the observer. The transformation equations applied to Newtonian mechanics:

$$x' = x - vt,$$

$$y' = y,$$

$$z' = z,$$

$$t' = t,$$

are now known as the 'Galilean transformations', a name given to them in 1908,[12] though the version now used, applying to rectilinear motion rather than Galileo's circular variety, was first stated by Newton.

The transformation is also accompanied by a principle of relativity. In Newtonian physics, states of rest and uniform motion in a straight line are equivalent, but nonuniform or accelerated motion is different in being

---

[12]Frank (1908, 898).

detectable by a field of force. Corollary 5 to the Axioms, or Laws of Motion is an explicit statement of the 'relativity' of uniform or unaccelerated motion 'The motions of bodies included in a given space are the same among themselves, whether that space is at rest, or moves uniformly forward in a right line without any circular motion'. This is the first statement of the 'principle of relativity'. It also incorporates the definition of an inertial frame of reference, a concept of immense importance to Newtonian physics. Newton further makes clear, at the end of *Principia*, Definition 3 (of the *vis insita*), that states of motion and rest can only be distinguished relatively by observation: 'but motion and rest, as commonly conceived, are only relatively distinguished; nor are those bodies always truly at rest, which commonly are taken to be so'.

The system of mechanics incorporated in Newton's laws of motion needed to be defined relative to an absolute standard of rest or inertial frame of reference. According to the principle of relativity, as defined in Corollary 5, the same laws will be true for all observers moving with *uniform* velocity with respect to this frame. The laws, however, will not be true for accelerating observers, except, as Newton recognised, in the case of observers with the same acceleration moving parallel to each other, where the bodies 'will continue to move among themselves, after the same manner as if they had been urged by no such forces'.[13] The discovery of a deviation from the laws can, therefore, be used to determine the observer's acceleration.

The 'relativity' of the measurement of natural phenomena was not itself an intrinsically revolutionary subject. The inherent 'relativity' of all motion had been a subject of scholastic debate at the end of the Middle Ages, and the fifteenth century German cardinal, Nicholas of Cusa, had presented many arguments in his famous work *On Learned Ignorance* against the traditional Aristotelian picture of the world system. Among many significant questions he asked: how could space have a boundary? how could time have a beginning? how could the universe have a centre? Though we might take the Earth as the centre in some respects, for our convenience, it was not the true centre, nor was the Sun, nor any other heavenly body. Observed motions were always relative, and it was arbitrary where we defined the origin. We could equally well take the Earth as moving as at rest. Newton's significant contribution is to incorporate the idea into a fully mathematical theory of dynamics.

---

[13] Axioms, or Laws of Motion, Corollary 6.

Ernst Mach, one of the later critics of Newton's absolute space and time, was enthusiastic about Corollary 5, where 'In spite of his metaphysical liking for the absolute Newton was correctly led by the tact of the natural investigator'.[14] But this had to be taken along with accepting the fixed stars as an inertial frame, as Newton had seemed to imply might be the case in contingent terms.

> It is a property of rest, that bodies at rest really do rest in respect to one another. And therefore as it is possible, that in the remote regions of the fixed stars, or perhaps beyond them, there may be some body absolutely at rest; but impossible to know, from the position of bodies to one another in our regions, whether any of these do keep the same position to that remote body, it follows that absolute rest cannot be determined from the position of bodies in our regions.

> ...it may be that there is no body really at rest, to which the places and motions of others may be referred.[15]

As he says elsewhere, even 'The fixt Starrs may move inter se by their mutual actions.[16]

In 1905, the Newtonian principle of relativity was used by Einstein in an entirely new context.[17] The Newtonian definition came before the discovery of electrodynamics; following this discovery, with its implication that electric and magnetic forces acted at the speed of light, rather than at the assumed infinite speed of Newtonian gravity, it was widely believed that motion with respect to an aether would be discovered in electrodynamic or optical systems. The failure of the Michelson-Morley experiment of 1887 to detect the motion of the Earth through the aether by observing a change in the velocity of light, led Einstein to extend the use of the relativity principle to electrodynamics, as well as mechanics. He also added a new principle, that the velocity of light remained the same in all frames of reference, and was unaffected by the motion of source or observer, though some commentators argued that this principle was a consequence of the first, and not an independent axiom. Einstein's use of the principle, however, led to him using a set of transformation equations for space and time coordinates, which Poincaré called the 'Lorentz transformations'[18] and which mainly

---

[14]Mach (1911, 36).

[15]*Principia*, Definitions, Scholium.

[16]Recorded by David Gregory as 'Mr Newton's exceptions ag[ains]t my book', RS Greg MS, f. 176ʳ; Hoskin (1985, 92).

[17]Einstein (1905b).

[18]Poincaré (1905).

differed from those in the Newtonian system by the appearance of the γ factor:

$$x' = \gamma(x - vt),$$
$$y' = y,$$
$$z' = z,$$
$$t' = \gamma(t - vx/c^2),$$

where $\gamma = (1 - v^2/c^2)^{-1/2}$. The dynamical consequences included the facts that lengths measured in a moving frame of reference would be contracted by the factor γ, while times would be dilated, and masses increased by the same factor.

As the name 'Lorentz transformations' would suggest, results of this kind had already started appearing, in one form or another, in electrodynamic theory, as a result of the electric and magnetic forces being transmitted at the velocity of light, with contributions supplied, variously, by Heaviside, Larmor and Lorentz. Einstein's key addition was to derive them purely from his new kinematics without any special reference to electrodynamics or the aethereal mechanisms of some of his contemporaries. Nevertheless, he never regarded his work in special relativity as 'revolutionary' in the way it is often seen now, treating it rather as the culmination of a long-term development in electromagnetic theory to which many had contributed. Much of the effectiveness of special relativity, also, comes from a subsequent development by Hermann Minkowski, Einstein's former teacher, in which a '4-vector' combination of space and time is treated as an invariant four-dimensional (4-D) space-time[19]:

$$s^2 = x^2 + y^2 + z^2 - c^2t^2 = x'^2 + y'^2 + z'^2 - c^2t'^2.$$

According to Robert DiSalle, because of his use of the 'Galilean transformation', 'Newton's laws presuppose absolute time, but not absolute space. In fact, rather than absolute space, Newton's laws require a 4-D structure known as 'Newtonian space-time'. A straight line of this structure represents uniform motion in a straight line, and therefore its physical counterpart is the notion of a body not subject to forces'. 'Spacetime is a 4-D affine space'.[20] For many, the special relativistic discovery of

---

[19]Minkowski (1909). The Minkowski representation is also more general, being applicable far beyond the special inertial motions treated by Einstein, and applying to all realistic cases.

[20]DiSalle (2002, 35).

time dilation has been presented as the revolutionary breakthrough in the 'challenge' to Newtonian physics, but it was, in fact, much less original and much less startling than is now generally recognised; for the fact that the involvement of the velocity of light in the process of measuring time could lead to distortions in the measured time scale did not depend on knowledge of the transformations used by Einstein and pre-dated special relativity by at least half a century.[21] It could even be said to have been implicit in the discovery of the finite velocity of light in Newton's own lifetime.

There are many problems with using the operational thinking of special relativity, aimed at finding a convenient derivation of the Lorentz transformations, to derive deep philosophical conclusions concerning the nature of space and time. For instance, since Einstein's conception of time required the abandonment of the idea of absolute simultaneity, he found it useful to define a new concept of simultaneity — a working definition aimed at deriving the required results by kinematics. According to Einstein's definition, a light-signal emitted from the mid-point between two observers would be received by them simultaneously. But it is a definition without fundamental validity, for, if there is no absolute simultaneity, there are no such things as simultaneous events. And, if causality is to be maintained, events have a definite order which cannot be changed, and this is what we mean by absolute time. The special relativistic emphasis on simultaneity is not fundamental, even if literally true. Both the Galilean and Lorentz transformations lead to cases where different observers appear to see the order of events reversed, for example, signals sent at the speeds of light and sound. The 1987 supernova produced both neutrinos and light, which arrived at slightly different times, the neutrinos arriving slightly earlier than the light, even though they must have been slower, as they had nonzero rest mass. No one believes these results proved that neutrinos travel faster, though appearances may suggest that they do. In fact, we know that neutrinos leave the core of the supernova immediately whereas the shock waves that

---

[21] Felix Eberty, for instance, a German professor of literature, explained how an observer, travelling at speeds comparable to c, would be able to magnify or reduce the scale of time (1846). In principle, an observer moving towards the Earth at a sufficiently high speed would be able to double the rate at which information was received, so doubling the apparent rate of events on Earth. Moving away from the Earth at the same speed would appear to halve the apparent rate. An observer travelling at the speed of light would see nothing. In 1883, the physicist J. H. Poynting showed how Eberty's conclusions followed from the Doppler effect, and explained that exceeding the speed of light would lead to events being seen in reverse order.

produce the light photons take time to reach the surface. Also, no one believes that jets emitted from quasars are actually moving apart at speeds that are many multiples of the speed of light, though line-of-sight observations from Earth suggest that they are. In such examples, as in a number of others, we find that careful attention to detail always leads to a resolution of the anomalies.

Special relativity also defines time as operational, but then proceeds to describe what happens when frames of reference are in *uniform* relative motion. Operationally, time measurement requires the use of acceleration — this is what is behind Newton's description of relative time via watches running at different rates, because they involve different frictions, etc. Relative time is always due to different accelerations. Now special relativity defines simultaneity using the one-way speed of light, but this cannot be used to *measure* time. Hence the twin paradox, where the assumption that two twins can be indistinguishable to each other in terms of relative uniform motion leads to the impossible conclusion that each would see the other as relatively aged. To measure time, you have to have a reversal, or acceleration.

Rapidly-moving muons entering the Earth's atmosphere certainly extend their lifetimes, and this can be measured using the 'time dilation' equation which comes from the Lorentz transformation for time, but it is the acceleration in the Earth's magnetic and gravitational fields and the reduction of the velocity to zero on detection which are relevant. To make the observation at all, we have to observe the trail of ions left by the particles in the detector, with the consequent loss of the particles' kinetic energy and velocity. The time dilation equation uses the velocity of the particle but the velocity is determined by acceleration mechanisms, not by uniform motion in a straight line. In effect, the Lorentz transformation equation holds for clocks of a particular kind operating under certain conditions, but it says nothing about the behaviour of time itself. It defines the effects of the forces which act to change the measured time signal from the clock on which they act, making it run slow in the same way as friction would slow down the time signal in the kind of classical clock known to Newton. The transformation equation is only different because different forces act differently on different clock mechanisms.

Again, the special theory of relativity is often said to have removed absolute time from mechanics, but this only applies to the measurement of time, not to time itself. Special relativity uses thought experiments in which time is the thing measured by clocks — an operational definition. But in theoretical physics, and especially in quantum mechanics, it is not. Measurements of time are unique; each has its own peculiarities, like years, tree rings, ticks on a clock, swings of a pendulum, orbits of a planet, and so on. No two of these are exactly the same, as Newton said for the orbits of a planet. Each swing of the pendulum is peculiar and special and could be labelled, just as we label our years in history; events either happen or they don't. But moments of time are not unique. True moments of time are all alike. Time has translation symmetry, and every moment of time is exactly like every other. This is the fact which is now known, via Noether's theorem, to be equivalent to the conservation of energy. In effect, time instants are all the same, while clock instants are all unique. Time is not the measurement of time, and measured moments of time are not the same as true moments of time. Newton knew this, as, in effect, did Einstein. Time, in fact, cannot be measured at all — in quantum mechanics, it is not an observable — neither, for that matter, can Newton's other fundamental quantity of mass. Only space can be measured, and it is measurements of space which, in clocks, are used in place of measurements of time.

Contrary to much philosophical reasoning since the early twentieth century, Einstein, in fact, also uses absolute time, just as Newton had, and for exactly the same reason. His system is based on causality. No event can precede its cause; and we can define the statement 'A causes B' as meaning no more and no less than A necessarily precedes B in time. And so, without making a direct statement, Einstein really introduces a third postulate: the fact that no information can be transmitted faster than the velocity of light. He also introduces a parameter which defines it, the proper time, which, in operational terms, becomes the time measured by a clock moving with the system measured. For Einstein, as for Newton, absolute time is the absolute order of physical events, determined by causality or exactly equivalent to causality, and not the *measurement* of time, which is the subject of ordinary observation. Newton's absolute time is no different, in principle, from the Einsteinian order determined by causality. Of course, a complete causal sequence can never be established, because we never have complete information about a physical situation through measurement, but such would certainly exist on a cosmic scale, and Einstein's general theory, which extended the special theory of relativity to incorporate gravity, also presupposes it.

Newton's absolute space and time are aspects of the abstract system, not the measurement process applied to individual events. The argument that special relativity does away with Newton's absolute time is fallacious because absolute time, in the same sense, is also required by Einstein's theory. Since Einstein's theory upholds the principle of causality, it must also allow an absolute sequence of events, which is the meaning of Newton's absolute time. Arguments about simultaneity are irrelevant because there is no absolute simultaneity in Einstein's system; the twin paradox is always solved by invoking a fundamental asymmetry between the two participants. Further justification for the Newtonian concepts is to be found in the concepts of quantum nonlocality and entangled quantum states; here, the absolute sequence of events is maintained by the indeterminacy caused by an instantaneous universal 'interaction' at a distance, just as in Newton's theory.

3.3. Acceleration and Rotation

With absolute space and absolute time defined, Newton was able to define true or absolute motion as motion through absolute space. Newton's opponents did not dispute that true or absolute motion must exist, believing rather that it must in some way be derivable from relative motions — a point he denied. Contrary to later opinion, they did not believe that all motion was relative. Newton's particular point was that absolute motion (which his opponents accepted) required absolute space (which they denied) and should be referred to absolute places.

Newton's view was that acceleration was fundamental, though velocity was not; and that absolute acceleration could be detected through rotational effects. His ideas were expressed in two famous thought experiments, the first of which involved a rotating water bucket:

> The effects which distinguish absolute from relative motion are, the forces of receding from the axis of circular motion. For there are no such forces in a circular motion purely relative, but in a true and absolute circular motion, they are greater or less, according to the quantity of the motion. If a vessel, hung by a long cord, is so often turned about that the cord is strongly twisted, then filled with water, and held at rest together with the water; after, by the sudden action of another force, it is whirled about the contrary way, and while the cord is untwisting itself, the vessel continues, for some time in this motion; the surface of the water will at first be plain, as before the vessel began to move: but the vessel, by gradually communicating its motion to the water, will make it begin sensibly to revolve, and recede by little and little, from the middle, and ascend to the sides of the vessel, forming itself into a concave figure (as I have experienced), and the swifter the motion becomes, the higher will the water rise, till at last, performing its revolutions in the same times with the vessel, it becomes relatively at rest in it.[22]

---

[22] *Principia*, Definitions, Scholium.

According to Newton's interpretation:

This ascent of the water shows its endeavour to recede from the axis of its motion; and the true and absolute circular motion of the water, which is here directly contrary to the relative, discovers itself, and may be measured by this endeavour. At first, when the relative motion of the water in the vessel was greatest, it produced no endeavour to recede from the axis; the water showed no tendency to the circumference, nor any ascent towards the sides of the vessel, but remained of a plain surface, and therefore its true circular motion had not yet begun. But afterwards, when the relative motion of the water had decreased, the ascent thereof towards the sides of the vessel proved its endeavour to recede from the axis; and this endeavour showed the real circular motion of the water perpetually increasing, till it had acquired its greatest quantity, when the water rested relatively in the vessel. And therefore this endeavour does not depend upon any translation of the water in respect of the ambient bodies, nor can any true circular motion defined by such translation. There is only one real circular motion of any revolving body, corresponding to only one power of endeavouring to recede from its axis of motion, as its proper and adequate effect; but relative motions, in one and the same body, are innumerable, according to the various relations it bears to external bodies, and, like other relations, are altogether destitute of any real effect, any otherwise than they may perhaps partake of that only one true motion.[23]

According to his account, Newton had actually performed the water bucket experiment, but the second thought experiment extends its context to the possibility of imagining rotation in an otherwise empty space:

For instance, if two globes, kept at a given distance one from the other by means of a cord that connects them, were revolved about their common centre of gravity, we might, from the tension of the cord, discover the endeavour of the globes to recede from the axis of their motion, and from thence we might compute the quantity of their circular motions. And then if any equal forces should be impressed at once on the alternate faces of the globe to augment or diminish their circular motions, from the increase or decrease of the tension of the cord, we might infer the increment or decrement of their motions; and thence would be found on what faces those forces ought to be impressed, that the motions of their globes might be most augmented; that is, we might discover their hindermost faces, or those which, in the circular motion, do follow. But the faces which follow being known, and consequently the opposite ones that precede, we should likewise know the determination of their motions. And thus we might find both the quantity and the determination of this circular motion, even in

---

[23] *Principia*, Definitions, Scholium.

an immense vacuum, where there was nothing external or sensible with which the globes could be compared. But now, if in that space some remote bodies were placed that kept always a given position to one another, as the fixed stars do in our regions, we could not indeed determine from the relative translation of the globes among those bodies, whether the motion did belong to the globes or to the bodies. But if we observed the cord, and found that its tension was that very tension which the motions of the globes required, we might conclude the motions to be in the globes, and the bodies to be at rest; and then, lastly, from the translation of the globes among the bodies, we should find the determination of their motions.[24]

The two experiments here described were not put forward as arguments for the existence of absolute space and absolute time, as defined earlier in the Scholium, but for the absoluteness of accelerative motion. The rotating bucket experiment is one of a series showing that absolute motion requires absolute space. The thought experiment of the revolving globes has a different purpose, being aimed at showing that one can determine the quantity of absolute motion of particular bodies, even though one cannot directly access absolute space.

Newton's absolute space and time, and his experiments indicating absolute rotation, were attacked within his lifetime, most famously by the philosopher George Berkeley, but there were numerous others who objected to his radical doctrine. Leibniz, for instance, wrote in a letter to Huygens of June 1694: 'Mr. Newton recognises the equivalence of hypotheses in the case of rectilinear motion, but with regard to circular motion he believes that the effort which revolving bodies make to recede from the axis of rotation enables one to know their absolute motion. But I have reasons for believing that nothing breaks this general law of equivalence'.[25] Leibniz has been portrayed as an early relativist, but there is no consistently relativist position in his writings; while some passages seem to suggest relativistic inclinations, others suggest quite opposite conclusions.

An idea now known as Mach's principle, first emerged during Newton's lifetime in Berkeley's *De Motu* (1721). Berkeley argued against Newton's absolute space and his interpretation of the water bucket experiment, stating that the inertial properties of matter with respect to the fixed stars (effectively, the rest of the universe) were a result of a long-range interaction between them. Einstein was influenced by Mach's revival and extension of

---

[24]*ibid.* Newton, of course, doesn't assume that the stars *are* fixed.
[25]Leibniz to Huygens, June 1694, C. J. Gerhardt (ed.), *Leibnizens mathematische Scriften*, 2: 184. Quoted Alexander (1956, xxvi).

the argument in 1872,[26] but was unable to incorporate the principle into general relativity.

As there is yet no established fundamental theory of physics it is premature to say we know how to explain space, time, and other fundamental quantities. Many people have nonetheless used a few comparisons between concepts defined according to the needs of different research programmes to make comments about their meaning in the entirety of physics. They have used theories of limited application to make general pronouncements on which aspects of space and time have survived into present-day physics, despite the fact that other aspects of physics would contradict their pronouncements. In fact, absolute space, time and motion, as well as relative space, time and motion, as defined by Newton, are very much alive and well at the present time, and Newton's belief that absolute rotation has a special significance appears to have held up as well.

Absolute space seems to correspond to something like aether in seventeenth-century language, or vacuum in our own. In this sense, the Higgs field, the zero-point energy and the cosmic microwave background radiation are all versions of absolute space. Ultimately, one of the most interesting things about Newton's discussion of fundamental quantities, at a more profound level, is that he finds physics needs *two* spaces, rather than one. One is fixed or 'conserved', the other is variable (translation-rotation symmetric and a source of gauge invariance according to modern ways of thinking), in effect acting like an infinite number of local spaces. Now, if we look at the way contemporary physics is constructed, say in relativistic quantum mechanics, we see that two spaces are needed to describe particles or singularities — space and vacuum space, one of which is relative and local, while the other is absolute and nonlocal.[27] We could quite reasonably call them relativistic space and absolute space. They may not correspond exactly with what Newton describes in the *Principia*, but they suggest that the Newtonian philosophical distinction still has validity in modern terms.

Absolute time is the sequence of events which remains unaltered in all physical theories, whether classical, relativistic or quantum. It is the time of Newton's calculus, and also that of quantum mechanics, where time is *not* an observable. Commentators have used special relativity as the theory which decides on the nature of space and time in a philosophical sense, but this can

---

[26]Berkeley (1721), Mach (1911, 79).

[27]The idea that there are two fundamental space-like parameters in nature has been a component of my own work from a very early period. Highly developed versions can be seen in Rowlands (2007, 2010).

only be done if we isolate it from the rest of physics, and take no account of such things as quantum mechanics and thermodynamics. However, at a more fundamental level, there is no special relativity, only relativistic quantum mechanics, and at this level the picture looks very different, with time taking on quantum characteristics, though the space-time connection remains. Robert DiSalle has, interestingly, argued that the space-time of *general* relativity is just as absolute as that of Newton. General relativity, he says, does not make space, time and motion 'generally relative' as Einstein thought, as it postulates a space-time which is an objective structure, every bit as absolute as Newton's.[28]

Newton seems to have been correct in saying that rotation was the aspect of motion that demanded an absolute condition. Even the rotating Kerr black hole of general relativity requires an absolute space to rotate in.[29] In special relativity, according to Brian Greene, Newton's water bucket would rotate and produce a concave water surface, even if the universe was otherwise completely empty, because it would be rotating with respect to *absolute space-time*. In general relativity, as normally interpreted, space-time is not absolute, as it is in special relativity, because it incorporates the effects of matter in the Einstein field equations, and the reference is instead made to observers in free fall. However, for a universe which is otherwise empty, there will be no gravitational field, and so the conditions will become the same as in special relativity; absolute rotation will still occur and, again, produce a concave water surface. In the 'Lense-Thirring', or frame-dragging, effect, first proposed by Einstein in 1912, in connection with his developing ideas of general relativity, a rotating massive object will drag space around with it, and, in the case of a bucket of water, drag the water, so, once again, making its surface concave.[30] On all counts, Newton seems to be correct against the objections of critics such as Ernst Mach.

Though the centuries-old privileged status of circular motion, based on everyday observation, had been replaced by rectilinear inertia in Newtonian dynamics, Newton here glimpses a new fundamental significance for rotation and angular momentum, which has persisted into quantum mechanics in the intrinsic significance of Planck's constant and the spin of fundamental particles. In general, the conservation of angular momentum does not seem

---

[28] DiSalle (2002, 34–35).

[29] The angular momentum of a black hole, like the mass and charge, must also be accessible instantaneously or we would never observe it. This is equivalent to defining it in an absolute space and time. The time-delayed effects occur with *changes* in the gravitational field, not with its static configuration.

[30] Greene (2004, pp. 51, 56–58, 72–74, 416–418).

to be affected by relativistic arguments, which makes angular momentum a more absolute quantity than either energy or momentum. Angular momentum is an axial or pseudovector, and its 4-D equivalent is not an axial 4-vector (a mathematical object which does not exist) but a rank 2 tensor or 2-form. The fourth component of this object is the product of $-ct$ and momentum, or the mass-moment of the system with respect to the centroid (mcr). For a single particle or for angular momentum observed about the centre of momentum or centroid of the system, this becomes 0. The quantum theory, of course, is based on single particles which absorb or emit angular momentum ($\hbar$) or action (h) in single units which are independent of any frame or relationship between space and time. We can in addition, detect absolute acceleration, in principle, via the Unruh effect (of which Hawking radiation is a special case),[31] and we can detect absolute motion via the cosmic microwave background radiation, if the frame of this radiation is the same as the aether frame. In this sense, it is interesting that temperature in effect determines time, as the absolute order of events is determined by increasing entropy.

---

[31]In this effect, the acceleration of an object through the vacuum creates a temperature difference between its two ends in the direction of the acceleration.

Chapter 4

Mass, Momentum and Energy

4.1. Mass and Momentum

If one concept stands out in the whole of Newton's work it is the invention of mass, or, more strictly, the introduction of mass as a parameter on the same level as the age-old principles of space and time. Simple yet profound, this is not at all an obvious thing to do, but it is the step which makes physics universal. Mass introduces the property of conservation, which can be described in absolute terms, leading to a whole series of conservation principles, which are invariably the bases of dynamical systems.[1] The conserved property of mass also shows up the contrasting, nonconserved natures of space and time, which are incorporated into physics in the differential forms with which these quantities are associated in fundamental equations. Newton's second law merges the contrasting properties of the fundamental parameters by defining force as a product of the conserved mass and the differential forms of the nonconserved space and time, and then establishes the conservation property of mass through the third law (which, through the notion of mass as 'impressed force' was directly equated by Newton with mass conservation).

All previous theories of mechanics had been kinematic only — they excluded the concept of mass and derived all physical laws only in terms of changes in measured values of space and time — Huygens, for example, solved the problems of impact by very skilfully *avoiding* the concept we now call momentum. Descartes, though he did have some idea of momentum, had made this approach inevitable by defining matter as extension only. Galileo and Kepler had discovered laws relating distances and times where

---

[1]See, for example, Rowlands (2007, 39–40), on the elemental significance of mass. The space, time, mass basis of physics was later expressed in dimensional analysis and the fundamental constants c, h and G.

the individual measurements could be expressed in terms of integers or rational numbers; they were unable to explain or extend these laws because they had no true idea of the agency of mass in producing the changes which they recorded in distance with respect to time in the particular cases they considered, and, even if they had found this, they would not have had the mathematics to develop it. Their laws were *kinematic*, rather than *dynamic*; they described *effects* but made no headway in explaining their *causes*. The effects were not universal, and depended on the particular local conditions which they described; as Newton showed, universality could only be attributed to the causes. To put it another way, the incorporation of mass as a fundamental parameter on the same level as space and time required the further realisation that the mathematics of variation or *real numbers* had also to be incorporated directly into physics.

The introduction of mass was also the introduction of *continuity* into mathematical physics; the *universal* interaction between masses and the universality of mass itself means that gravity is a continuum process. Though Newton, like many contemporaries, was a committed atomist, his interactions between masses are between point-elements of mass, infinitesimally small mathematical constructs in a continuum field, and not between discrete particles. It is noticeable that the version of calculus developed by Newton to apply to his physical theory is a continuum one, as opposed to the discrete one normally associated with Leibniz.

No one had previously used the concept of mass in this fundamental way, and it proved to be crucial, for many concepts and laws depend on it. Descartes' extension was clearly not enough. In *De gravitatione et aequipondio fluidorum*, a work directed against Descartes, Newton says: '...spaces are not the very bodies themselves but are only the places in which bodies exist and move...'[2]

> ...we may imagine that there are empty spaces scattered throughout the world, one of which, defined by certain limits, happens by divine power to be impervious to bodies, and *ex hypothesi* it is manifest that this would resist the motions of bodies and perhaps reflect them, and assume all the properties of a corporeal particle, except that it will be motionless. If we may further imagine that that impenetrability is not always maintained in the same part of space but can be transferred hither and thither according to certain laws, yet so that the amount and shape of that impenetrable

---

[2] *Unp.*, 148. Janiak (2008, 102–129), characterizes mass, rather than force, as the crucial, 'revolutionary', concept in Newton's philosophy because it means that mechanism can be transcended. From my own point of view as a physicist, it is crucial because it allows the purely abstract mode of reasoning that alone has been successful to be universalized.

space are not changed, there will be no property of body which this does not possess. It would have shape, be tangible and mobile, and be capable of reflecting and being reflected, and no less constitute a part of the structure of things than any other corpuscle.... In the same way if several spaces of this kind should be impervious to bodies and to each other, they would all sustain the vicissitudes of corpuscles and exhibit the same phenomena. And so if all this world were constituted of this kind of being, it would seem hardly any different.[3]

In the *Principia* he stated: 'All spaces are not equally full.... And if the quantity of matter in any given space can, by any rarefaction, be diminished, what should hinder a diminution to infinity?'[4]

Newton's definition of mass as quantity of matter is fundamental, as it requires the concept of density, the quantity of matter in a given volume, to relate it to space: 'The quantity of matter is the measure of the same, arising from its density and bulk conjunctly'.[5] It arises very early in Newton's work, in *Quaestiones quaedam philosophicae*, along with the distinction between mass and weight, which subsequently led to the dependence of gravity upon mass.[6] Earlier ideas of density had treated it as a purely relative quantity, not an absolute one. Several other concepts were defined by Newton for the first time with the new understanding of mass playing a crucial role in their construction. The first was the quantity now known as momentum, measured as the product of the mass of a body and its velocity ($\mathbf{p} = m\mathbf{v}$). 'The quantity of motion is that which arises from the velocity and quantity of matter conjointly'.[7] Momentum would play a significant role in Newtonian dynamics, as, together with position, it allowed a complete specification of a dynamic system, the combination of the two quantities now being described as 'phase space'.[8]

---

[3] *Unp.*, 139. It is interesting that Newton here conceives (as General Relativity would do later) that space *itself* could be so constructed as to have the attributes that we associate with material bodies — principally because all physical observation is channeled through the medium of space.

[4] III, Proposition 6, Corollary 3.

[5] *Principia*, Definition 1.

[6] QQP, MS Add. 3996, 'Of Motion','Of Attomes'. 'The gravity of bodys is as their solidity, because all body{s}descend equall spaces in equal {times}consideration being had to the Resistance of the aire & c.' The concept of mass could not emerge until it was separated from weight, under which it had been subsumed by earlier authorities such as Galileo and Descartes, and weight recognised as a force.

[7] *De motu corporum*, c 1684 (MS A), Definition 2; *Unp.*, 241.

[8] Gutzwiller (1992) says of phase space: 'Quite amazingly this concept, which is now so widely exploited by experts in the field of dynamic systems, dates back to Newton'. Phase

4.2. The Laws of Motion

Classical mechanics is famously founded by Newton on the basis of three laws of motion coupled with the inverse-square law of gravitation. Again, the laws of motion arise very early in his work, as though they represent a fundamental level in his thinking from which, once established, he would never deviate.[9] Laws, in fact, are 'established' rather than 'discovered'. Law means statement as an axiom or fundamental principle; a previous statement, whether in full or in part, cannot be considered an anticipation unless specifically stated as a fundamental principle. Though they are of great importance, Kepler's 'laws' of planetary motion are of a quite different kind, phenomenological in origin rather than fundamental, an accidental result of special conditions and derived empirically rather than from basic principles. Newton defined three axioms which codified dynamics, in the same way as Maxwell would produce a series of equations to codify electromagnetism. The definition of such principles as axioms is a separate exercise from discovering them individually.

Interestingly, Newton's laws of motion were not quite used by Newton in the way we would use them today, as foundational axioms on which he would develop a rigidly deductive science of mechanics. They were rather a codification of a practice with which he was already familiar. The same also applied to his versions of the differential and integral calculus (or, in his terminology, the method of series and fluxions).[10] In contrast with the work of Leibniz, the practice, with Newton, usually long preceded the codifications. This has led to a great deal of confusion in subsequent commentators who have been baffled by the existence of two versions of Newton's second law of motion, only one of which is formally stated in an explicit way, and by his rather cavalier attitude to his own ownership of them. Clearly, for Newton, the first two laws, at least, were almost a statement of the obvious; he was able to do calculations without needing to keep referring to statements of them.

Newton's first law of motion is a version of the principle of inertia, which, as he recognised, had ancient antecedents. Here, a body continues in its state of rest or uniform motion in a straight line unless acted on by an external force, or, as Newton himself expresses it: 'Every body perseveres in its state of rest, or of uniform motion in a right line, unless it is compelled

---

space is also implicit in the phase plane diagram of *Principia*, Book I, Propositions 30 and 31.

[9] *The Lawes of Motion*, Herivel (1965, 208–215).

[10] See Guicciardini (2009, 239, 379–380), on laws of motion, *passim* on calculus.

to change that state by forces impressed thereon'.[11] Newton says that: 'All those ancients... attributed to atoms in an infinite vacuum a motion which was rectilinear, extremely swift and perpetual because of the lack of resistance'.[12] The law was preceded by many statements of the principle of inertia, of varying degrees of accuracy, but Newton was the first to establish it as an axiomatic law of motion. An intuitive understanding of inertia may be ancient; Newton claimed Lucretius and other classical authors as predecessors, and earlier in the seventeenth century, as is well known, Galileo and Descartes had made their contributions. Newton's version, however, differed from his predecessors' because the Newtonian understanding of forces is needed for a true statement. The first law of motion cannot be stated without the second — and this specifically as a momentum law. To understand what happens in the absence of force, you have to have a definition of force.

Newton describes inertia as a principle in itself: 'Matter is a passive principle and cannot move itself. It continues in a state of moving or resting unless disturbed. It receives motion proportional to the force impressing it, and resists as much as it is resisted'.[13] 'A body, from the inactivity of matter, is not without difficulty put out of its state of rest or motion'.[14] 'Inertia is force within a body, lest its state should be easily changed by an external exciting force'.[15] 'The innate force of matter is the power of resisting by which a body continues, as much as in it lies, in its normal state, either of resting or of moving uniformly in a straight line....a body only exerts this force in the change of its state made by means of another force impressed upon it...'[16]

The second law of motion provides the definition of force. In *The Lawes of Motion*, his earliest manuscript on dynamics, Newton says that '...force is equivalent to that quantity of motion wch it is able to beget or destroy'.[17] And in *De gravitatione et aequipondio fluidorum*, he explains that: 'Force is the causal principle of motion and rest. And it is either an external one that generates or destroys or otherwise changes impressed motion in some body; or it is an internal principle by which existing motion or rest is conserved

---

[11] *Principia*, Law I. 'Impressed forces' were described, in Definition 4, as the ones causing such changes.

[12] Manuscript fragment on Law of Inertia for *Principia*; *Unp.*, 310.

[13] Draft Q 23 for *Opticks*; McGuire (1968), 171.

[14] *Principia*, Definition 3.

[15] *De gravitatione*, Definition 8; *Unp.*, 148.

[16] *De motu corporum*, c 1684 (MS A), Definition 3 (Law I); *Unp.*, 241.

[17] *Unp.*, 157.

in a body, and by which any being endeavours to continue in its state and opposes resistance'.[18]

In our language, and in the bulk of Newton's usage in the *Principia*, force is the rate of change of momentum:

$$F = \frac{d\mathbf{p}}{dt}.$$

This simple algebraic definition of force postdates Newton but is equivalent to his usages. According to the official statement of Law II in the *Principia*: 'The alteration of motion is ever proportional to the motive force impressed; and is made in the direction of the right line in which that force is impressed'. '... and the exercise of this force may be considered as both resistance and impulse...'[19] Here, the force is defined as an impulse ($\Delta(mv)$ or $d(mv)$), with an implied infinitesimal time interval (Δt or dt) to convert it to continuous force ($\Delta(mv)/\Delta t$ or $d(mv)/dt$). In Definition 8 of the *Principia*, for example, he writes: 'The motive quantity of a centripetal force, is the measure of the same, proportional to the motion which it generates in a given time', which specifies the definition as we now use it; and also: 'the motive force arises from' the product of the 'accelerative force' and the 'quantity of matter'.[20] In effect, Newton considered a continuous force to be equivalent to an infinite succession of infinitesimal impulses,[21] which is interestingly close to the view of contemporary quantum physics.

In proving Proposition 24 of *Principia*, Book II, Newton explicitly uses a continuously acting force in writing that 'the velocity that a given force can generate in a given [quantity of] matter in a given time is as the force and the time directly, and the matter inversely', with the specific statement that this is 'manifest from the second Law of Motion'.[22] There is also a manuscript in which he comes close to the modern definition of force in symbolic calculus as ma or $m\ddot{x}$. In considering a case of unresisted linear motion of a body (presumably of unit mass)[23] at a vertical distance y from a central force of gravitational attraction, he says that, 'if the body ascends or descends straight up or down its speed will be $\dot{y}$ and gravity $\ddot{y}$. For the

---

[18]Definition 5; *Unp.*, 148.

[19]Law II; Definition 3.

[20]Definition 8.

[21]*MP* VI, 98. Guicciardini (2009, 239) considers that it may derive from his 'endorsement of atomism', with 'impacts between hard atoms' causing 'instantaneous changes of velocity'.

[22]Book II, Proposition 24. Cohen (1999, 112).

[23]An often unstated assumption in Newton's work.

fluxion of the height is the body's speed and the fluxion of the speed is as the body's gravity'.[24] Seventeenth century scientists were seldom concerned about dimensional exactitude in the way we are today, and, though Newton's definition of force implied an impulsive or instantaneous quantity, he also used it in the infinitesimally equivalent sense of a quantity acting over time, or, as we should say, a quantity proportional to rate of change of momentum.

In effect, Law I, as stated, refers to continuous forces, and Law II to impulsive or instantaneous forces. Both laws are needed for the complete picture; and both definitions are assumed to be valid in both laws; the easy mathematical transition from instantaneous to continuous forces means that this is not a fundamental distinction. In addition, forces, whether treated as impulses $F \times dt$ or as continuous quantities F, usually involve ratios in specific calculations in the *Principia*, so no dimensional problems emerge. In fact, the forces as defined in Definition 8 and Law II are *dual* quantities, one of many dualities to be found in Newtonian physics. Ultimately, the first is used to explain how a causal change can be made from one physical state to another, while the second explains how any change can occur at all. The duality extends to the quantities which encompass the entire system by combining momentum and position — energy for the continuous and action for the impulsive, or torque for the continuous and angular momentum change for the impulsive — and the two possibilities for doing this, which we now call the vector and scalar products, form another duality with a significant role in Newtonian physics. I. Bernard Cohen believed that emphasis was given to impulsive forces in the *Principia* to make them acceptable to contemporaries, who would not accept action at a distance,[25] but the Newtonian definition of an independent free-flowing time parameter means that the two force definitions can be used interchangeably in all physical situations.

The second law, in either statement, introduces calculus into physics. Space and time need to be differentials because they are variables; time being the independent variable, as implied by Newton's preferred version of calculus, it finds itself in the denominator. (Newton had, of course, developed calculus, largely to understand the idea of motion, his first tract on calculus being entitled 'An attempt to measure by motion'.) In addition, the two definitions of force, discrete and continuous, also reflect the two methods

---

[24] Add MS 395.6 ff. 38^r–39^r, *MP* VII, 128 (Latin), 129 (translation). Guicciardini (2009, 240) calls it 'a prototype of $F = ma$'. Guicciardini (1999) discusses this and related manuscripts on pp. 108–112.

[25] Cohen (1999, 111–113). Pourciau (2016, 94), quotes a manuscript of the 1690s (MP VI, 538–609) which he says makes clear the fact that Newton's statement of the second law requires both impulsive and continuous forces.

of calculus which are specifically outlined in Newtonian mathematics, the discrete algebraic method of infinitesimals (moments or fluxions), and the continuous geometrical method of limits (prime and ultimate ratios). In either method, the Newtonian definition of force means that physics is henceforth described primarily by differential equations.

In the particular case where mass is a constant, force is mass $\times$ acceleration:

$$\mathbf{F} = m\frac{d\mathbf{v}}{dt} = m\mathbf{a}.$$

Definition 7, which is concerned with the accelerative quantity of centripetal force, requires the continuous force definition of the second law ($\mathbf{a} = \mathbf{F}/m$), rather than the impulsive, while gravitational force is taken as proportional to the quantity of motion in a given time, $d(m\mathbf{v})/dt$.[26] While Sections I-X of *Principia*, Book I are kinematical, requiring the accelerative quantity, Section XI changes to dynamics as the accelerative force becomes the motive force. (Newton also has a third measure of force, the absolute value, which is significant in those cases, such as magnetism, where the force is not a function of mass.)[27] However, by defining force primarily in terms of the momentum equation, Newton actually avoided the limitations now associated with its formulation as the product of mass and acceleration. He also considered cases where the force involves a changing mass, for example, in reaction propulsion and water coming from a tank, which is either described by the formula $\mathbf{v}\,dm/dt$, or has an additional component of this nature.[28]

Newton was also precise in defining force, displacement, velocity, momentum and acceleration as *vector* quantities, composed or added by describing the diagonal of a parallelogram, with the first formal statement of the parallelogram law in Corollary 1 to the Axioms, or Laws of Motion in the *Principia*, and, in resolving forces into components along arbitrary directions, in Corollary 2, was the first to recognise the affine structure of space. The space he used became specifically three-dimensional in his proof in the third book of the *Principia* that the gravitational action of a sphere

---

[26]Cohen (1999, 104).

[27]Definition 6. Cohen (1999, 105). The use of the accelerative quantity in I–X also reflects the fact that, as Densmore emphasizes (2003, 5–6 and passim), Newton tended to write his results in *proportions* rather than equations.

[28]Pierre Varignon, who attributed Newton's second law to *Principia*, I, Proposition 39, stated it in the form $F = m\,dv/dt$ in 1700. This formula was subsequently given again by Bernoulli in 1710 and 1727, where it was described as 'well-known', by Jakob Hermann in 1716, and by Euler in 1736 and 1750, where it was claimed as 'new' because it was being applied in a new context (Harman, 1988, 84).

acts as though all its mass were concentrated at the centre, and Immanuel Kant subsequently showed that inverse-square laws necessarily presupposed a space of dimension 3.

The first part of Corollary 2 states that any two mechanical forces will produce a resultant force by vector addition or composition. The fact that a load pulled by two individuals in different directions would move in a direction between them (which, in the theory presented in Corollary 2, can be constructed from the diagonal of the parallelogram representing them) must have been known to many before the seventeenth century, no doubt including Aristotle. The second part states the far from obvious converse, that the force in any plane can be resolved into non-parallel component forces. Though the original force may be a real one, along an observed line of action, the components can be purely 'fictitious' ones, set along arbitrarily-selected lines of action, an idea without precedent. The same, of course, applies to velocity, momentum, acceleration and displacement. It is clear here that Newton is the originator of the mathematical idea of the vector, though the version found in modern textbooks may date from as late as the twentieth century.

In 'the short space' of this corollary, Newton, according to John Roche, 'dealt with virtually the whole science of statics'. It is, he says, 'an astonishingly condensed piece of analysis which uses a single diagram with multiple interpretations. One is immediately struck by Newton's casual mastery of his medium, and by the density and economy of his style'.[29] The properties of the balance, the lever, and the wheel, and also the screw, are described, and, using the resolution of forces into components, the law of the lever or principle of moments is seen to be a consequence of the parallelogram of forces. Newton concludes that:

> the use of this Corollary spreads far and wide, and by that diffusive extent the truth thereof is further confirmed. For on what has been said depends the whole doctrine of mechanics variously demonstrated by different authors. For from hence are easily deduced the forces of machines, which are compounded of wheels, pullies, levers, cords, and weights, ascending directly or obliquely, and other mechanical powers; as also the force of the tendons to move the bones of animals.[30]

---

[29] Roche (1988, 52).
[30] Axioms, or Laws of Motion, Corollary 2. The Corollary gives a lengthy description of the device for illustrating the composition and resolution of force now known as 'Newton's wheel'.

It was not his first venture into statics. Perhaps about a decade earlier, he already had enough information to resolve the general problem of a string under arbitrary loading, for in a lecture supposedly given about October 1678, he had shown that a string carrying a load and supported at each end could be explained by Stevin's principle of equilibrium in static tensions.[31]

Newton recognised that the parallelogram or vector addition law was not a direct consequence of the laws of motion, though it was 'abundantly confirmed from mechanics'.[32] In fact, it is nothing to do with forces as such, as it applies also to space itself and to all quantities incorporating vector space in their definition. The law also implied that forces could be added according to a principle of linear superposition, a principle which became particularly significant in deriving the total gravitational effect of large-scale bodies and systems.[33]

While Newton's second law is original in providing the first valid mathematical definition of force, and his first law is original in requiring this definition for its complete statement, his third law is original in an entirely new way. It is a universal law, which uniquely defines the nature of a physical system. Law III in the *Principia* states that: 'To every action there is always opposed an equal reaction; or the mutual actions of two bodies upon each other are always equal, and directed to contrary parts'. And explains that: 'This law takes place also in attractions...', that is, in actions at a distance where there is no contact. Newton clarifies this in *The System of the World*: '...all action is mutual, and (... by the third law of motion)...must be the same in both bodies. It is true that we may consider one body as attracting, another as attracted; but this distinction is more mathematical than natural. The attraction resides really in each body towards the other, and is therefore of the same kind in both'.[34] 'The iron draws the loadstone, as well as the loadstone draws the iron; for all iron in the neighborhood of the loadstone draws other iron. But the action between the loadstone and iron is single, and is considered single by philosophers'.[35] Another way of stating the principle

---

[31]Lectures on algebra, 1673–1683, 1, 2, §1, *MP* V, 282.

[32]*Principia*, Book I, Axioms, or Laws of Motion, Corollary 2. This is a particularly interesting observation in relation to the simultaneous existence of an infinite number of centres of mass in the universe.

[33]Pask (2013, 127–128), discusses this, noting that Sommerfeld (1942) wanted to define the parallelogram law as a fourth law of motion.

[34]§20.

[35]§21.

is given in *The Elements of Mechanicks*: 'The actions of two bodies upon one another are always equal and directly contrary'.[36]

The third law of motion implies that the action and reaction are part of a single action. This is related to the fact that potential energy is twice kinetic in an inverse-square law system. It is also connected to the fourth Corollary to the Laws of Motion, which defines the common centre of gravity of a system of interacting bodies, which is not acted upon by an external force, as being at rest or in uniform motion in a straight line. 'The common centre of gravity of two or more bodies does not alter its state of motion or rest by the actions of bodies among themselves; and therefore the common centre of gravity of all bodies acting upon each other (excluding outward [i.e. external] actions and impediments) is either at rest, or moves uniformly in a right line'. The proof of this theorem, which was vital to the development of Newtonian dynamics, required Lemma 23, defined later in Book I, and was completed by induction.[37]

Despite statements to the contrary, the third law still holds even in the case of the Lorentz force of electrodynamics, if the force is defined as being between an individual current element and a closed current loop external to it. The third law is manifestly universal, as it acts on all bodies equally. Newton realised that it must apply to *every* action and reaction. This was a far from obvious concept; even Newton's brilliant disciple Roger Cotes found it hard to accept. There are many common occurrences, of course, such as colliding bodies, in which action and reaction can be observed in particular cases, but only Newton made it a universal principle. The laws of motion, taken together and, in effect, considered as a *single* law, define an absolute or *extremum* condition: there is no force in a system. Such null principles are among the most powerful in physics, because they can be defined absolutely.

It would seem that, in addition to requiring a universal law of gravitation, Newton's system of dynamics is also universal, in being composed of a law defining a quantity, force, which he then shows must be zero within a system. There is necessarily, therefore, a zero force totality in the entire universe. We can even consider the law as referring to the reciprocal action and reaction between any defined body or system and *the rest of the universe*, and, only by approximation, referring to the action and reaction between two bodies. The fact that it actually applies to finite systems as well as universal ones, and to systems within systems, could be taken as an indication that Nature repeats similar patterns at all scales, as Newton himself believed. The universal

---

[36] *Unp.*, 165.
[37] Axioms, or Laws of Motion, Corollary 4.

validity of the third law at all scales, in fact, appears to be one of the most significant organizing principles in the whole of Nature.

The law also meant that the centre of gravity of mutually attracting bodies would constitute an inertial frame, either at rest or moving with uniform motion in a straight line:

I have hitherto been treating of the attractions of bodies towards an immovable centre; though very probably there is no such thing existent in nature. For attractions are made towards bodies, and the actions of the bodies attracted and attracting are always reciprocal and equal, by Law III; so that if there are two bodies; neither the attracted nor the attracting body is truly at rest, but both (by Cor. 4, of the Laws of Motion), being as it were mutually attracted, revolve about a common centre of gravity. And if there be more bodies, which either are attracted by one single one, which is attracted by them again, or which all of them, attract each other mutually, these bodies will be so moved among themselves, as that their common centre of gravity will either be at rest, or move uniformly forward in a right line'.[38]

In the system of the world, this meant that the correct theory of the Solar System was neither geocentric nor heliocentric, but one in which the revolution of the heavenly bodies was about the centre of gravity of the entire system. Even the Copernican theory had now been corrected as the result of an even more profound revolution. The Sun itself was 'agitated by a perpetual motion', though it never receded 'far from the common centre of gravity of all the planets',[39] and it was only a working hypothesis that this centre was itself immovable. The real 'centre of the world' (i.e the Solar System) was 'the common centre of gravity of the Earth, the Sun, and all the planets'.[40]

A practical aspect of the third law was reaction propulsion, as applied to rockets and to vessels propelled by water. Newton explained this in Proposition 37 of Book II of the first edition of *Principia*:

And further if a [tall] vessel [filled with fluid] be suspended by a very long thread from a nail, like a pendulum, when water flows out from it in any horizontal direction the vessel will always recoil from the perpendicular in the opposite direction. And the cause of the motion of those darts [rockets] that are filled with damp gunpowder is the same: gradually expelling the

---

[38]I, Section XI, preliminary words.
[39]*Principia*, III, Proposition 12.
[40]*Principia*, III, Hypothesis 1, Proposition 11, Corollary to Proposition 12.

material through a hole in the form of a flame, they recoil in the opposite direction from the flame and are violently propelled away from it.[41]

Though he showed no particular interest in the steam engine, developed late in his lifetime, Newton proceeded to create the theoretical possibility of steam reaction propulsion somewhat by accident, after Denis Papin had proposed to the Royal Society that a boat weighing 80 tons could be built 'to be rowed by oars [i.e. paddle wheel] moved with heat' (11 February 1708), based on an idea he had published in the *Acta Eruditorum* for 1690.[42] In the following month he proposed to install an engine for use in 'the moving of ships' but without any suggestion that it was based on 'a reaction principle'. Newton, the President, was asked to comment and his draft (in the Cambridge University Library) discusses the 'towing & moving of ships & Galleys by the recoil of the engine & force of the Steam duly applied'. In Rupert Hall's opinion, Newton had 'misinterpreted Papin's proposal and himself introduced the idea that the boat was to propelled by reaction, thus inventing reaction propulsion almost by inadvertance'.[43] 'sGravesande subsequently extended the idea further to a road carriage, in *Physices Elementa Mathematica* ... (1720–1721), leading to the idea that Newton had invented such a vehicle.[44] Hall speculates that, if Newton really did contribute to this idea, it would have been when 'sGravesande was in England, between March 1715 and February 1716,[45] during which period he was elected to the Royal Society and met the President, and involved Newton in another thermodynamic area — the discussion of perpetual motion.

The third law led to an extremely profound insight by Newton, which has not been understood up to the present day, that mass was actually also a *force*. Definition 3 of the *Principia* introduces *vis insita* ('inherent force'). 'The *vis insita*, or innate force of matter, is a power of resisting, by which every body, as much as in it lies, continues in its present state, whether it be of rest, or of moving uniformly forwards in a right line'. Newton wrote that the '*vis insita*' may be known by 'a most significant name, *vis inertiæ*'. In explaining it, he says:

This force is ever proportional to the body whose force it is and differs nothing from the inactivity of the mass, but in our manner of conceiving it.

---

[41] *Principia*, III, Proposition 37, first edition; Hall (1985). Reaction propulsion requires the force $dp/dt = v \, dm/dt$.

[42] Papin, Royal Society, 11 February 1708; Hall (1985).

[43] Hall (1985).

[44] Book IV, Ch. X.

[45] Hall (1985).

A body, from the inert nature of matter, is not without difficulty put out of its state of rest or motion. Upon which account, this *vis insita* may, by a most significant name, be called inertia (*vis inertiae*) or force of inactivity. But a body only exerts this force when another force impressed upon it, endeavours to change its condition; and the exercise of this force may be considered as both resistance and impulse....

Mass, in Newton's theory, is clearly something other than the mere quantity of matter, and here he gives it force-like properties, creating an intrinsic 'reaction' concept from his third law of motion, with the 'inherent force' representing the other half of the system to the 'impressed force'. The relation between matter and force in the theory allows something outside of matter and opposing it, which can be treated abstractly. In later physics, there is always something other than matter, which has this characteristic. In the nineteenth century, it might be the field concept or aether; in the twentieth- and twenty-first centuries it might be energy or vacuum. A version of it comes into particle physics with the distinction between fermions and bosons. The ultimate origin is probably the distinction between mass and charge or gravity and gauge theory (which roughly correspond to the nineteenth-century aether and matter), but it is important that physicists have always found the need for something in opposition to the matter concept — even Descartes has to have an aether, though it is a material one. Newton's creation of an abstract principle of gravitation inherent in the mass concept first quantifies the necessary opposition in physics.

All of Newton's laws of motion, even taken individually, were original in significant ways, but, taken together, they constituted a structure with a power never previously seen in natural philosophy. Despite their apparent simplicity, they are far from obvious and are even counter-intuitive. They are one of many examples of Newton's special ability at creating systematic structures based on a prior identification of the most secure foundations. In relation to all previous attempts at understanding dynamical processes, which included work on falling bodies and mechanical impact, they indicated at once that all physical occurrences could ultimately be explained on the basis of relatively simple principles. No predecessor had developed anything resembling a universal system of dynamics, and no successor would create one which could not be referred back to Newton's laws.

The laws of motion were, of course, not the only way of doing 'Newtonian mechanics'. New versions of dynamics were developed in the century after Newton. These were based on new quantities, rather than on Newton's force, though he had already used several of them in more restricted senses; one was *action*, the product of momentum and distance *mvs*.

Following earlier considerations of action by Leibniz, Maupertuis and Euler both developed a Principle of Least Action, in 1744.[46] This was applied in both optics and dynamics, and effectively took the place of Newton's third law. Another such concept was energy; special aspects of the conservation of energy, were understood in the seventeenth century by such as physicists and mathematicians as Galileo, Huygens, and Leibniz, while Newton included general mathematical energy conservation theorems in the *Principia*. Gradually, it came to be recognised that this conservation principle, like the conservation of momentum, was another alternative to the third law. In each example of the new dynamics, in virtually exact repetition of the Newtonian pattern, there was always a law which defined a quantity (force, momentum, energy, action) and another which defined a universal condition to which the quantity applied, setting it to zero, or a constant value, or a maximum or minimum. In all applications, a further ('physical') law specified the operation of the source of the defined quantity, as Newton's law of gravitation had defined the operation in his system of the quantity force.

4.3. Centripetal and Centrifugal Force

Newton held force, as he had defined it mathematically, to be responsible for all physical effects. What he called impressed or inertial force, or resistance of a body to any change in its state of motion, was a passive principle to be differentiated from the various active principles, or fundamental forces of attraction and repulsion, which were the real causes for such changes. The programme of Newtonian physics was to apply the second and third laws of motion (or principles which are equivalent) to the particular laws that are relevant to these individual forces. The physicist, he said, was to use 'the phænomena of motions to investigate the forces of nature, and then from these forces to demonstrate the other phænomena'.[47] Physics has never deviated from this programme. Just four fundamental forces, of which gravity is one, are held to be responsible for the entire range of phenomena in the

---

[46]These, respectively, minimised the integral of *vis viva* (or $2 \times$ kinetic energy T) over time, and momentum p over spatial coordinate q. The modern definition, due to Hamilton, minimises or maximises, i.e. finds the stationary value of, the integral of the Lagrangian function $(T - V)$ over time. The claim that Leibniz preceded Maupertuis and Euler in the discovery of the principle, based on a non-autograph copy of a letter produced long after his death, but shortly after Maupertuis and Euler had made their pronouncements, still causes controversy (see Ramm, 2011; Breger, 1999).

[47]*Principia*, Preface.

universe; and our understanding of these forces is still just as abstract and non-mechanistic as was Newton's own.

In Newton's mature physics, active principles act by generating centripetal forces, or forces which act to accelerate a body towards (or, in some cases, away from) a point acting as a centre: 'I call that by which a body is impelled or attracted towards some point which is regarded as a centre centripetal force'.[48] (As so often with Newton, this definition was more general than would at first appear. The centre, for example, did not need to be the location of a physical source, just the point towards which it was directed.) This approach began with an early result for what was then regarded as *centrifugal* force in circular motion. This is described in *The Lawes of Motion*: 'If a body be moved in the perimeter of a given conic...' '...the real quantity of its circular motion is more or less as the body hath...power and force to preserve it in that motion; which motion divided by the body's bulk is the real quantity of its circular velocity'.[49]

Newton had quantified the force, on the first page of his Waste Book (Add. 4004), by imagining a body moving inside a square with a circumscribed circle, and then generating impacts at each reflection from the sides, which could be added to find the total force. He then imagined increasing the number of sides, until, in the limit, the body is 'reflected by the sides of an equilateral circumscribed polygon of an infinite number of sides (i.e. by the circle it selfe)'.[50] The same calculation would be recalled by Newton in the *Principia*, in the Scholium to Book I, Proposition 4.

Huygens had obtained a similar expression for the centrifugal acceleration at an earlier date, which he would publish in 1673, but Newton's early ideas on action and reaction were already leading him to think of such motion in terms of centripetal, as well as centrifugal, forces. In the Waste Book Axiom 20, Newton investigates the path of a ball moving on the inside surface of a hollow sphere. He finds there is equal pressure (the later action and reaction) between the ball and surface in both directions. 'Hence it appears that all bodies moved circularly have an endeavour from the center about which they move, otherwise the body...would not continually press upon ...' the sphere.[51] Equal and opposite 'centripetal' and 'centrifugal'

---

[48] *De motu* 1685, Herivel (1965, 277), Westfall (1980, 411).

[49] Manuscript on centripetal force, after 1687, Scholium, *Unp.*, 68; *The Lawes of Motion*, 1666, *ibid.*, 158.

[50] Waste Book, Add. 4004, f. 1, Herivel (1965, 129–131), quoting 130.

[51] Waste Book, 'Axiomes. And Propositions', Add. 4004, ff. 10ᵛ–11ᵛ, Herivel (1965, 141–159), f. 1, Axiom 20, *op. cit.*, 146–147, Axiom 21, quoting 147.

forces apply in the *special case* of circular motion, where the force is directed along the radius of the circle of curvature.

Very early in his work, and much earlier than scholars used to think, Newton, as a result of his development of calculus techniques, had a method of using the mathematical concept of 'curvature' to deal with curved orbits. His first steps to finding the curvature of a conic had been made as early as December 1664.[52] He had three approximate curvature methods — polygonal, parabolic and circular — and obtained solutions by graphical or iterative numerical analysis.[53] In the polygonal approximation, as already described, the impulse moves in a square, then an n sided polygon; n is finally extended to infinity. An important statement is in the Scholium to *Principia*, I, Proposition 4.[54] In the parabolic approximation, the force is taken over a small region, considered constant; the orbiting body is assumed to have a uniform tangential component of motion, on which the force acts with an inward acceleration, directed towards a centre, producing a parabolic path, which becomes circular if there is constant force or acceleration in that region. The parabolic approximation is used in *Principia*, I, 4, 6, and 10–11, where it is extended to any curved path. In the circular approximation, the curvature method solves elliptic motion by means of a tangent circle, or circle of curvature. The circular approximation is used in the first edition of *Principia*, in Book III, Proposition 28; it is also used in some parts of the second edition.[55]

Centrifugal force remained important to Newton, even after he had fully developed his idea of 'centripetal' forces. He used his old ball within a hollow sphere argument in the Scholium to Book I, Proposition 4, in the second edition of the *Principia*: 'This is the centrifugal force, with which the body impels the circle: and to which the contrary force, wherewith the circle continually repels the body towards the centre, is equal'. But he only ever considered it for circular orbits, and at maximum or minimum radius in others — the only points in a noncircular orbit where the force is directed along the radius of the circle of curvature.

Newton also used the term 'centrifugal force' for the inertial (tangential) component of curved motion, in the Scholium to *Principia*, III, Proposition 4; here the inertial component is just the body's inertia, not a force. Though it has often been assumed that Newton's early treatment of orbital motion

---

[52] *MP* I, 252–255.
[53] Brackenridge and Nauenberg (2002).
[54] Second edition, I, Proposition 4, Scholium.
[55] Cohen (1999, 73–74).

was based on a 'centrifugal force' outwards or Descartes' *conatus*, or that it was based on a balance of forces inwards and outwards, the outward force, such as it was, gradually changed into an inertial force (*vis inertiae*), or the inertia to motion, rather than a real force. The use of 'centrifugal force', rather than 'centripetal', has been misunderstood as a 'wrong' way of looking at the data rather than as an *alternative*. In Newton's case the centrifugal force becomes the *vis inertiae*.

There is, of course, another context in which the term 'centrifugal force' is used by contemporary authors. Since all real physical observations involve accelerating or noninertial frames of reference, it has often been found convenient to introduce 'fictitious' inertial forces, which enable us to preserve the form of Newton's laws, even in noninertial frames. The Earth, for example, as a rotating body, with angular velocity ω, is supposedly acted on by a fictitious centrifugal force ($m\omega v$ or mv^2/r), which causes it to become an oblate spheroid, a fact of which Newton was well aware. The force is fictitious because the conservation of angular momentum enables a frictionless body to continually maintain its rotation even without the action of any external agent. A second fictitious force, the Coriolis force ($2m\omega \times \mathbf{v}$), which is discussed in the Scholium to Book I, Proposition 2, is also caused by the Earth's rotation and is well-known as the cause of the deflection of the trade winds. It was predicted by Newton as the force which caused a falling body to be deflected to the east.[56] Newton thought that the effect might prove the rotation of the Earth if it could be observed, as was later accomplished with the Foucault pendulum.

Inspired by the apparent discovery of stellar parallax, reported in Hooke's *An Attempt to Prove the Motion of the Earth by Observations* of 1674, which would have provided a direct observation of the Earth's annual motion through space, Newton wrote to Hooke on 28 November 1679, proposing that the Earth's *diurnal* rotation could also be discovered by dropping a bullet from a high church tower or down a well.[57] The top of the tower would have a higher tangential speed than the base and so the effect would be measurable. Newton gave precise details for its observation. Classically, the object would have fallen to the west, as the Earth moved under it, but this had not been observed. Newton, however, believed (correctly) that it would be deflected

---

[56] *Principia*, Book I, Proposition 2, Scholium: 'if any force acts perpetually in the direction of lines perpendicular to the described surface, this force will make the body to deviate from the plane of its motion'. Also, Newton to Hooke, 28 November 1679, *Corr.* II, 301–302.

[57] Hooke (1674), Newton to Hooke, 24 November 1679, *Corr.* II, 301–302.

to the east. Hooke claimed a few weeks later that he had been successful with Newton's experiment, but it seems he was mistaken.[58]

Newton sought to extend the curvature argument from circular to elliptical motion, and the expression v^2/r was extended to elliptical orbits early on.[59] He wrote in his Waste Book in 1664: 'If the body... [be] moved in an Ellipsis, then its force in each point (if its motion in that point bee given) may bee found by a tangent circle of Equall crookednesse with that point of the Ellipsis'.[60] That is by the curvature (Leibniz's osculating circle) at that point. The radius of the acceleration is the radius of curvature at the point. If the motion is not circular, there is a perpendicular component of acceleration but the centre of force and centre of curvature do not coincide.

Newton already had a method for computing the orbit, and the shape of orbits for central forces before he wrote his letter to Hooke on 13 December 1679.[61] The diagram in the letter shows that he had already found by approximation the orbit of a body under a constant central force and knew how it would change if the force increased towards the centre. The rotating oval figure in the diagram 'shows an orbit for the case of a constant central force, one which shows the symmetries expected from energy conservation and the reversal of the direction of time'; 'the initial segment of the orbit... is almost exactly symmetric under reflection — a symmetry that Newton must have discovered and we now know is due to time-reversal invariance'.[62] He also gave the result, which could have only been developed by analysis, that one particular force increasing towards the centre (an inverse-cube force) would produce an orbit that was an equiangular spiral; his analytic solution showed that there would be an infinite number of spiral revolutions.

The early work shows that Newton was already used to dealing with scalar dynamical quantities, such as those now known as action and energy, as well as with vector ones, such as momentum and force. His early derivation of the force for circular motion involves treating impulse or momentum as a scalar quantity. Of course, momentum is a vector, as Newton well knew, and in the inscribed polygon argument is directed in different directions at different times, and so cannot be added as magnitudes. However, what *can* be added is a scalar product of momentum and displacement in the same

---

[58]Hooke to Newton, 17 January 1680, *Corr.* II, 313.

[59]Cohen (1999, 80).

[60]Waste Book, 1664, Add. 4004, 1, Herivel (1965, 130), The Waste Book (f. 1191ʳ) also has the polar equation for an ellipse referred to a focus (*MP* VI, 148, and Whiteside, 1964, 123).

[61]*Corr.* II, 307–308.

[62]Nauenberg (2005, 24).

direction, and, if the polygon has sides of equal length, then the information about the individual impulses is retained. The scalar quantity defined by momentum × displacement is now called *action* ($S = \mathbf{p} \cdot \mathbf{r}$), and the rate of change of action with time is a measure of the potential energy ($E = dS/dt$) in the system. If the circular motion is uniform then the potential energy or action per unit time will remain constant. So there will be the same amount of action in each cycle.

In the calculation, the scalar value of momentum or impulse, mv or $\Delta(mv)$, is multiplied by the scalar value of the total displacement around the square or polygon L, determined as n units of the side length l, which is itself proportional to the radius $r(L = nl = 2nr\sin(\pi/n))$. The action is then $2nrmv\sin(\pi/n)$. This is the same as if we had taken n *changes* of momentum of value $2mv\sin(\pi/n)$ at the vertices of the polygon directed along the radius to the centre of motion. With time period equal to $2nr\sin(\pi/n)/v$, the total potential energy is mv^2 and the force along the radius to the centre at each vertex is mv^2/r. In the limit, when $n \to \infty$, $2n\sin(\pi/n) \to 2\pi$, L becomes $2\pi r$ and the action sum becomes $2\pi mvr$, with the change at each point on the circumference now infinitesimally small.

Essentially the calculation is about identifying the two components of phase space, momentum and displacement, as the two pieces of information needed to specify the dynamics of the system. Because of the symmetry of the problem, they exhibit a proportionality, when we take the limit, so effectively we can treat the result as giving us momentum squared or even L^2. It is, of course, physically meaningless to square a vector with itself, but the momentum and displacement are different physical quantities, though numerically proportional. While phase space, mvL (not mvr), mv and r are the information, mv and L (consisting of n units of the fixed length l) being the independent units in the general case, in the limit, when $n \to \infty$, mv, L and r are proportional. So we get $(mv)^2$ or L^2, and, for a fixed mass and radius, mv^2/r.

4.4. Conservation of Momentum and Angular Momentum

Several important conservation principles are first stated or first stated with mathematical and physical correctness in Newton's work. The most fundamental is the conservation of mass or quantity of matter, which is implied by everything which follows. There is also a kind of conservation principle assumed for the fundamental particles of which matter is assumed to be composed. Newton defines the vector concept of linear momentum ('quantity of motion') in *Principia*, Definition 2, and then proceeds, in

Corollary 3 to the Laws of Motion, to the first precise and correct statement of the law of conservation of momentum, that the vector sum of momentum in a closed system remains the same if there is no external force acting on the system. 'The quantity of motion, which is collected by taking the sum of the motions directed towards the same parts, and the difference of those that are directed to contrary parts, suffers no change from the action of bodies among themselves'. This is true, even when the collision is oblique, in which case the vector properties must be invoked and resolution and composition be carried out as in Corollary 2.

According to an entry in the Waste Book, Newton was already investigating inelastic collisions quantitatively in January 1665,[63] and he soon extended Descartes' scalar law of conservation, which took into account the product of the size and speed of the interacting bodies, by showing that the conservation was only valid when it included the direction of motion in a vector definition of momentum as mass × velocity or speed in a particular direction. Newton relates the principle to that of equal and opposite forces, for the conservation of momentum, in particular, implies the third law and vice versa. As Newton shows, it follows from a combination of Laws II and III. 'For action and its opposite re-action are equal, by Law III, and therefore, by Law II, they produce in the motions equal changes towards opposite parts. . . . '[64] Though Descartes had defined momentum as mass (or rather size) times velocity, his principles of motion had failed because he did not recognise the vector nature of momentum in elastic collision. Huygens solved the problem by carefully avoiding any use of the momentum concept at all and pitching his discussion in purely kinematic terms. Newton's solution, unlike Huygens', was dynamic. Nevertheless, though he used vector momentum in definitions and in clarifying the problem of inelastic collision, Newton also liked using a scalar version for specific purposes, for example, in his derivation of the law of centrifugal force in circular motion; the scalar momentum, in effect, behaved similarly to a kind of energy term and led to important conclusions about dissipative systems.

Another very important principle, the conservation of angular momentum (a skew-tensor or pseudovector), finds its first statement in *The Lawes of Motion* of c 1666, 'here enunciated for the first time in the history of mechanics':[65] 'Every body keepes the same real quantity of circular motion

---

[63] Waste Book, 'Axiomes and Propositions', Add. 4004, 13^v–39, January 1664/65, Herivel (1965) 162–179).

[64] *Principia*, Axioms, or Laws of Motion, Corollary 3.

[65] Westfall (1971, 360), Herivel (1965, 211).

and velocity so long as tis not opposed by other bodys. And it keeps the same axis too . . . '[66] This was defined in the context of collisions, but Newton also later used the concept to describe the dynamics of orbital motion.

In Proposition 1 of the first book of the *Principia*, Newton uses second-order infinitesimals to show that inertial motion of and by itself implies an area-conservation law, with equal areas swept out in equal times. Unlike Kepler's empirically-derived similar area law for planetary motion, 'Newton's law of areas'[67] is a fundamental law of physics, fully equivalent to the conservation of angular momentum, and applicable to *any* system in which the force on a body is directed towards a centre, irrespective of the kind of force law involved, and whether or not the force is directed towards a central body, as also whether it is the product of a single source or many. Considering its momentous importance, Newton's proof is remarkably simple mathematically, requiring only equal-area triangles in addition to the limit concept.[68] Nauenberg also considers it to be a statement of Newton's second law of motion which Leibniz was able to translate almost immediately into his form of calculus, though he failed to give it a correct physical interpretation. He says that Newton, using Lemma 3, Corollary 4, gave a proof that, 'in the limit of an infinite sequence of impulses at equal but vanishingly small time intervals, these impulses became equivalent to a continuous force'.[69]

Following this significant opening Proposition, Corollary 1 is an implicit introduction of the conservation of angular momentum (for a unit mass object) in general central forces. 'The velocity of a body attracted towards an immovable centre, in spaces void of resistance, is reciprocally [i.e. inversely] as the perpendicular let fall from that centre on the right line that touches the orbit'. The product of the velocity and the perpendicular distance from the centre of rotation (in modern terms $\mathbf{v} \times \mathbf{r}$, where $\mathbf{r}$ is measured between the Sun at the focus and the tangent to the ellipse at the position of the body) is, therefore, a constant. For a body of mass m, the conserved quantity becomes $m\mathbf{v} \times \mathbf{r}$ or $\mathbf{p} \times \mathbf{r}$, the angular momentum. Newton was also able to show in the main Proposition that equal areas were swept out in equal times

---

[66] *The Lawes of Motion*, Herivel (1965, 211), *Unp.*, 160.

[67] So designated in www.cleonis.nl/physics/phys256/angular_momentum.php.

[68] Nauenberg (2003) and Pourciau (2003) demonstrate the essential validity of Newton's proof of Proposition 1. Propositions 1–10 constitute Section II of Book I on 'The invention [determination] of centripetal forces'.

[69] Nauenberg (2012), referring to Leibniz (1689), and citing Nauenberg (2003) and (2010).

by a line which moved from a body in the direction of any point not on the line of the body's motion. This meant that the area law was simply a particular aspect of the law of inertia for linear motion.

As in his earlier, more explicit statement, the conservation law here includes both magnitude and direction, for Proposition 1 required the orbits to 'lie in the same immovable planes'. In fact, the law of conservation of angular momentum is really a combination of three separate conservation laws — of magnitude, direction and left- or right-handed rotation — a fact of fundamental importance to physics.[70] This is determined by the expression $\mathbf{p} \times \mathbf{r}$, where the magnitude is fixed at $pr \sin \theta$, for an angle θ between $\mathbf{p}$ and $\mathbf{r}$, and the direction is that of the fixed axis of rotation, perpendicular to both $\mathbf{p}$ and $\mathbf{r}$. For a planar orbit, there is no absolute right- or left-handedness, but the relative handedness (which depends on the direction assigned to $\mathbf{p}$) must remain fixed, as would have been evident from the Solar System, where the planets all orbit in the same direction, while Halley's comet is one of a number of bodies orbiting in the opposite direction.

Corollary 1 is followed by a number of others which give technical results important for later Propositions, but Corollary 6 has an additional significance in stating that the same results apply when the entire system is not at rest but in uniform motion in a straight line, that is, in an inertial frame of reference. Proposition 2 is the converse of Proposition 1, a demonstration of the logical connection between the law of inertia and Kepler's law of areas, generalised to hold for an arbitrary central orbit. According to Corollary 1, if areas described in equal times continually increase (or decrease), then the additional force acting on the particle must accelerate (or decelerate) it in the direction of motion. Corollary 2 shows that this is also true if the orbit is in a resisting medium. Newton sees the physical significance of the law of areas as a necessary and sufficient condition for a central force; for each conic there exists a toward-one-focus force law under whose influence and with suitable initial conditions a particle must traverse the conic as orbit.[71]

A 'truly remarkable' Scholium shows that motion in the orbital plane is unaffected by any force acting normally to the plane. If, in addition to the centripetal force, there is a force perpendicular to the orbital plane, the radius vector connecting the body to the focus will continue to generate

---

[70]Rowlands (2007), where it is related to $SU(3) \times SU(2) \times U(1)$ symmetry-breaking.

[71]Pask (2013, 180), describes the combination of Propositions 1 and 2 as an 'if-and-only-if' (iff) result, saying that the equal area law only applies iff the force creating the motion is centripetal. He regards such iff results as one of Newton's major innovations (510).

equal areas in equal times, irrespective of any motion outside the plane. The angular momentum due to the orbital motion will be parallel to this additional force, and the moment of the force will be perpendicular to this direction, so the rate of change of angular momentum will also be perpendicular, which means that the magnitude of angular momentum (or the length of its vector), and the rate at which area is swept out, $1/2$ $r^2 d\phi/dt = h/2$, will be unchanged. Donald Lynden-Bell says: 'This is all so obvious to Newton that he does not stoop to prove it — he simply states it!'[72]

According to Lynden-Bell, Newton inquires about the circumstances under which the magnitude of a vector which is a constant of the motion can be conserved without also 'requiring each component to be separately conserved'. The three such quantities in classical mechanics are: the angular momentum, 'the linear momentum $\mathbf{p}$ and $\mathbf{r} - \mathbf{v}t = r_0$, the position of the barycentre at $t = 0$'. To preserve the magnitude of $\mathbf{p}$ constant without keeping $\mathbf{p}$ constant itself, we need 'a force $d\mathbf{p}/dt = m\mathbf{v} \times \mathbf{B}$ that is always perpendicular to $\mathbf{p}$'. This requires 'forces of gravo-magnetic type of which Coriolis force is a special example'. Newton had, as we have seen, previously invoked such a force, in 1679, in an attempt to detect the rotation of the Earth. In the case of $\mathbf{r} - \mathbf{v}t$, the force which will preserve its magnitude is $\mathbf{r} \times \mathbf{v}t \cdot d(\mathbf{r} - \mathbf{v}t)/dt = 0$ which means that '$-\dot{\mathbf{v}}t$ must be of the form $(\mathbf{r} - \mathbf{v}t) \times \mathbf{B}$', which means that we need a force per unit mass 'of the form $\mathbf{F} = \mathbf{v} \times \mathbf{B} - \mathbf{r} \times \mathbf{B}/t$, where $\mathbf{B}$ may be any function of position, time and velocity'.[73]

Proposition 3 gives the case of a body moving around a moving, rather than a stationary, centre. A Scholium shows that equal area motion indicates a centre, about which all circular [circulatory] motion is performed in free spaces. Proposition 4 concerns uniform circular motion, with the forces shown not only to be directed to the centres of the circles, but also to each other as the squares of the arcs described in equal times divided respectively by the radii of the circles. Corollary 1 to this Proposition demonstrates that the centripetal forces are as the squares of the velocities divided by the radii. Further corollaries (2-9) show that, for forces F and f, on bodies with tangential velocities V and v, and period of revolution T and t, then F: $f = V^2/R : v^2/r = R/T^2 : r/t^2$. Universally, also, for $T \propto R^n, V \propto 1/R^{n-1}$, then $F \propto 1/R^{2n-1}$, and conversely. In the special case in Corollary 6, where

---

[72]Lynden-Bell (2000, 132).
[73]Lynden-Bell (2000, 131–132).

$T \propto R^{3/2}$, then $F \propto 1/R^2$.[74] And, even if the orbits are not circular, the equal area law means that the same relations will apply if we use the distances of the bodies from the centre of the orbit in place of the radii.[75]

In Proposition 6 Newton provides 'a general concept of instantaneous measure of force, for a body revolving in any curve about a fixed center of force'.[76] He shows that it is directly proportional to the versed sine (the perpendicular distance between the mid-point of a chord and its arc) and indirectly to the time squared, and then proceeds to translate this into a purely geometric construction in a series of Corollaries. Following this, he writes in Corollary 5: 'Hence if any curvilinear figure APQ is given, and therein a point S is also given, to which a centripetal force is perpetually directed, that law of centripetal force may be found, by which the body P will be continually drawn back from a rectilinear course, and, being detained in the perimeter of that figure, will describe the same by a perpetual revolution'. This applies to *any* plane curve. Any such curve will be 'an orbit for some central force, relative to an arbitrary force center', because such a force conserves angular momentum. Newton also gives the formula by which the force can be determined, which is identical to the one found in the previous four Corollaries: 'find, by computation, either the solid $SP^2.QT^2/QR$ or the solid $SY^2.PV$, reciprocally proportional to this force'.[77]

In rigid body dynamics, Newton's early work, in the manuscript *The Lawes of Motion*, includes an equivalent of the concept of moment of inertia, the 'real quantity of circular motion'.[78] Having largely concentrated on translational (or 'progressive') motion in his consideration of impact in the Waste Book, Newton now extended his dynamical analysis to include rotational (or 'circular') motion as well. He considered a body rotating on an axis defined through its centre of gravity which impacts on another body of the same mass in such a way that the first transfers all its rotational motion to the second. In this case, '...the reall quantity of its circular motion is more or lesse accordingly as the body hath more or lesse power and force to preserve it in that motion; which motion divided by the bodys bulke is the reall quantity of its circular velocity'.[79]

---

[74] Corollaries 2–9. Among the other special cases are T constant, $F \propto R$ (Corollary 3); $T \propto \sqrt{R}, F$ constant (Corollary 4); $T \propto R, F \propto 1/R$ (Corollary 5).

[75] Corollary 8.

[76] Cohen (1999, 72).

[77] See Ehrlichson (1990, 882).

[78] *The Lawes of Motion*, NP; Herivel (1965, 209), Westfall (1971, 360–362).

[79] *The Lawes of Motion*, NP; Herivel (1965, 210), *Unp.*, 158.

To describe a body's equator of circulation, he defined a 'radius of circular motion', which is equivalent to the modern radius of gyration (k). The real quantity of the body's circular motion was then found by multiplying the body's 'bulke' (or mass) M and the velocity of a point on its 'Equator of circulation'. Effectively it becomes $Mk\omega$, where ω is the angular velocity.

Let us suppose two bodies, A and a, collide, each of which has a component of translational and one of rotational motion. After a perfectly elastic collision, the relative velocity of separation of the bodies may be assumed to equal the relative velocity at which they approached each other before collision, say $Q/2$. Then the total change in velocity at the points of collision is Q. Now, another quantity P represents the sum of the 'smallnesse of resistance' to the change of motion which will occur. This will be the sum of the reciprocal masses and the reciprocal of the 'real quantity of circular motion'. Q/P is now equivalent to the total change of motion (momentum and angular momentum). It is to be divided between the four terms in the equation in proportion to 'the easinesse (or smallnesse of resistance) with which those velocitys are changed...' The total change in the 'progressive motion' of A then becomes equal to Q/AP, and so on. Essentially, this is equivalent to the modern analysis of the collision of rotating bodies, combining the conservation of momentum and angular momentum, and using Newton's 'real quantity of circular motion' in place of the modern 'moment of inertia'.[80] Newton's real quantity of circular motion differs from the modern definition of angular momentum $(Mk^2\omega)$ by a factor k, while Mk differs from the modern moment of inertia (Mk^2) by the same factor. The effect of the quantities is essentially the same as the modern equivalents but Newton defines them to be dimensionally equivalent to the linear terms in the collision problem.[81]

How much Newton understood about rigid body dynamics has been the source of a considerable amount of debate. In the discussion following Law I in the *Principia*, he makes a point of applying the concept of inertia to the spinning motion of a 'top, whose parts by their cohesion are perpetually drawn aside from rectilinear motion', equating this with the linear motion of projectiles, and the motions of the planets and comets, 'both progressive and circular', in exact agreement with the definition of angular momentum as $\mathbf{p} \times \mathbf{r}$. The examples are clearly chosen to emphasise that both linear and orbital motion of material objects and the spinning motion of a rigid body

---

[80] *The Lawes of Motion*, NP; Herivel (1965, 209).
[81] Herivel (1965, 216).

follow exactly the same process of conservation unless acted on by an external force, which changes either the state of linear or rotational motion. The top is described in the precise terms needed to identify it as a rigid body.[82]

One of the reasons why interpretations of some aspects of his work have led to controversy, however, is that Newton often uses language that is very different from current practice, and frequently without the precision now demanded for scientific terms. In Book III, Lemma 2, for example, which is a key part of his treatment of the precession of the equinoxes, he refers to 'the total force or power of all the particles situated every where about the sphere to turn the Earth' about its axis, and this can be taken, quite reasonably, to be the torque or moment which the tidal forces due to the gravitational attraction of the Sun cause to be exerted on the Earth. Chandrasekhar, however, also took it to mean the angular momentum or moment of momentum, and produced a reconstruction of Newton's calculation of precession on this basis.[83]

Now, one of the strongest areas of criticism of Newton's treatment of precession and other complex gravitational effects has been that his 'geometrical' mathematical procedures were unable to accommodate the extensive developments in dynamics that were produced by his eighteenth-century successors, using the more analytical procedures of the 'Leibnizian' calculus. The fact that Newton had no true understanding of rigid-body dynamics, based on angular momentum and moment of inertia, has been a very considerable component of this case.[84] Clearly, he did have a degree of understanding in his early work on collisions, and the qualitative aspect of this is seen in his commentary on Law I, but the mathematical development does not seem to have formed a significant part of his mature dynamics. However, this does not decide the issue of whether Chandrasekhar is justified in using angular momentum rather than torque in interpreting Lemma 2, for Newton, in the *Principia*, has a dual concept of force, defined by Law II and Definition 8, with two separate measures, the impulsive and the continuous,

---

[82]Whiteside prefers 'hoop' to 'top', suggesting that it means something like 'roundabout' (*MP* VI, 97). Cohen and Whitman (1999, 416), also translate the Latin '*trochus*' as 'hoop', while saying that it means 'a top or some kind of spinner'. Whatever the specific meaning, the physics is unchanged.

[83]Chandrasekhar (1995, 466–470).

[84]Dobson (1998, 1999, 2001). It is important to point out that, while there are significant errors in Chandrasekhar's work, there are also significant insights. While some historians have been quick to point out that some of Chandrasekhar's claims are anachronistic, it is important to separate out such instances from those where his mathematical insight has observed the kind of deep connections which are of special interest in this study.

each of which can be used in physical examples, and these are distinguished only by invoking a nonobservable free-flowing time in the second. If we apply this to the creation of a turning effect, the impulsive measure will imply an angular momentum change when the continuous implies torque. (The same duality also applies to what we would call a change of action and the work done.) This means that it is certainly possible to interpret the torque in Lemma 2 as an angular momentum change. In effect, a distinction that we make in rigorous terms is meaningless to Newton because of the difference in his use of terminology.

Of course, angular momentum, whether implicit or explicit, is a major aspect of the whole of Newton's argument in the *Principia*, and there is a very striking discussion of the conservation of angular momentum based on a symmetry argument in Corollary 22 to Book I, Proposition 66. Newton imagines a perfectly spherical uniformly dense body to be at rest in free space, and then to receive an impulse, which is partly rotational and partly translational. He says: 'Because this globe is perfectly indifferent to all the axes that pass through its centre, nor has a greater propensity to one axis or to one situation of the axis then to any other, it is manifest that by its own force it will never change its axis, or the inclination of it'. If then a new impulse is applied to the globe on any part of its surface, it will compound with the first as if they had been applied at the same time, and produce 'a simple and uniform motion about one single given axis, with an inclination perpetually invariable'. Here, we have an argument that conserves angular momentum because of the rotation symmetry of free space, exactly as we now deduce as a particular consequence of Noether's theorem.

There is also an important aspect of reflection symmetry, for, if the globe is divided into two hemispheres by any plane passing through its centre, 'and the centre to which the force is directed', the force will apply equally to each hemisphere, and 'will not incline the globe any way as to its motion round its own axis'. However, if there is a mountain, or any such matter between the pole and equator, breaking the uniformity of the globe, 'this, by its perpetual endeavour to recede from the centre of its motion, will disturb the motion of the globe, and cause its poles to wander about its superficies describing circles about themselves and their opposite points'. If the mountain is placed at a pole, the nodes at the equator will move forwards; if placed in the equatorial regions, the nodes will go backwards. If a new quantity of matter is added on the other side of the axis, the nodes will go either backwards or forwards depending on whether the new matter is nearer the pole or equator. Chandrasekhar comments that the arguments 'are remarkable for the use that Newton makes of the geometrical *symmetry* of the sphere to reflection and to rotation about its centre to draw *physical conclusions*.[85] He relates them to arguments made by L. Lictenstein more than two hundred years later 'to prove that a fluid mass in equilibrium under its own gravitation must necessarily be spherically symmetric'.[86]

4.5. Conservation of Energy

From the third law, combined with the second law, it is possible also to derive the conservation of mechanical energy, that the total energy in a conservative system, or sum of potential and kinetic energies, remains constant, following any change within the system. The modern concept of energy really begins only in the nineteenth century. Thomas Young was the first to use the word in the modern sense in 1807,[87] but the idea was not fully adopted until the middle of the nineteenth century, when physicists like Joule and Helmholtz began to consider a principle of the conservation of energy in all interactions. Newton used a variety of force-related quantities, especially in his early work, sometimes distinguishing them with particular names such as 'total force', a term which occurs in his first analysis of circular motion and represents 'action' or the product of momentum and distance; but, in the *Principia*, he avoided creating new terminologies even though he clearly used the quantities which we now describe as work, potential energy, kinetic energy and power. He does not use any term which represents energy, or any form of it (if we except his 'active principles'). He does not treat it as an entity with an independent existence, like force or 'quantity of motion' (his name for momentum).[88]

---

[85]Chandrasekhar (1995, 264).

[86]Lictenstein (1918).

[87]Young (1807, I, 44, 78). Terms *dimensionally* equivalent to energy, such as *vis viva* (mv^2), were, of course, widely used by Leibniz and others from as early as the seventeenth century.

[88]Stein (2002) quotes a passage from Definition 8 in the *Principia*, which, in his translation, reads: 'I refer the Motive force to the Body, as an endeavour and propensity of the whole towards a centre, arising from the propensities of the several parts taken together; the Accelerative force to the Place of the body as a certain power or energy diffused from the centre to all places around to move the bodies that are in them; and the Absolute force to the Centre, as indued with some cause, without which those motive forces would not be propagated through the spaces round about; whether that cause is some central body, (such as is the Load-stone, in the centre of the force of Magnetism, or the Earth in the centre of gravitating force) or any thing else that does not yet appear'. (286) The Latin original of 'a certain power or energy' is *efficaciam quondam* (literally, 'a certain efficacy') (*ibid.*, 305 n 64). Stein, who had previously written 'On the Notion of Field in Newton, Maxwell, and Beyond' (1970), states that the passage is concerned

The dynamics in the *Principia* is often very different to that suggested by its popular reputation. It was not until well into the twentieth century that scholars began to realise that Newton had produced an extensive investigation into the relation between energy and work in Sections VII and VIII, and that Section VIII, which considers the general case of 'orbits in which bodies will revolve, being acted upon by any sort of centripetal force', was concerned in its entirety with equations involving energy. No longer concerned only with motions produced by inverse-square forces, he now extends his analysis to 'the motions of bodies tending to centres by means of any forces whatsoever ... '.[89] The theorems in this section (Propositions 40–42) are clear exercises in the calculus, referring to integration methods published only in *De quadratura* in 1704 ('granting the quadratures of curvilinear figures'). Newton's work here extends into a dynamics far beyond that commonly associated with the *Principia*.

Galileo and Huygens had previously established relations that would now be regarded as requiring special cases of the conservation of energy, though they were always structured in kinematic terms, specifically excluding the concept of mass, which is required for a definition of the concept, for the idea of conservation as a fundamental property, and for a dynamic treatment, and which Newton incorporated from the beginning. Galileo had derived the acquisition of velocity by a falling body due to uniform acceleration, while Huygens applied similar reasoning to the velocity acquired by a cycloidal pendulum. With the mass included, these would become energy calculations, and Newton has an independent calculation for the cycloidal case, from about 1671, which incorporates his specific understanding of mass and weight.[90] The connection with conservation of momentum and angular momentum ensured that the Newtonian view of energy was always dynamic, and his early work on collisions cannot be understood without a simultaneous application of all three conservation principles.[91] Propositions 39–41 of *Principia*, Sections VII–VIII, extend the specific uses of the conservation of energy to a general treatment.[92]

---

with the idea of a 'field of force' directed towards a centre (286). Interestingly, it concerns magnetic and other possible types of field yet unknown, as well as the gravitational.

[89] *MP* VI, 380; *Principia*, I, Proposition 45, statement after Corollary 2.

[90] *On Motion in a Cycloid*, Herivel (1965, 199–203), translation 203–206; *Unp.*, 170–180; *MP* III, 420–430. 'While stressing its several parallels with Huygens' kinematical proof', Whiteside insists 'on the originality of its dynmical basis' (*MP* VI, 402–403).

[91] Herivel (1965, 217–218).

[92] *MP* VI, 336–355. Aiton (1965).

In Proposition 39, which considers the velocity of a rising or falling body produced by the action of an arbitrary force ('a centripetal force of any kind') varying in any manner with distance, differential equations of motion are set up and solved for mechanical systems in the modern way. Though the earlier draft version of the *Principia* known as the *Lectiones* had found the solution in the case of an inverse-square force by 'an ingenious use of conics',[93] Newton now made no attempt to confine himself to the standard mathematical approaches of the *Principia*, but relied on a direct application of the fluxional calculus, and specifically on a double integration.

In the proposition, Newton finds both the instantaneous velocity at any point on the trajectory of a body ascending or descending in a straight line and the time taken for the body to arrive at any point. According to Richard S. Westfall, the proposition is 'the mechanical equivalent of the work-energy equation, expressed indeed in a more general form than Leibniz ever achieved' (as well as at an earlier date),[94] and in more or less exactly the same terms as we use it today. Newton shows that the work done, the integral of force over distance (expressed as the area of the figure ABGE in Newton's diagram), is equal to the change in kinetic energy:

$$\Delta W = \Delta(mv^2/2).$$

The third corollary to the proposition states the same fact even more precisely, while the first corollary relates the kinetic energy to the change in kinetic energy as $W/\Delta W = v^2/2v\Delta v$ and the second gives the case when there is a nonzero initial velocity. In modern notation, Proposition 39 is equivalent to

$$\mathbf{F} \cdot \mathbf{ds} \propto \int v dv \propto v^2$$

or

$$\mathbf{F} \cdot \mathbf{ds} = \int mv dv = 1/2mv^2.$$

The quantity mv^2 is the one that Leibniz would soon afterwards call *vis viva*, and use as the basis of his own rival and differently-conceived system of dynamics.

Proposition 39 also demonstrates that the time of fall is proportional to the area under the curve ('granting the quadratures of curvilinear figures').

---

[93]Westfall (1980, 438).
[94]Westfall (1971, 487).

Proposition 16 of the *Lectiones de Motu*, an earlier version of the *Principia*, Book I, had demonstrated how the orbit of a body in a conic section originating from a central inverse-square force could always be constructed from a given initial starting-point and initial velocity.[95] On the basis of Proposition 39, Section VIII universalised the result, showing how, for any force defined as a function of distance, the elements of the orbit can be determined.[96]

Proposition 39 became a major breakthrough in Newton's dynamics. Everything that he discovered after this followed directly from it. Along with Propositions 40 and 41, it provided a clear demonstration that Newton knew how to write down differential equations of motion for applications in mechanics, and to make direct use of the concept of energy in his calculations. Proposition 40, again using an arbitrary force, and for a body describing any trajectory, finds the velocity at any point on the trajectory. Expressed in modern terms, the Proposition finds an integral which is equivalent to the sum of the kinetic and gravitational potential energies over the trajectory remaining invariant or constant over time.[97] In the nineteenth century, Hamilton would invent a new function, the Hamiltonian, which would use this summation, $\mathcal{H} = T + V$, as the basis of a new way of structuring dynamics.

Proposition 40 is followed by two very remarkable corollaries. In the first, the kinetic energy acquired by a body in ascending or descending a certain distance under a centripetal force of any kind is seen to be the same irrespective of whether the ascent or descent is under the action of a harmonic oscillator in a straight line or along a curved line, by, for example, being constrained to move along a curved smooth surface. In other words, the potential energy applied is independent of the path taken, the condition for a conservative force. The second corollary, which requires the result of the first, uses the procedure of Proposition 39 to integrate a centripetal force proportional to r^{n-1} over distance for a body moving between points at distances P and A to show that the difference in the squared velocities $(v_A^2 - v_P^2)$ is proportional to $(P^n - A^n)$. If we take the change of kinetic energy (T) as $1/2(v_A^2 - v_P^2)$ and the change in potential energy (V), by integration as $(P^n - A^n)/n$, then we have an equivalence between kinetic and potential

---

[95] *Lectiones de Motu*, Proposition 16, *MP* VI, 156–160 (Proposition 17 of *Principia*, Book I); Westfall (1980, 439).

[96] *MP* VI, 344–388; Proposition 41 and Corollaries.

[97] Aiton (1965).

energies of $T = nV/2$, exactly as is found for the time-averaged versions for large-scale systems in the virial theorem.[98]

Of the cases of particular interest, an inverse-square law force $(n = -1)$ and a constant force $(n = 1)$ are dual in providing the same *numerical* relation between T and V, respectively $2T = -V$ and $2T = V$, while the harmonic oscillator $(n = 2)$ has numerically equal time-averaged kinetic and potential energies, each switching between maximum and minimum over the cycle. The dual case to this is the inverse cube relation $(n = -2)$, where $T = -V$ and the time-averaged total energy is zero. Newton would famously prove that an inverse cube orbit would be unstable with the satellite spiralling into the centre. Ultimately, such relationships as $2T = V$, applied to impulse forces, would lead to the general principle of least action,[99] but this way of thinking appears to be a long way from that used by Newton in this corollary.

Proposition 41, another obvious exercise in the calculus, which has been called 'the central dynamic theorem of the *Principia*',[100] gives the orbital equation for a body acted upon by any central force ('a centripetal force of any kind', varying 'in its recess from the centre according to some law, which anyone may imagine at pleasure, but [which] at equal distances from the centre [is taken] to be everywhere the same'), given the speed and direction of motion at any point. In principle, this is a version of the well-known '*vis viva*' integral, a conservation of energy equation used for finding the paths of bodies subject to any type of force, which Johann Bernoulli would put into algebraic form in 1710:[101]

$$v^2 = u^2 + 2 \int f(r)dr.$$

Proposition 41 has been rightly praised. Though presented in a superficially geometrical form, its infinitesimal moments could be easily translated

---

[98] A disguised form is also present in a manuscript calculation, using both geometry and fluxions, which was printed as 'Calculations of Centripetal Forces' in *Unp.*, 65–68, translation, 68–71. The usual way to state the virial theorem is for a force proportional to r^{-n}. Then we have $T = (1-n)V/2$. Newton, however, in expressing the theorem for a force proportional to r^{n-1}, is clearly conscious of the fact that this will lead to a particularly neat expression for the relationship between T and V, and, in fact, between $2T$ (the *vis viva*) and V.

[99] As stated by Maupertuis (1744, 1746).

[100] Bechler (1973).

[101] Referring to Aiton (1965), Whiteside says that Varignon, Hermann and Bernoulli 'groped their way to equivalent representations' of Propositions 39–41 'in terms of Leibnizian calculus'. He notes Bernoulli's 'grudging' acknowledgement of Newton and 'the essential justice of Newton's claim to have had complete mastery of the inverse problem' of deriving the force law from the orbit 'long before Bernoulli' (*MP* VI, 349–350).

into either Newtonian fluxions or Leibniz's differentials,[102] as was later done by several mathematicians (including Newton himself in Corollary 3), and it could be said to have been the true foundation and *direct source* for *all* the dynamics which followed in the eighteenth century. I. Bernard Cohen said that nothing in any of the previous literature on dynamics had 'the same magnitude or importance' and that, with it, the whole subject 'achieved its modern maturity for the first time'.[103]

Under the general conditions specified, Newton determined both 'the curves in which bodies will move' and 'the times of their motions in the curves found'. In other words, he presented to his readers a truly general resolution of the inverse problem of finding an orbit from a given law of force.[104] Two corollaries give the method of finding the line of apsides by reducing the kinetic energy to zero, while the third uses Proposition 41 to solve the case for an inverse-cube force. For seemingly unknown reasons,[105] the inverse-square law case was not singled out for special treatment, though it would have been well within Newton's mathematical capabilities. Proposition 41 gives a complete solution of the problem of determining the resulting orbit in a given central force-field when the velocity and direction of initial motion at a point are known. In a steady state orbit, an orbital term equivalent to h^2/r^2 substitutes for $-u^2$ on the left-hand side, h being the angular momentum constant for the orbit, and typically doubling the v^2 term to give a direct equation for potential energy, as opposed to a typical value for Proposition 39 of $u^2 = 0$, which gives a direct expression for the kinetic energy gained starting from rest.

Propositions 39 and 41 provide the two main ways of representing conservation of energy in dynamics. The first is the kinetic energy equation which applies under changing conditions; the second is the potential energy equation which applies to a steady state. There are yet further conservation of energy calculations in the *Principia*, dealing with specific cases. Newton, for example, also extends energy considerations to resisted motion, by investigating the motion of pendulums in resisting media, particularly

---

[102] Guicciardini (2009, 248–250).

[103] Cohen (1999, 141–142). Cohen points out that it uses both integral and differential calculus, and says that it 'represents a climax, a high peak' in the dynamics of the *Principia*. Pask (2013, 213), says that 'it surely represents one of science's greatest achievements'.

[104] This was fully recognised by Hermann (1711), who also said (1716, 73) that Bernoulli had used the same method and that Varignon had found the same result by a different method (Guicciardini, 1999).

[105] But see Smith (2012, 371–373).

involving cycloidal oscillations, and shows how we can find 'the resistance of mediums by pendulums oscillating therein', which means that it provides a theoretical structure for using pendulum experiments to find fluid resistance. Book II, Proposition 30 shows that, in the case of a pendulum moving through a resisting medium, 'the loss of potential energy due to resistance is equal to the work done against resistance'.[106]

Newton's analysis of machines in the Scholium to the Laws of Motion introduces the concept of power or rate of working, suggesting that the rate of energy consumption, calculated as the product of force and velocity, is equal to the sum of the useful power generated and the power dissipated by friction of the parts.[107] Here, he writes:

> For if we estimate the action of the agent from the product of its force and velocity conjunctly, and likewise the reaction of the impediment conjunctly from the product of the velocities of its several parts, and from the forces of the resistance arising from the attrition [friction], cohesion, weight, and acceleration of those parts, the action and reaction in the use of all sorts of machines will be found always equal to one another. And so far as the action is propagated by the intervening instruments, and at last impressed upon the resisting body, the ultimate determination of the action will be always contrary to determination of the reaction.

The treatment of machines, based on the idea of turning effects, also connects with the dynamics of orbital motion, and circular motion in particular. Here, the key concept is the torque or moment, the 'vector product' of force × perpendicular distance about the fulcrum or axis of rotation, whereas the energy or work done is the 'scalar product' of force times distance moved along the line of action. The two distances are related for machines such as the lever. Using impulse instead of continuous force gives the respective quantities as angular momentum and action, as in circular motion (which, in effect, it is).

In the late nineteenth century, Peter Guthrie Tait, who never realised the fundamental significance of Propositions 39–41, made large claims for the significance of Newton's statement about power in machines (closely related as it is to the Scholium to the third law of motion) as a conservation of energy proposition.[108] In Tait's interpretation:

> The complete theory of all such cases was enuntiated [*sic*] in a perfect form by Newton in the *Principia* as a scholium to his Third Law of

---

[106] See Westfall (1971, 487).
[107] See also *Elements of Mechanicks*, 3, *Unp.*, 165.
[108] Tait (1877), quoted in quoted in *Corr.* III, 354.

Motion; in which he not only laid down the so-called Principle of Vis-viva, and D'Alembert's Principle, for which others long afterwards gained great credit; but stated, so far as the development of experimental science in his time permitted, the great Law of Conservation of Energy.

By the context it is easy to see that the *actio* here spoken by Newton is precisely what is now called rate of doing work or horse-power. Also the *re-actio*, as far as acceleration is concerned, is precisely what is now known as the *rate of increase of kinetic energy*. Newton's statement is, therefore, in modern phraseology, equivalent to this:

'Work done on any system of bodies has its equivalent in the form of work done against friction, molecular forces, or gravity, if there be no acceleration; but if there be acceleration, part of the work is expended in overcoming resistance to acceleration, and the additional kinetic energy developed is equivalent to the work so spent'.

Tait's argument was at one time considered to be extravagant — a desperate attempt to find an indication that the conservation of energy had its foundations in the *Principia*. While the treatment of machines alone would not justify such assertions, the application of energy principles to projectiles and pendulums, and the general energy theorems incorporated into Propositions 39–41, show that Newton was well aware of the nature of the integral of force over distance and of its relationship with terms involving mass and velocity squared; and though he expressed it in Proposition 39 in terms of the area of a geometrical diagram, his use of the phrase 'granting the quadrature of curvilinear figures' indicates that this particular theorem was completely analytic in origin and that the relationship between work and energy was derived by him in essentially the form we derive it today. It is no longer possible to doubt that Newton understood the relation between work done and increase of kinetic energy, without having an explicit terminology. In principle, he recognised that, for purposes of *calculation*, it was as valid to state equations using the mathematical construct which we now describe as 'energy' as it was to work from the laws of motion based on the concept which he had defined as 'force'.

It is not always realised that the treatment of energy implicit in the Propositions of the *Principia*, and made explicit later by the more developed systems of dynamics of Euler, Lagrange and Hamilton, allowed Newtonian mechanics to survive the challenge introduced with Einstein's special theory of relativity in 1905, and even that of quantum mechanics twenty years later. Special relativity is constructed to preserve classical energy relations, and does so by redefining dynamic quantities, such as force and mass, to preserve this compatibility. Consequently, many Newtonian equations survive, even where objects are travelling with 'relativistic' velocities (close to c) because the quantities have been redefined to absorb the change. Some of these redefined quantities, for example, 'relativistic mass' have not proved to be universally popular, but they are important in showing that the history of physics is mainly one of continuity rather than revolutionary change, and they are also consistent with Newton's own practice, embryonic aspects of them appearing in Newton's own works. (They are also important in relation to the fact that all mass is dynamic in origin, and it makes no sense in fundamental terms to make arbitrary separations between different forms of a fundamental quantity appearing in different manifestations.) The same is true of quantum mechanics, where the conservation principles survive intact — the most famous quantum mechanical equations, for example, those of Schrödinger, Klein and Gordon, and Dirac, are all conservation of energy equations. The conservation principles that appear in Newtonian dynamics, based ultimately on the definition of mass as a conserved quantity, are still the most unchallengeable facts on which all physics is founded, and are likely to remain so for any future that we can envisage.

4.6. *Vis Viva* and Collisions

The 'power' quantity in Newton's statement on machines (force × velocity) (also found in simpler form in the *Principles of Mechanicks*, 3) becomes *vis viva* (or momentum times velocity) if we take the impulsive definition of force, $\Delta(mv)$, rather than the continuous one, $\Delta(mv)/\Delta t$, since these quantities play identical roles in Newtonian mechanics. In fact, in Newtonian physics, all fundamental dynamical quantities obey the same conservation law or equivalent extremum principle as they all assume the same time variable in differentiation. Of course, the quantity mv^2 never had the heuristic role in Newton's work that it had for Leibniz in his vision of a 'living force' or '*vis viva*', Cartesian in origin, passing from one body to another in interactions but never lost as a result of them. Something like this entered into the mainstream of physics, somewhat transformed, as the main aspect of the physical concept of energy, which provided a simple mechanistic principle as a unified way of describing physical interactions, and satisfying the desire, we might suspect, to find some single tangible entity responsible for all the signs of life induced in otherwise 'inert' matter. The nearest equivalent for Newton might be his association of the 'life-principle' with light, or something which causes it, and perhaps even responsible for the creation of inert matter itself.

Newton's dynamics, however, actually incorporates a principle equivalent to the conservation of *vis viva* for perfectly elastic collisions. Let us suppose we have a collision between bodies of mass m_1 and m_2, with initial velocities u_1 and u_2 and final velocities v_1 and v_2. Here u_1, u_2, v_1, v_2 may be positive or negative, and have either the same linear direction, or may be components of the velocity in any given direction. Using conservation of momentum

$$m_1 u_1 + m_2 u_2 = m_1 v_1 + m_2 v_2.$$

But there is also another law that applies — Newton's law of restitution or Newton's experimental law — which says that the ratio of the relative velocities before collision to the relative velocities after collision is given by a constant $-e$, where e, the coefficient of restitution, has a value between 0 and 1, and is 1 for a perfectly elastic collision.[109] Here, we have

$$(v_1 - v_2) = -e(u_1 - u_2).$$

If the collision is perfectly elastic, then the equations may be written

$$m_1(u_1 - v_1) = m_2(v_2 - u_2),$$

$$(u_1 + v_1) = (v_2 + u_2)$$

from which it is easy to derive the 'conservation of *vis viva*' equation

$$m_1 u_1^2 + m_2 u_2^2 = m_1 v_1^2 + m_2 v_2^2$$

with the consequence that the total amount of *vis viva* (mass $\times$ velocity^2) is conserved in a perfectly elastic collision. In more modern terms, the conserved quantity is the kinetic energy, which is half the *vis viva*. So the conservation of kinetic energy in elastic collisions is given by

$$1/2\, m_1 u_1^2 + 1/2\, m_2 u_2^2 = 1/2\, m_1 v_1^2 + 1/2\, m_2 v_2^2.$$

Equally, if the collision is inelastic ($e \neq 1$), some *vis viva* or kinetic energy will be lost. In addition, if the source of momentum is kinetic (i.e. not generated directly by a potential energy, as in, say, an explosion), then the sum of the *scalar* values of momentum after collision will necessarily be less than that before collision, unless the collision is elastic and all the momentum is directed along the same straight line.

Characteristically, the law was established on the basis of an extensive series of experiments, meticulously carried out, which at the same time verified the conservation of linear momentum and the third law of motion. Newton measured the 'motions' (momenta) of colliding pendulums, using hard and soft spheres, with both equal and unequal masses, while taking

---

[109] *Principia*, Axioms, or Laws of Motion, Scholium.

into account the air resistance.[110] He verified that the third law applied for 'attractions', or actions at a distance, as well as for collisions, by observing that the mutual attractions of a loadstone and a piece of iron, made to float on water, would not displace a nonmagnetic obstacle placed between them; and inferred from this that, if the Earth was divided into two parts by any plane drawn through it, then the two parts would attract each other with an equal force, because a nonzero resultant force in any direction would propel the whole Earth to infinity through the non-resistant aether.[111]

Newton's subscription to the kinetic theories of heat and matter shows that he knew about heat motion inside bodies made of smaller particles, while his analysis of power in machines, his work on resistance in fluids, and various statements on heat and friction, may suggest that he was not unaware of the transfer of motion to the internal particles of bodies in inelastic collisions, as advocated by Leibniz, and the role of heat in this process. Roger Cotes expressed what seems to be the Newtonian viewpoint in his Preface to the second edition of the *Principia* (1713): 'Bodies in going on through a fluid communicate their motion to the ambient fluid by little and little, and by that communication lose their own motion, and by losing it are retarded. Therefore the retardation is proportional to the motion communicated...' At a later date, Newton's close disciple, Samuel Clarke, working probably under his direct inspiration and certainly with his complete approval, showed a clear understanding of the Leibnizian concept of *vis viva* (mv^2) and how it related to the Newtonian concept of force.

It could, of course, be argued that it is impossible to have a fully-functioning dynamics or a theory of matter without a fully explicit concept of energy or something equivalent; this is, after all, the fundamental quantity responsible for all physical changes, 'the capacity to do work'. However, despite its considerable heuristic value, Leibniz's *vis viva* is only a metaphysical notion, and 'energy' is not a concept that is privileged in the way that Leibniz seems to have believed. Taken as a mechanical idea, energy is no more 'meaningful' a concept in physics than 'action' or 'momentum' or 'entropy'. It is simply a mathematical construct with certain predictable properties, and it is only 'conserved' in the universe by

---

[110] *Principia*, Axioms, or Laws of Motion, Scholium. The apparatus for the pendulum experiments, which was no doubt similar to that which Christopher Wren had used in demonstrating the kinematic laws of impact before the Royal Society in 1668, has a generic similarity to that used in the modern multi-pendulum device known as 'Newton's cradle'.

[111] Possibly, here, we have a hint that the third law really involves the entire Universe, not just two isolated bodies.

incorporating potential energy, which is certainly not an example of 'living force', and heat, which cannot be equated with the 'capacity to do work', as energy terms, and by making it interconvertible with mass. In other words, while the mathematical principle of conservation of energy is essential to physics, the metaphysical connotations are, in the strictest terms, an unnecessary accretion, and have to a considerable degree obscured the true significance of the concept.

The question that needs to be asked is: is it possible that an 'alternative' system of dynamics could have been developed on Leibniz's principles — a system of the world which could not have been more mechanistic and which was opposite in every respect to everything in which Newton believed — without reference to the concepts of Newton? It is a question to which there is a relatively simple answer. Leibniz did not have a consistent 'system' of dynamics, based on his conservation principle, whether or not we regard it as correct; Newton, on the other hand, had a completely consistent system, even without a separate concept of energy. The very success of this system shows that energy is *not* the fundamental quantity in physics and that the conservation of energy is *not* the most important conservation principle. The historical question in this instance shows how a consideration of the circumstances in which a modern idea has developed may lead to a better understanding of how we now use that idea.

As we have already implied, there is in fact a problem with the very nature of energy. It is impossible to define in physical terms, for only 'useful' energy is the 'capacity to do work', and not all energy is 'useful'; nor are all types of energy interconvertible with each other, so even the 'conservation' of energy has no satisfactory physical description. Energy, in this sense, is, therefore, a less important quantity than force which is always the cause of physical change. Energy terms frequently become separated from their physical origin, but force terms can always be traced to one of four types of physical source; the behaviour of physical systems is determined by the nature of these physical sources and not from the total amount of energy available. The conservation of energy, therefore, is a necessary condition for any physical process but is not a sufficient one. Like the conservation of momentum, it is really only a version of Newton's equal and opposite forces, and is ultimately derived, via Newton's 'inertial force' from the conservation of inertia or mass itself.

Nevertheless, energy has tremendous importance as a *mathematical* concept, and we could hardly imagine quantum or particle physics without it. This is because it is a scalar, unlike force and momentum, and so is ideal for the treatment of many-body systems in which knowledge of sources is not required or even possible. It is the establishment of energy as a mathematical quantity derived from the primary physical concept of force that justifies our discussion of Newton's use of the concept, for as a mathematical term of this nature it occurs in many places in his writings on mechanics; it reappears, again as an unnamed quantity, in some of his writings on optics. The 'absence' of the concept from Newton's work is therefore more apparent than real, and merely reflects the fact that he did not regard it as an independent entity.

In all of these writings Newton shows a modernity and sophistication which has not yet been fully appreciated, especially in those writings on optics with close parallels to modern ideas on mass-energy.[112] The controversy with Leibniz, which followed Newton's publication of the *Principia* in 1687 and Leibniz's publication of the *Tentamen* in 1689, is only one aspect of this work, and it is one in which Newton made sure that he did not become personally involved. It is important, however, in that it enables us to fill out the details of the Newtonian position with some confidence in the absence of more detailed primary sources.

Thus, while Leibniz's metaphysical principle of the conservation of *vis viva* is important in historical terms, Newton's mathematical treatment of energy is much closer to the modern conception. Leibniz, as Westfall points out,[113] had no notion of work or potential energy; his *vis viva* was mv^2, not $1/2\,mv^2$, like our kinetic energy. He added nothing, mathematically, to the achievements in dynamics of Galileo, Descartes and Huygens, and was unable to proceed beyond the kinematics of free fall. Even though his orbital dynamics was taken directly from Newton, it was still based on the concept of centrifugal force. Though he was successful in cases where forces could be integrated over time, he was unable to integrate force over distance to calculate work done, or even to admit that this integral had significance, and so failed in his analysis of circular motion where the work integral must be equated to zero. Though he rightly emphasised the conservation of *vis viva* in elastic collisions, he was unable to give a valid derivation.

Newton's dynamics was, of course, capable of responding to these challenges and few would doubt that his mathematical treatment of problems involving energy terms was vastly superior to that of Leibniz; but, despite the intrinsic appeal of Leibniz's universal principle of the conservation of living force, it is not unreasonable to argue that Newton's fundamental or 'philosophical' understanding of the basic concept was also superior. There

---

[112]These are discussed in *Newton and Modern Physics*.
[113]Westfall (1971, 283–322).

were two vital aspects of the concept of energy which Leibniz, with his Cartesian vision of the universe as a machine kept in perpetual motion by the transfer of 'living force' between its parts, was unable to comprehend.

The first was that *vis viva* or kinetic energy is not a conserved quantity in the ordinary sense. For example, it is not conserved in interactions where there is no collision, as when a projectile rises in the Earth's gravitational field, or, in general, in any interaction where there is an interconversion between kinetic and potential energies. It is true, of course, that electromagnetic potential energy can be regarded as due to the 'kinetic' energy of the exchange photons, but this follows the potential rather than the kinetic energy equation, potential energy being twice the value of kinetic, and it is by no means certain that gravitational potential requires a similar exchange. Also, the concept of 'exchange mechanism' is itself only a heuristic substitute for more abstract modern notions, such as gauge invariance.

For Leibniz there could be no transfer of energy without collision because there could be no action at a distance. For Newton, however, the 'active principles' causing such effects as gravity and fermentation, or, as we should say, gravitational and chemical potential energies, were responsible for keeping the universe in motion, and these were due to forces acting at a distance; mv^2 was not conserved in itself, but could be restored by the equivalent degree of active principle. His mathematical analyses in the *Principia* presumably explained how this would happen. Though Newton believed in the conservation of momentum or 'quantity of motion' as a *vector*, he gave examples of how the *scalar* magnitudes of momentum (and, by implication, the total scalar kinetic energy) would increase or decrease under the action of forces, even without impact. As Samuel Clarke, Newton's brilliant defender and the principal source for his views on the energy question, wrote in his celebrated correspondence with Leibniz: 'Action, is the beginning of a motion where there was none before, from a principle of life or activity: and if God or man, or any living or active power, ever influences anything in the material world; there must be a continual increase and decrease in the whole quantity of motion in the universe'.[114]

Leibniz, and some other commentators, assumed that Newton's active principles, stemming, as he implied, directly from God, were therefore supernatural rather than natural, but Clarke denied that there was any meaningful distinction: 'a *natural* and a *super-natural* *action* of God, are terms whose signification is only relative to us';[115] and he emphasised

---

[114] Alexander (1956, 110).
[115] *ibid.*, 115.

Newton's view that no changes in motion took place without the mediation of some active principle:

> Every action is (in the nature of things) the giving of a new force to the thing acted upon. Otherwise 'tis not really action, but mere passiveness; as in the case of all mechanical and inanimate communications of motion. If therefore the giving of a new force, be supernatural; then every action of God is supernatural and he is quite excluded from the government of the natural world; and every action of man, is either supernatural, or else man is as mere a machine as a clock.[116]

Even supposed 'mechanical communications of motion' were actually the result of new action:

> I alleged [wrote Clarke] that every action is the giving of a new force to the thing acted upon. To this it is objected, that two equal hard bodies striking each other, return with the same force; and that therefore their action upon each other, gives no new force. It might be sufficient to reply, that the bodies do neither of them return with their own force, but each of them loses its own force, and each returns with a new force impressed by the other's elasticity: for if they are not elastic, they return not at all'.[117]

Clarke thus showed that bodies do not conserve motions in collisions, or when any force acts, but acquire new ones. This is because there really is no such thing as a mechanical collision, or a complete coincidence in space of two material particles.

Newton and Clarke seem to have deliberately blurred the distinction between the natural and supernatural actions of God in the Universe. No doubt this was partly a result of Newton's typical process of keeping his options on facts which were currently inexplicable, but it probably also reflects a genuine Newtonian belief that the actions of God could not be fathomed to the extent that some could be described as 'natural' and others as 'supernatural'. In other words, it would seem that the drift of the argument presented by Clarke to Leibniz was that all the actions of God in the Universe would *appear to man* as though they were natural.

They would, in fact, follow what could be described as mathematically predictable 'laws of nature', but these laws would themselves be unfathomable, or, as we would say, *abstract* in principle. Possibly gravity followed such a law, possibly it could be explained mechanically by some other cause which itself was nonmechanical; but the final cause would most certainly

---

[116] *ibid.*, 51.
[117] *ibid.*, 110.

not be 'mechanical'. The ordered Universe could not have arisen only from matter and motion; something else must have been responsible for creating the systems of stars and planets. This was attributed by Newton to his set of 'active principles', but it seems that he believed that, once these were taken into account, it would then be possible to explain all natural phenomena, whether terrestial or cosmological, using the concept of matter in motion.

Newton was never sure whether his fundamental principles had reached the ultimate level of abstraction but in his most mature thinking he never sought a non-natural explanation for a phenomenon if a natural one was available. He was prone to claiming that various aspects of the Solar System were a result of divine action rather than 'natural' causes but, wherever he actually specified the nature of such action, he always did so in terms which we would regard as natural. His 'active principles', whether mechanical or abstract, were undoubtedly natural. As he wrote in the Latin version of Query 23/31: 'natural phenomena show that such principles really exist: though it has not yet been explained what their causes are.[118] It is hardly possible to doubt, therefore, that Newton's 'active principles' followed mathematical laws of nature and did not represent an arbitrary kind of supernatural interference as implied by Leibniz; and, in separating them from the mechanical motions which were their effects, he was able to provide a much subtler and more powerful version of dynamics than his mechanistic opponent.

The abstract nature of the active principles ensured that potential energy remained differentiated from kinetic, whereas, in Leibniz's system all energy terms were subsumed under a metaphysically exciting, but not mathematically helpful, concept of 'living force', whose ultimate origin was the Cartesian principle of explaining the universe by 'matter in motion'. It was an exact parallel to the distinction which Newton had made between active and passive forces, and, indeed, was a direct result of that distinction. The evidence suggests that Newton, or at least his closest supporters, fully understood the concepts involved in Leibniz's principle of the conservation of *vis viva*, and understood them better than Leibniz himself. Samuel Clarke, in his correspondence with Leibniz, described, for the case of a projectile, how the concept of kinetic energy or *vis viva* results from considering variation of the body's motion with respect to displacement, while that of momentum results from considering the variation with respect to time. That is, if one considers only the displacement of a body acted on by a force, the concept

---

[118]Latin version of Q 23/31, McGuire (1968, 195).

of kinetic energy will be taken as fundamental, and if one considers only the time during which the force acts, the concept of momentum. Though stated for a constant force, the point is clearly made, and could be extended by integration to any force. He subsequently extended these ideas in a separate paper.[119]

Newton did not choose to base a dynamical theory on the concept of *vis viva* because he did not believe that it was fundamental. It was mechanistic rather than abstract, and, apart from making everything depend on an impact mechanism which did not really exist, it assumed perpetual motion. A dynamics based on *vis viva* emphasised a unity in nature but did not allow scope for future development because it did not distinguish between the different sources of active principles and did not distinguish between usable and dissipated energy. Finally, it made the concept of conservation of energy seem to be a more important principle than the conservation of mass, an emphasis which has continued to this day, even though mass (with its additional gravitational and inertial implications) is really the more fundamental concept. Newton, however, never doubted that the conservation of mass or, as he called it, the force of inertia, was one of the most fundamental principles in the whole of physics, and he would never have contemplated the possibility of making a merely mathematical construct, such as energy, into a concept with more fundamental significance. The power of Newton's methodology, which is apparent from his success in applying it to the great world system, was ultimately due to the fact that he incorporated mass into his system as a fundamental parameter on the same level as space and time, and this is reflected in his priorities.

A footnote by Clarke to his fifth letter to Leibniz has a striking passage which seems to show the conservation properties of 'energy', 'momentum' and other mathematical constructs are fundamentally dependent on that of mass:

> The *vis inertiae* ... is always proportional to the quantity of matter; and therefore continues invariably the same, in all possible states of matter, whether at rest or in motion; and is never transferred from one body to another. Without this *vis*, the least force would give any velocity to the greatest quantity of matter at rest; and the greatest quantity of matter in any velocity of motion would be stopped by the least force, without any the least shock at all. So that, properly and indeed, all force in matter either at

---

[119] Alexander (1956, xxxf), 121–125.

rest or in motion, all its action and reaction, all impulse and all resistance, is nothing but this *vis inertiae* in different circumstances.[120]

This extremely interesting passage illustrates better than anything else the true sophistication of the Newtonian approach to 'energy' as a mathematical construct related to mass. It is the clearest possible statement of the conservation of mass implied in Newton's description of matter as composed of ultimately indestructible particles, with the even more important implication that the conservation laws involving mass- or force-based quantities are ultimately nothing other than the conservation of mass itself.[121]

---

[120] *ibid.*, 111–112.

[121] It could even be said further that, rather than being seen as anti-Newtonian, as it frequently is, *Mach's principle* could actually be regarded as the logical consequence of Newton's belief in the inertial mass of a body as a *force* (*vis inertiæ*), effectively the reaction force to the accumulated gravitational interaction of all the other bodies in the universe. It is conceivable that the *seeming* inexplicability and pointlessness of the concept of *vis inertiæ* is simply a result of the failure of current advanced thinking to create a structure to match that of Newton. However, in the Newtonian case, the real inertial force implied by Berkeley and Mach would need to be defined as fictitious to avoid postulating a geocentric universe (see 5.8), and the same would be true of Einstein's general relativity.

Chapter 5

Gravity

5.1. The Universal Law of Gravitation

The law of gravity proved to be a major watershed in the history of physics, not because it was an inverse square law or because it applied to the celestial bodies, but because it applied to every particle of matter in the universe acting on every other, and, in conjunction with Newton's system of dynamics, in particular the third law of motion, this meant that physics became truly universal. The law states that there is a force of attraction between all particles of matter or elements of mass distributions in the universe, which is proportional to the masses of the particles and inversely proportional to the square of the distance between them. The attraction is mutual, as required by the third law of motion: 'the equal attraction of bodies is a branch of the third law of motion'.[1] It is now most conveniently expressed by the equation

$$F = -G\frac{m_1 m_2}{r^2},$$

where m_1 and m_2 represent the masses of the interacting particles, and r their distance of separation, while the negative sign indicates a force of attraction; G, the universal gravitation constant, or Newton's gravitational constant, is there to relate the units defined for force with those defined for mass and distance. This was the first universal law ever proposed, precisely because it was about mass, which Newton recognised as a universal source of interaction in nature.

There is a version of the law in the third Book of the *Principia* (Proposition 7), but the best statements in Newton's own writings come from a proposed preface for the *Opticks* of 1704: 'all bodies in ye Universe have a tendency towards one another proportional to ye quantity of matter

---

[1] Newton to Cotes, 28 March, 1713; *Corr.* V, 396–399, 397.

in them & ... this tendency in receding from the body decreases & is reciprocally proportional to ye square of ye distance from ye body'.[2] Another version can be found in *The Elements of Mechanicks*, a manuscript dating from some time after 1687: 'All bodies are impenetrable and have a force of gravity towards them proportional to their matter and this force in receding from the body decreases in the same proportion that the square of the distance increases and by means of this force the Earth Sun planets and comets are round'.[3]

The universality can be assumed by its conforming with experimental observations: 'Universally, all bodies about the Earth gravitate towards the Earth; and the weights of all, at equal distances from the Earth's centre, are as the quantities of matter which they severally contain. This is the quality of all bodies within the reach of our experiments; and therefore (by Rule III) to be affirmed of all bodies whatsoever'.[4] It means that even the individual particles of bodies attract those of other bodies, however distant:

> It proceeds from some cause that penetrates to the very centres of the Sun and planets without any diminution of its virtue, and which acts not on the surface of particles alone, but on all matter to the very centre since its action is proportional to the quantity of matter in all bodies. It proceeds from a cause whereby the single particles of bodies act at immense distances with a virtue decreasing in the duplicate ratio of the distances ... [5]

The gravity of any body or system arises from the sum of the gravities of its component parts, however small. 'Gravitation towards the Sun is made up out of the gravitations towards the several particles out of which the Sun is composed; and in receding from the Sun decreases accurately as the inverse square of the distances ... even to the remotest aphelion of the comets' ... [6] 'But we have before proved that these forces arise from the universal nature of matter, and that, therefore, the force of any whole globe is made up of the several forces of all its parts'.[7] 'Therefore the force of gravity towards any whole planet arises from, and is compounded of, the force of gravity towards all its parts. Magnetic and electric attractions afford us examples of this;

---

[2] Cambridge University Library, MS Add. 3970.3, ff. 475ʳ–482ᵛ, f. 479ᵛ, NP.
[3] *The Elements of Mechanicks*, Unp., 167.
[4] *Principia*, III, Proposition 6, Corollary 2.
[5] draft General Scholium (MS A), 1713; *Unp.*, 353.
[6] *Principia*, 1713, General Scholium.
[7] *De Systemate Mundi*, §25; 571.

for all attraction towards the whole arises from the attractions towards the several parts'.[8]

In the *Principia*, Newton built a system based on an abstract concept of action at a distance between point particles of matter. The results of the actions could be described even if the causes were not understood. Gravity was one particular force, acting between all particles with the property *mass*, and, as such, could be shown to be responsible for all the larger-scale actions in the universe: the motions of the heavenly bodies, the tides, the precession of the equinoxes, and the laws of falling bodies. The three 'laws of motion' which he had described as applying to all mechanical motions (only the last two of which were really necessary), could now be used, in conjunction with the additional gravitational principle, to explain such large-scale events with mathematical precision.

The law of gravity, because it was so unprecedented in its universality and abstractness, was only slowly developed into its final form, and it was only given in the *Principia* after a long build-up in which the case was made with intense mathematical rigour. After the fundamental principles of centripetal force laws and the resulting orbits were established with great thoroughness in the first book, the third book applied the inverse-square relation, successively, for Jupiter's satellites, for the planets moving round the Sun, and for the Moon as a satellite of the Earth,[9] and then asserted that, in each of these cases, a centripetal force drew the satellite body from a rectilinear path and kept it in orbit.[10] He noted that even a small deviation from an inverse-square law in the case of the planet would produce a notable precession of the line of apsides.[11] Then, after a demonstration that 'all bodies gravitate towards every planet', with a weight that is proportional to the quantity of matter,[12] he comes to the conclusion that an inverse-square law gravitational principle must be universal.[13]

---

[8]*Principia*, III, Proposition 7, Corollary 1.

[9]III, Propositions 1–3, assuming in the last case that observed deviations in the lunar motion were due to an additional attraction by the Sun (Proposition 3 and Corollary).

[10]III, Propositions 4–5 and Corollaries 1–2 to Proposition 5.

[11]III, Proposition 2.

[12]III, Proposition 6.

[13]III, Proposition 7 and Corollary 2. In the *Principia*, these propositions had been preceded by Phaenomena 1–6, which stated the correctness of the Copernican system, the area law and Kepler's third law for the 'primary planets' and for the satellites of Jupiter and Saturn, the third law for Earth, and the area law for the Moon, with the statement that the law does not apply exactly 'since the motion of the moon is a little disturbed by the action of the sun'. The term 'Phaenomena' was first introduced in the second edition,

Newton also argued for the universal principle on more general grounds, as in the Rule III he added to the Rules of Philosophical reasoning in the 1713 edition of the *Principia*:

> ... if it universally appears, by experiments and astronomical observations, that all bodies about the Earth gravitate towards the Earth, and that in proportion to the quantity of matter which they severally contain; that the Moon likewise, according to the quantity of its matter, gravitates towards the Earth; that, on the other hand, our sea gravitates towards the Moon; and all the planets mutually one towards another; and the comets in like manner towards the Sun; we must, in consequence ..., universally allow that all bodies whatsoever are endowed with a principle of mutual gravitation. For the argument from the appearances concludes with more force for the universal gravitation of all bodies than for their impenetrability; of which, among those in the celestial regions, we have no experiments, nor any manner of observation.

The constant G was defined after Newton's time, but knowledge of its value is entirely equivalent in Newtonian theory to knowledge of the mean density of the Earth (ρ). If g is the acceleration of free fall and R the radius of the Earth, both known at the time, then G can be found from $3R^3/4\pi\rho g$. The mean density of the Earth had not been measured at the time, but Newton guessed, with remarkable accuracy that it was between 5 and 6 times that of water, and so the constant is correctly named after him. (The accepted value is 5.46.)[14] Experimental values for the mean density were found by the Astronomer Royal, Nevil Maskelyne, in 1774, from measurements of the deflection of a plumb line on Mount Schiehallion in Scotland, and, then, more accurately, by Henry Cavendish, in 1798, using a torsion balance in his own laboratory.[15] Proposition 92 of *Principia*, Book I, gives the basis of the Cavendish experiment. 'An attracting body being given, it is required to find the ratio of the decrease of the centripetal forces tending to its several points'. Newton proposes that the experimenter should take a regular body, such as a sphere or cylinder (Cavendish used spheres), and then find the force of attraction at several distances: 'the law of attraction towards the whole made known by that means, will give the ratio of the decrease of the forces of the several parts'.

---

of 1713. In the original edition, Phaenomena 1, 3–6 were termed Hypotheses 5–8 and 9, and Phaenomenon 2 was absent.

[14] *Principia*, III, Proposition 10.

[15] Maskelyne (1775), Cavendish (1798).

5.2. Early Thoughts on Gravity

Though it had a long gestation, Newton, in later life, remembered that the idea of universal gravitation had come to him as a young man when, in his garden at Woolsthorpe, in 1666, he saw an apple fall. This account has created a degree of scepticism, some authors even stating it as a 'fact' that the incident never happened. Originally, it was claimed that the story came from Voltaire.[16] However, it was subsequently discovered that Newton had spoken about it himself in conversations recorded by John Conduitt[17] and William Stukeley.[18] He explained that his insight, at that moment, was that he saw that the force that kept the Moon in its orbit was the same as that which caused the apple to fall, but attenuated by the inverse square principle. He immediately tried out a calculation which gave the answer 'pretty nearly' but not exactly, with perhaps a 10% discrepancy, mainly because he used an inaccurate value of the radius of the lunar orbit. He suspected the influence of Descartes' vortices, which he had yet to eradicate from his thinking, though an obvious other source of discrepancy would be that the Moon's orbit was elliptical, rather than circular, and with many irregularities already known.[19]

For a long time, it was doubted that this test had ever been made, until the paper was discovered in the middle of the twentieth century.[20] The document was written in Latin, suggesting that it was intended as part of a more formal presentation rather than being simply a draft of his first thoughts on the subject. With its discovery, the scepticism shifted to what was actually meant by the calculation. A number of authors, beginning with Whiteside, and without the knowledge of crucial aspects of Newton's early work on the dynamics of orbits that were only established later, have emphasised its supposed limitations, in the process offering the perfect opportunity to the conspiracy theorists who have always operated on the periphery of Newton studies.[21] Certainly, this is not Newton's final law of

---

[16] Voltaire (1727, 1733/1980, 75).

[17] KCC, MS Keynes 130.4, 10–11, Westfall (1980, 154).

[18] Stukeley (1752, 19–20), recalling a conversation on 15 April 1726.

[19] An annotation Newton made c 1670 in his copy of Vincent Wing's *Astronomia Britannica* (1669) discusses the possibility that the solar vortex might press on the terrestrial vortex carrying the Moon to reduce the width of its elliptical orbit by 1/43 (Gjertsen, 1986, 613).

[20] *On Circular Motion*, ULC, Add MS 3958 ff 87, 89, Hall (1957); also in *Corr.* I, 297–303, Herivel (1965, 192–198).

[21] Whiteside's monumental edition of Newton's *Mathematical Papers* is one of the greatest feats of intellectual scholarship performed during the twentieth century, and is an indispensable guide to Newtonian mathematics, which I have found especially valuable in chapter 2 of this book. The monumentality of the work has led to it being taken, with some justification, as an almost final authority on most matters. However, discoveries made

gravity of 1687, conceived in the most austere and abstract terms that any physical law has been conceived before or since. Newton would only establish that completely abstract way of thinking after a long and gruelling struggle against both himself and virtually all his contemporaries. But it is a good deal closer than their descriptions would suggest.

Inevitably, there was a reaction by certain scholars against the mythology associated with the 'annus mirabilis' of 1665–1666, and a desire to establish more precisely the exact historical process by which the law of gravity was finally established. A problem with this approach, however, has been an overemphasis on the limitations of what Newton discovered in his early work, and a lack of realisation of the fact that his approach to the mathematical treatment of dynamics was abstracted at a very early stage from any consideration of 'physical' hypotheses that might be needed to explain the results. Newton certainly investigated such hypotheses but his mathematical work in dynamics was, from the beginning, a quite separate enterprise. This dualistic way of thinking was characteristic of many aspects of his work — for example, in the simultaneous use of figurative language in his investigation of alchemical authors and of sober chemical language in his experimental notebooks and his published works, or in the simultaneous pursuit of rational dynamics and prophetic interpretation, or in the use of action-at-a-distance forces alongside aethereal mechanisms. Newton was quite capable of looking at the same problem simultaneously from quite different viewpoints in a way that Robert Hooke, for instance, never could. It eventually led to his establishment of a purely abstract approach to physics, quite separate from physical hypotheses, *whether or not these were true.*

One consequence has been that such scholars have almost overlooked the fact that Newton's few pages worked on in the autumn of 1666 or around that time, possibly soon after his return to Cambridge in 1667, contain nearly all the vital ingredients that went into his later, more precise, definition of

---

since Whiteside completed his work have led to some of his opinions needing to be revised. Particular examples from the *Principia* where this is desirable include his comments on the solid of least resistance, the integrability of oval curves, the status of the lunar theory, and the sine-squared law of resistance. The work of Brackenridge and Nauenberg on Newton's curvature method, in particular, has led to significant developments in our understanding of Newton's work on orbital theory, and this, I believe, makes a revision in our understanding of Newton's early gravitational theory an absolute necessity also. Whiteside's emphasis, though influential on scholars of his generation such as Westfall, Cohen and Hall, seems odd even in the context in which he first discussed it; however, in the light of subsequent developments, it appears to be completely untenable.

the law: it used an inverse-square law, derived from an orbital mathematical equation that he had found for himself; it applied to Sun and planets, Earth and Moon, and defined the orbits of satellites to agree with Kepler's third law; it assumed that the orbital motion was produced in the same way as the free fall gravitation or weight that acted on bodies at the Earth's surface, with an acceleration that was independent of the size or mass of the body; in being related to weight, it depended on mass, but the acceleration was the same on all bodies in the same place; at the Earth's surface it was clearly attractive, and, in being of the same form as weight at a further distance, it was also attractive at all distances, whether or not it was 'caused' by some kind of 'endeavour' outwards.[22] All this was *completely independent of whatever hypothetical mechanism he might have imagined could have been the physical cause.* In addition, a second calculation assumed that, with respect to the Moon, there was competition between the attractions by the Earth and Sun, which meant that it was already seen as a multibody process.

In the sense that this was not yet the abstract law of attraction without physical explanation between all particles of matter in the universe, which would revolutionise the *meaning* of physics as well as its application, the law of 1666 was not yet universal, but it is difficult to see how one can equate terrestrial and lunar gravitation by the same law without having *some* idea that it is universal, and another kind of universality is implicit in the fact that it affects all bodies on Earth, and, so potentially implies an interaction between all bodies. Newton's manuscript also contains a statement of yet another kind of universality in its proposal of an inverse square relationship between the Sun and all the known planets (Mercury, Venus, Earth, Mars, Jupiter and Saturn) from Kepler's third law,[23] and a second, less stringent, test of the inverse square dependency from a calculation that the gravitational force between Earth and Moon must be greater than that between Sun and Moon to ensure that the Moon always faces the Earth rather than the Sun. A connection of this interplanetary force specifically with *gravity* is impossible to deny if the calculation is made with reference to the acceleration of free fall, whatever additional reference is made to Cartesian vortices or streams of aether. That the force must be dependent on mass is obvious from Newton's very early understanding of

---

[22]Westfall (1971, 358–359), is one of the few authorities who have come near to this realisation. Others have denied that the calculation refers to gravity at all (which it clearly does at the Earth's surface) or that there is any concept of attraction, which, as Westfall realised, is unavoidable if we invoke gravity.

[23]Here the relationship was notably exact (Westfall, 1971, 359), suggesting that any deviation in the case of the Moon might mean that the lunar orbit was a special case.

the relationship between mass and weight. The extension of the force to a *mutual* interaction might not have been so obvious, but Newton seems to have been familiar with the third law of motion long before its first formal statement in *De motu* and the *Principia*.[24] Treating astronomical bodies on the same basis as falling objects suggests that no distinction was made between them.

Chandrasekhar has made the point that, when Newton equates the ratio of the acceleration at the radius of the Moon's orbit to that at the surface of the Earth, to the squared ratio of the radius of the Earth to the radius of the Moon's orbit, he necessarily implies that the attraction at the Earth's surface requires its entire mass to act as though concentrated at the centre, which is definitely against 'common sense'.[25] Newton said later that he had long believed that the force inside the Earth would not be inverse-square. Either gravity did not extend to the centre of a solid body, or, if it did, the mass of the body would have an effect on the relationship between force and distance. A linear dependence on the distance would have been a relatively straightforward consequence on the assumption of uniform density, but Newton, dissatisfied with a '*quam proximé*' result, required the theorem of 1685 that all the mass of the sphere acted as though at its centre before he could prove it.[26]

The assumption has always been made that the late 1660s calculation used a very restricted understanding of the inverse-square law force, but it is not obvious where the restriction actually begins. Though Newton took a great deal of effort in convincing himself later of the universality of the gravitational law, this doesn't necessarily imply that he hadn't previously considered its wide application. If the inspiration came from analytical reasoning, as the suddenness of the revelation would seem to suggest, then the idea would have been present from the beginning, even if not fully recognised. The lack of understanding of Newton's curvature method of calculation has led to erroneous conclusions about the Moon test and what he achieved in 1666; it may also be that the idea on which it was based included a closer approximation to universal gravitation than has yet

---

[24]Whiteside, *MP* V, 148–149, note 152 and VI, 98–99, note 16. Apart from the version in his conservation of momentum calculations for oblique collisions in the *Waste Book* (Ax. 119, 121, MS IIe, Herivel, 1967/1970, 129), his lectures on algebra include a statement which is almost identical to that in the *De motu corporum, Definitiones*.

[25]Chandrasekhar (1995, 6).

[26]Newton to Halley, 20 June 1686, *Corr.* II, 435–436, 435: 'I never extended the duplicate proportion lower then to the superficies of the earth & before a certain demonstration I found the last year have suspected it did not reach accurately enough down so low'.

been appreciated. The fact that it didn't match up yet to the perfectionist standards that Newton himself would introduce into physics does not excuse some of the more extreme exercises in revisionism that became fashionable in the second half of the twentieth century.[27]

While it is not at all obvious what relation the force that Newton imagined has to the endeavour outwards, implied by Descartes' *conatus*, the most available physical model at the time, a comparison of the lunar orbit with the gravity of a terrestrial object implies that each is concerned with a force towards the centre of the Earth, the kind of inward force that Newton already knew was required by a circular orbit, regardless of whether he also believed there was a balancing force outwards, as in the hypothesis of Giovanni Borelli, published in 1666. It is clear, also, that a force $\propto 1/R^2$ could be easily derived for a *circular* orbit, from Kepler's third law, $T^2 \propto R^3$, once Newton knew the law of centrifugal force, mv^2/R (as this is also equivalent to $m\omega^2 R = 4\pi^2 mR/T^2$). Richard S. Westfall, who was writing before a more extensive knowledge of Newton's early work on orbital dynamics became known and when many people doubted that he could have been inspired by the fall of an apple, wrote that 'Newton must have had something in mind when he compared the moon's centrifugal force with gravity, and there is every reason to believe that the fall of an apple gave rise to it. Though he did not name the force explicitly, something had to press back on the moon if it remained in orbit. Something had to press back on the planets'.[28]

A question that has yet to be answered, of course, concerns the use that Newton made of aether theories in his early 'physical' understanding of gravitation. The early *Quaestiones quaedam philosophicae* of c 1664 has an entry 'Of Gravity and Levity', preceding the date of the Moon test, that uses ideas very similar to those discussed in the later *Hypothesis of Light*, seemingly (at least, in the latter case) with an inverse square law in mind.[29] Effectively, Newton put forward a purely mechanical impact theory, with aether particles descending towards the Earth's centre, and forcing material particles down with them towards the Earth's surface, the downward aether stream becoming more concentrated as it came closer to

---

[27]Remarkably, the special pleading that was once attributed to those who claimed that Newton had not made the 1666 test before the paper was found, has been transferred to those who claim in support of Hooke that the mathematical calculation cannot mean what it actually says it does. It permeates many popular accounts of Hooke, and even one or two scholarly ones. It is important to point out the superiority of Newton's work, even as early as 1666, over that of Hooke later in being founded on valid dynamical principles.

[28]Westfall (1980, 155).

[29]QQP, 'Of Gravity and Levity, 97ʳ, NP.

the Earth and occupied a decreasing surface area. A continuing circulation was then maintained by an upward stream of aether particles rising in a transformed state from the centre of the Earth.

There is, however, no mention of such ideas in the paper of the late 1660s. With the rapid advancement of his mathematical skills, Newton could certainly have derived the mathematical results and completed the Moon test without considering the full implications of the aether interpretation; alternatively, he could have had the aether interpretation in mind from the beginning, linking the areal dependence of the intensity at the Earth's surface with the inverse square relation he had now derived mathematically. As yet we simply don't know, but in connection with the purely mathematically-derived law he was using, it is of no significance. Unlike Hooke, who never worked out a mathematical theory of any physical process independently of an assumed physical mechanism, and, in fact, never worked out any such theory entirely on his own account, Newton was perfectly capable of developing a mathematical theory of a dynamical system irrespective of the 'physical' description of the cause that produced it, and he was already doing this with rigid body dynamics in his early manuscripts. Whatever the details of the explanation, it is simply preposterous to claim that Newton knew nothing of an inverse square law involving gravity before 1679.

The main problem with the 1660s calculation had been that the value of the Earth's radius Newton had used was inaccurate and not the best then available. By 1672, having edited the *Geographia generalis* of Bernhard Varenius,[30] he had good measurements of the Earth's radius, for the book contained Snell's 'nearly exact' value for the Earth's circumference of 1617.[31] Newton actually used it to correct Varenius's erroneous estimate of the distance at which a mountain of 1 German mile in altitude could be seen at sea, from 29.25 to 41.5 German miles.[32] In 1672, therefore, he had every opportunity to revisit his calculation and find that it gave much better agreement with his theoretical postulate. We know that, by 1673, Newton had prepared 'divers astronomical exercises, which are to be subjoined to *Mr Nicholas Mercator's Epitome of Astronomy*, and to be printed at Cambridge'.[33] The only one of which we have any details is a theory of

---

[30] According to Whiston (1749), Newton (at some date) used Picard's value of the radius from *La mesure de la terre* (Paris, 1671, 1676, referred to in *Phil. Trans.*, 10, 261, 1675 and 11, 591, 1676) to correct his 'old imperfect Calculation'.

[31] Hall (1992, 110).

[32] Cajori (1934, 664), note 40 to *Principia*, III, Proposition 4. Cajori cites *Mathematical Gazette*, 14, 415, 1929.

[33] Sir Edward Sherburne, *The Sphere of Marcus Manilius*, 1675, 116.

the libration in longitude of the Moon, which was discussed in lost letters to Mercator, but was included in Mercator's *Institutiones astronomicae* in 1676, and later referred to in *Principia*, Book III, Proposition 17.[34] This may well connect with the second test of the inverse-square law which Newton included in his paper of the 1660s, for it was at exactly this moment that Newton quoted the precise calculation from this document in a letter intended for Christiaan Huygens, which highlighted the fact that he had considered the possibility that the Moon did not appear to rotate with the Sun because the Earth's force on it was greater than that of the Sun.[35] Mercator's text had been intended for use at Cambridge, and Newton annotated it, possibly becoming properly familiar for the first time with Kepler's area law, which was discussed on p. 145 of Book II, though Mercator himself rejected it in favour of a hypothesis of his own.[36]

On 20 June 1674, at the prompting of John Collins, Newton discussed parabolic projectile motion under the uniform gravity at the Earth's surface, suggesting that air resistance would mean the horizontal components of a bullet's speed decreasing in equal time intervals in something like a geometrical progression, a principle which he may have seen as equally applicable to orbital velocities.[37] At about the same time, or soon after, he wrote the first version of the work that was subsequently sent to the Royal Society, in 1675, as *An Hypothesis Explaining the Properties of Light*.[38] Here, also, he revisited his early speculations on gravity from *Quaestiones quaedam philosophicae*, and also some from the seemingly more 'alchemical' text, *Of Natures Obvious Laws & Processes in Vegetation*, of c 1672.[39] The aether theory in both these works, had a number of similarities to the one described in the entry on 'Gravity and Levity' in the *Quaestiones*. There were again streams of aether particles directed downward towards the centre of the Earth and upwards from it. Terrestial gravity was still seen as the mechanical effect of the pressure produced by the bombardment of aether particles, but now more specifically attributed to the existence of an aethereal density gradient, the aether in the confined spaces being less dense than that in the freer spaces outside. Unlike the aether described in the 1660s, however, this one was not purely mechanical. *Of Natures Obvious Laws* spoke about a

---

[34] III, Proposition 17.

[35] 23 June 1673, *Corr.* I, 300. Hall (1957, 68). The passage was not included in the final text transmitted to Huygens.

[36] It was also discussed in a paper by Mercator in the *Philosophical Transactions* of 1670.

[37] *Corr.* I, 309.

[38] draft for *An Hypothesis Explaining the Properties of Light*, possibly early 1675, NP.

[39] NPA.

very active 'subtil spirit' within the aether, while the *Hypothesis* attributed gravity to a specific and specially active component, 'something very thinly and subtilly difused through it', rather to than the entire medium.

It seems also that the aether theory in the *Hypothesis*, at least, if not in the earlier works, implied an inverse-square law force for gravity, though this was never explicitly stated, and in this case it extended beyond the terrestrial example. According to the *Hypothesis*, the Earth and Sun and other celestial bodies are places at the centre of a continual cycle of transformations of the aether from the spaces surrounding them. A stream of aether from the 'aetheral spaces' moves continually through all parts of the Earth, and loses momentum and increases in density as it interacts with the Earth's solid matter; the condensed aether then escapes to create the Earth's atmosphere, before proceeding further into the 'aethereal spaces', where it takes on its original form, ready to repeat the cycle. In Rosenfeld's modern reconstruction, S aether particles per unit time moving in a constant inward stream with radial velocity v, have a density $Sm/4\pi R^2 v$ at a distance R from the centre of the Earth, and this will increase in inverse ratio in proportion to the velocity. The stream exerts a pressure $Smv/4\pi R^2$ towards the centre of the Earth's solid matter. If there is no significant change in the velocity, this force will be inversely proportional to the square of the distance, exactly as Newton had calculated for the attractions of the Sun on the planets and the Earth on the Moon by Kepler's third law. In this way, gravitation would emerge as a universal force with an inverse-square attraction, produced by mechanical contact between aether and particles of solid matter.[40]

In the *Hypothesis*, Newton specifically postulates that the Sun absorbs the universal aethereal matter to keep the planets in their orbits in the same way as the Earth absorbs it to make objects within its vicinity gravitate towards its surface. He specifically invokes a universal gravitating mechanism, the Sun acting in exactly the same way as the Earth. In 1686 he would identify this as implying that the force would be constant at the Earth's surface, and, because of the areal dependence of the pressure, be subject to an inverse square diminution upwards, but not downwards, from the surface.[41] An aether-density theory of gravity would necessarily exclude the interior parts of solid matter from an inverse-square law dependence, which explains why, as Newton stated on a number of occasions, he suspected that the force relation below the Earth's surface would be different, and also

---

[40]Rosenfeld (1962–1965, 373). See also Aiton (1969).

[41] *Corr.* I, 365–366; II, 439–440. The logic of the aethereal hypothesis would imply that, within dense matter, the aethereal pressure, and so the gravity, would decrease.

why his subsequent proof that the entire mass of a body acts as though at its centre was such a breakthrough. There is notably no discussion of vortices in the *Hypothesis*. Newton may have recognised this as the 'standard' theory of the planetary motions then available, but it was not one to which he was obliged to commit himself. In this work, and in other places, he is clearly presenting a different theory of his own, one based on a relatively minimalist treatment of forces and aethereal pressure. If he thought that Cartesian vortices might conceivably have an effect on the orbits of the planets, it was clearly not the main cause of their motions.

A slightly later version of this theory appears in a letter to Boyle, written on 28 February 1679. Here, Newton gives only the mechanical aspect of the aether density gradient of 1675, while making it responsible for cohesion, surface tension, capillarity, optical refraction, diffraction and various chemical phenomena. There is no mention of the 'vital spirit' proposed in *Of Natures Obvious Laws*, nor of the active aethereal ingredient in the *Hypothesis*, but this may be because Newton is writing specifically about the mechanical aspect. There is, however, a refinement on the density gradient theory which Newton seems to have conjured up in the act of writing the letter. He imagines that the aether particles between the pores of bodies are 'finer' or more 'subtle' than the 'grosser' ones in the free spaces outside, and that the aether particles in a body resting on the Earth's surface or suspended above it will be finer nearer the surface and grosser further away.[42] A body descending to the Earth would gradually replace its grosser particles with ones of successively finer texture, so producing the effect of gravitational attraction. Direct impact is no longer required. Despite the subtle differences between them, all four of Newton's aethereal explanations of gravitational attraction, whether terrestrial or solar, are broadly similar. Though they are all indebted to Descartes for the aethereal concept, they show no sign of interest in other Cartesian hypotheses.

Newton never gave up trying to find an aethereal explanation for the gravitational interaction, but it is clear that the explanation from 1675 (like those from c 1664, c 1672 and 1679) is a 'hypothesis', not to be regarded as on the same level of scientific explanation as a mathematical theory derived directly from phenomena such as the dynamics of orbital motion. It was not the *source* of his gravitational theory any more than it was the source of the optics it also contained; it was merely one possible physical explanation.

---

[42]There is a bizarre similarity here to the way in which the mass-energy of photons (or, indeed, of any particle) decreases and the photon frequency redshifts nearer to the Earth's surface due to the increase in gravitational force.

Much of the misunderstanding of Newton's early understanding of gravity is due to a lack of appreciation of this very important distinction. To have an aethereal explanation of gravity, or any other hypothetical one, as Newton showed in his later optical Queries, does not in the least exclude a mathematical one. We have to remember that the ideas presented in *An Hypothesis Explaining the Properties of Light* came many years after he had produced a number of documents establishing dynamical laws of motion and of circular orbits, based on largely abstract mathematically-based reasoning. Because his most fundamental work, unlike that of many of his contemporaries, was not grounded on hypotheses, Newton could simultaneously imagine speculations of various degrees of extravagance without compromising his more completely assured mathematical derivations and the results he derived directly from experiment.

In addition, Newton, again unlike many of his contemporaries, never assumed the validity of a single hypothesis while possibilities of others still existed, and his explorations of the available options and refusal to accept as *necessarily* true the one that later authors would favour has again been misunderstood. He didn't finally accept the inverse square law because it *could* explain the planetary motions but because he had proved that *no other law* was possible in relation to the phenomena. In the period 1666 to 1684 he had many options for explaining the planetary motions and his explorations of some particular ones doesn't imply lack of understanding of others. The extreme precision of his scientific methodology and his relentlessly Ockhamist search for final explanations was not, as he pointed out in other contexts, due to a lack of understanding of possible explanations, but due to the desire to establish the one that came nearest to the final fundamental truth. In the end, this approach proved itself by creating a scientific methodology that not only surpassed all others at the time, but still surpasses any alternatives today.

5.3. A New Direction

The possibility of deriving an inverse-square law of planetary motion, for a circular or near-circular orbit, became available to many after Christiaan Huygens published the law of centrifugal acceleration in 1673.[43] Newton later claimed that he had discussed the possibility of such a law of gravity with Christopher Wren, in 1677 during the months when Wren was at Cambridge

---

[43]Huygens (1673).

building the library at Trinity College.[44] Wren may well have made his own independent discovery of the inverse-square law by this date, and he had already developed an understanding of the dynamics of orbits on which Hooke seems to have drawn.[45] Newton later made the claim that: 'By the inverse Method of fluxions I found in the year 1677 the demonstration of Kepler's Astronomical Proposition, *viz* that the Planets move in ellipses, which is the eleventh Proposition of the first book of Principles ... '[46] It is by no means impossible that, under the stimulus of the discussion with Wren, he could have worked this out by the curvature method, as subsequent correspondence seems to suggest.[47] We have no justification for assuming that he made this up; all the other claims that he made have been substantiated, despite earlier scepticism.[48] And this would suggest that, rather than there being a hiatus between the revelation of 1666 and the work that led to the *Principia*, the search for a law of gravity was a relatively continuous enterprise. There are many indications that he remained actively concerned with dynamics, and didn't simply withdraw into alchemical studies as has sometimes been assumed.

The *Hypothesis*, though highly speculative, contained possibilities which Newton clearly thought to be of major importance; otherwise, he would never have risked sending it to the Royal Society. However, the work on gravity certainly took a new direction when, in late 1679 and early 1680, Newton received letters from Hooke, asking him to consider the problem of the force keeping the planets in orbit, and stating his belief that the force must involve an inverse-square law relation.[49] Hooke first wrote to Newton on 24 November stating his hypothesis 'of compounding the celestial motions of the planetts of a direct motion by the tangent & an attractive motion towards the central body' and asking Newton to state any objections he had towards it. He also asked for Newton's opinion on his newly published law for the spiral spring.[50] Newton wrote on 28 November, not replying to

---

[44]Newton to Halley, 27 May and 20 June 1686, *Corr.* II, 433–434, 435.

[45] Halley to Newton, June 1686, *Corr.* II, 441–442. Bennett (1982, 60–62). Wren had the idea but never made any claim that he had actually solved the problem, and he was certainly clear that Hooke hadn't either.

[46]draft letter to Des Maizeaux, c 1718, Cohen (1971, 295).

[47] Newton to Hooke, 13 December 1679, *Corr.* II, 307–308.

[48]Nauenberg (1994, 245), has said that there is no reason to believe that the claim for a demonstration in 1677 is 'bogus history'. There is, in fact, no general evidence of Newton actually falsifying evidence in other cases.

[49]Hooke to Newton, 24 November 1679, *Corr.* II, 297–298.

[50]*ibid.*

Hooke's enquiry, but providing, with an accompanying diagram, 'a fansy of my own about discovering the Earth's diurnal motion, a spiral path that a freely falling body would follow as it supposedly fell to Earth, moved through the Earth's surface into the interior without material resistance, and eventually spiralled to (or very near to) the Earth's centre, after a few revolutions'.[51] Clearly, this was for an Earth imagined as a centre of force rather than as a material body, and it assumed that the observer rotated with the Earth.[52] Hooke, however, who effectively assumed the observer was at rest and not situated on the Earth, considered that Newton had made a significant error and replied on 9 December with his own diagram showing that, in the absence of resistance, except at the equator, the path would be an ellipse, though, in a resistive medium, it would spiral. Hooke also proposed a 'correction' of Newton's suggested experiment for observing the body's easterly deflection, saying that the body would fall as much to the south as to the east.[53]

Despite the fact that he had promised to keep the correspondence confidential, Hooke could not prevent himself from reporting on his correction of Newton's 'error' to the Royal Society, a fact which made Newton so angry that he could hardly bring himself round to make a reply. However tempting it may be to present Hooke as a 'victim' in the subsequent quarrel with Newton, it would be foolish to deny that Hooke's frequent desire to score points over his rivals was an unattractive aspect of his character, and that it is not surprising that it finally led to his undoing.

Newton's original carelessly-produced diagram (which does not match his text, as it shows the spiral reaching the Earth's centre before the completion of the first revolution) may have been drawn for a resistive medium, as Newton himself later told Halley,[54] the lack of resistance implied by the text being from the Earth's material body. He would define a spiral trajectory in *Principia*, Book II, Propositions 15 and 16 for a body in a resisting medium moving under the influence of a centripetal force inversely proportional to some power of the distance; the trajectory in his diagram actually looks like the one he had apparently discovered for an inverse-cube force.

---

[51] *Corr.* II, 300–303.

[52] Whiteside (1991, 22–23), comments that 'when it is still rare for modern commentators to notice this nicety, we should perhaps not take Hooke too much to task for his blindness to it'.

[53] *Corr.* II, 304–306.

[54] Newton to Halley, 27 May 1686, *Corr.* II, 433; Newton says of his spiral, 'which is true in a resisting medium such as our air is'.

Newton assumed that Hooke had based his trajectory on a calculation. He had no reason to suppose at this time that Hooke's grasp of dynamics was as limited as it subsequently turned out to be. He agreed with Hooke that his falling body would 'not descend in a spiral to the very centre', and would go in a direction more south-east than east, but he also considered that Hooke had himself made an error in assuming that the path would be elliptical, and 'would not descend to the center, but at a certaine limit returne upwards againe', as the gravitational force at the Earth's surface was very nearly constant and would produce a quite different orbit for a falling body. He would take 'the simplest case for computation, which was that of a Gravity uniform in a medium not Resisting', and, he says, 'in this case I granted him what he contended for, and stated the Limit as nearly as I could', which implies that he used an iterative, rather than analytic method.[55]

In a second letter to Hooke, on 13 December,[56] he included another diagram, this time for the motion of an object under a constant central force, which implies that he had a method of computing the orbit, probably based on his curvature procedure, and that he also understood the orbit's reflection symmetry, based on the conservation of energy and the symmetry under reversal of time,[57] a 'fundamental dynamic symmetry' in such cases.[58] The calculation appears to have been sound though the faulty drawing suggests the wrong angles; the correct angles are worked out, under exactly the same suppositions, in the *De motu* of 1684.[59] Hooke must have been astonished that Newton could arrive at such an orbit by calculation, for he recognised immediately that it was identical to the trajectory of a ball rolling inside an inverted 'Concave Cone', and this he knew from experiments required a constant force.[60] Newton said his method involved 'innumerable & infinitely little motions (for I here consider motion according to the method of indivisibles) continually generated by gravity...'[61]

---

[55] *ibid.*

[56] *Corr.* II, 307–308.

[57] Nauenberg (1999, 79).

[58] Nauenberg (1994, 221). The connection between symmetry principles and classical dynamics, which Newton seems to have prefigured in a number of particular instances, would become a major aspect of twentieth century physics. It is a classic example of his tendency to abstraction.

[59] *De motu*, revised version, 1684–1685, *MP* VI, 149–153. Whiteside (1989, 89–91). Nauenberg (1994, 238, 245–246).

[60] Hooke to Newton, 6 January 1680, *Corr.* II, 308.

[61] Newton to Hooke, 13 December 1679, *Corr.* II, 307–308.

But Newton also indicated that he could compute by iterative processes the orbits of bodies under central forces that increased with decreasing distance from the centre, and inferentially diverge there. His letter indicates that his solutions included the orbits for inverse-square and inverse-cube forces.

> Thus I conceive it would be if gravity were the same at all distances from the center. But if it be supposed greater nearer the center the point *O* may fall in the line *CD* [inverse square] or in the angle *DCE* or in other angles that follow, or even nowhere. For the increase of gravity in the descent may be supposed such that the body shall by an infinite number of spiral revolutions descend continually till it cross the center by motion transcendently swift [inverse cube] ... [62]

In effect, the force directed to a focus is inverse-square and that producing a logarithmic spiral must be inverse cube, and Newton could well have had *analytical* solutions in these cases, using the equivalent of a conservation of energy equation, such as he later included as Propositions 39 and 40 of *Principia*, I. In fact, the inverse-cube solution, with 'an infinite number of spiral revolutions' into the centre, can only be solved analytically and is not amenable to an iterative approach.[63]

Newton says that Hooke 'replyed that gravity was not uniform, but increased in descent to the center in A Reciprocall Duplicate proportion of the distance from it, and thus the Limit would be otherwise than as I had stated it, namely at the end of every intire Revolution, and added that according to this Duplicate proportion, the motions of the planets might be explained, and their orbs defined'. What Hooke actually wrote was: 'But my supposition is that the Attraction always is in a duplicate proportion to the Distance from the Center Reciprocall, and Consequently that the Velocity will be in a subduplicate proportion to the Attraction and Consequently as Kepler Supposes Reciprocall unto the Distance'. He also supposed that, below the surface of the earth, 'the more the body approaches the Center, the lesse it will be Urged by the attraction — possibly somewhat like the Gravitation on a pendulum or a body moved in a Concave sphere where the power Continually Decreases the nearer the body inclines to a horizontal motion, which it hath ... in the lowest point'.[64]

---

[62] *ibid.*

[63] Nauenberg (1994, 234).

[64] Newton to Halley, 27 May 1686, *Corr.* II, 433; Hooke to Newton, 6 January 1680, *Corr.* II, 309.

If Newton ever studied it carefully at all, he would immediately have seen that Hooke's proposal of an attraction 'in a duplicate proportion to the Distance from the Center Reciprocall' was not, in fact, an inverse-square law of gravitation at all. It was based, not on dynamics, but on an analogy with light and sound, and depended on an idea that gravitating sources, like light, emitted pulses. Even this analogy between the powers of light and gravity, though it alluded to Kepler's inverse-square law of light attenuation, was not founded, as we might expect, on spherical symmetry or the 3-dimensionality of space, but on a much vaguer notion of their intrinsic connectedness. It was, in many ways, less advanced than Newton's aethereal theory of 1675. There was no true understanding of what *quantity* was involved in the relation. Hooke had not, at this time, used the combination of Kepler's third law and the law of centrifugal force which had led Newton, and would subsequently lead others, to the inverse square relation for the circular case.[65] With Huygens' centrifugal equation available from 1673, this was the obvious route to follow for anyone versed in dynamics, such as Wren or Halley, but Hooke was seemingly not equipped to pursue it.

The speculations recorded in his letter show his lack of any insight at this stage into the basic principles of Newtonian dynamics or any viable alternative. His use of Kepler's erroneous version of orbital kinematics suggest that the quantity subjected to an inverse-square law attenuation was not Newtonian force, and so the idea could not have led to a physically valid law. An argument based on *vis viva*, momentum, velocity or action, for example, would be erroneous because it would lead to the wrong force relation. In the case of planetary motions, Hooke combined a faulty law of velocities from Kepler, that the speed was inversely proportional to the distance from the focus, and a faulty law, from Galileo's $v^2 = 2as$, relating force to velocity squared. Hooke seems to have based his argument on an idea similar to *vis viva*, or 'work',[66] which would have led to an inverse cube relationship between force and distance, and a completely unstable non-conic orbit. In addition, the decreasing 'power' towards the centre supposed to act below the Earth's surface could not have been gravitational force or acceleration, which would have been constant in the analogies he used, but again something like velocity squared. Essentially, Hooke's idea is not

---

[65]*MP* VI, 199.

[66]Westfall noted (1971, 272) that Hooke's definition of 'force' is 'similar to our concept of work'. We should also realise that Hooke's proposals can only be made to sound like our abstract physical laws by selective quotation. Much more than Newton, he was tied to 'physical' models which seldom coincide with the kind of physical explanations that we would find acceptable today.

merely mathematically inadequate, as is frequently made out (which merely highlights the incorrect Keplerian relation), but completely wrong *physically* (because it uses the wrong physical concept), a rather more serious failing. The inverse-square principle he derived from Kepler's law of light intensity is not applied to anything that makes physical sense.

Hall considered Newton's account of the correspondence 'quite fair to Hooke'.[67] It was even generous as it did not continue with Hooke's attempted justification based on Kepler's erroneous reciprocal relation between velocity and distance or his use of a 'force' which was proportional to velocity squared and so not equivalent to force in the Newtonian sense at all. He also said that the correspondence with Hooke had put him on to the method of explaining planetary orbits using the law of areas, an admission he need never have made if he had wanted to deprive Hooke of having had any influence over him.[68] There is no reason to suppose that any aspect of Newton's recollection was anything but the truth exactly as it had happened. Newton, however, did not reply to Hooke's letter of 6 January 1680, in which the proposal of investigating an inverse-square relation was made. He had no desire to be Hooke's mathematical 'drudge'.[69] It is arguable whether a statement from Newton at this stage of his own work on an inverse-square law would have prevented the controversy that ultimately erupted six years later or whether it would simply have shifted it forward.

After the correspondence, Newton seemingly discovered the significance of Kepler's area law for the first time, and immediately applied it to the demonstration that an inverse-square law would produce an elliptical orbit.[70] After correspondence with Flamsteed on the comet of 1680–1681, he also seems to have established a theory for a comet in orbit round the Sun by

---

[67] Hall (1992, 206).

[68] Newton to Halley, 14 July 1686, *Corr.* II, 444–445: 'his Letters occasioned my finding the method of determining Figures, which when I had tried in the Ellipsis, I threw the calculation by being upon other studies. & so it rested for about 5 years till upon your request I sought for that paper, & not finding it did it again & reduced it into the Propositions shewed you by Mr Paget; but ... the duplicate proportion ... I gathered ... from Keplers [third] Theorem about 20 yeares ago ...' Also: 'In the end of the year 1679 in answer to a letter from Dr Hook ... I computed what would be the Orb in described by the Planets ... & I found now that whatsoever was the law of the forces wch kept the Planets in their Orbs, the areas described by a Radius drawn from them to the Sun would be proportional to the times in wch they are described'. (Add.3 968.9, 101r; Cohen, 1971, 293). It is notable that Newton's accounts at all times tell a *consistent* story concerning all the incidents.

[69] Newton to Halley, 20 June 1686, *Corr.* II, 438.

[70] This is possibly recorded in 'A Demonstration that the Planets, by their Gravity towards the Sun, may move in Ellipses', *Corr.* III, 71–77.

'... the *vis centrifuga* at C [perihelion] overpow'ring the attraction & forcing the Comet there notwithstanding the attraction, to begin to recede from ye Sun'.[71] A passage deleted from a letter to Flamsteed of 16 April 1681 additionally states that 'I think I have a way of determining ye line of a Comets motion (what ever that line be) almost to as great exactness as the orbit of ye Planets...'.[72]

By early 1684, Hooke, Wren and Halley had all independently decided that an inverse-square law was the most likely explanation for the motions of the planets, but none had found a mathematical proof. A meeting between the three men and Wren's wager led to Halley's visit to Newton at Cambridge, and, ultimately, the *Principia*, during the composition of which Newton came to his final, and most profound, understanding of the law of gravitation, one which provided the culmination, as well as justification, of his unique process of developing a quasi-mathematical style of qualitative reasoning. The process by which this occurred cannot now be totally recovered, though some of the steps in the development are known. A number of scholars have sought to attribute the germ of the concept to the influence of alchemy or the idea of 'attraction', but this seems a hardly adequate explanation for what really happened. Whatever factors contributed, the idea was an inductive development in the recursive style which Newton had been slowly developing as he left behind the hypothetical reasoning he had inherited from his early reading, and the final presentation came from Newton's process of analytical abstraction in which all accretions were gradually pared away, with only the pure abstract idea left behind.

5.4. Angular Momentum and Orbits

Everyone agrees that the correspondence with Hooke in 1679–1680 completely changed Newton's view of planetary orbits, apparently almost immediately after the awkward exchanges with his rival. Various assumptions have been made about what Newton learned as a result of this correspondence. Once the original source of doubt about the paper of the 1660s had been cleared, the ground shifted to Newton's supposed view of orbital motion as centrifugal. But Newton's change of direction after the letters he received from Hooke was not about the difference between the traditional view of centrifugal *conatus* being replaced by a force directed towards the centre

---

[71]Newton to ? Crompton, ? March 1681, *Corr.* II, 358–362, 360. Cited in Nauenberg (1994), note added in proof, 252.

[72]Passage deleted from Newton to Crompton, 16 April 1681, *Corr.* II, 366. Cited in Nauenberg (1994), note added in proof, 252.

of the orbit,[73] as some have thought, or to a force producing an orbit in a body attempting to continue in rectilinear motion. Though he wrote in a letter to Crompton, in April 1681, that the centrifugal force at perihelion made the comet overcome the attraction of the Sun, this does not mean that Newton needed to convert from outward to inward forces for closed orbits, for Newton's use of the word 'centrifugal' does not usually mean an endeavour outwards as it did for Huygens.[74] In addition, comets were not even established as planet-like at the time. There was no sudden switch from outward to inward, only a gradual clarification, with the 'centrifugal force' (*conatus*/endeavour) becoming the straight-line inertial motion.

In Definition 5 of the *Principia*, Newton writes:

> A stone, whirled about in a sling, endeavours to recede from the hand that turns it; and by that endeavour, distends the sling, and that with so much the greater force, as it is revolved with the greater velocity, and as soon as ever it is let go, flies away. That force which opposes itself to this endeavour, and by which the sling continually draws back the stone towards the hand, and retains it in its orbit, because it is directed to the hand as the centre of the orbit, I call the centripetal force. And the same is to be understood of all bodies, revolved in any orbits. They all endeavour to recede from the centres of their orbits; and were it not for the opposition of a contrary force which restrains them to, and detains them in their orbits, which I therefore call centripetal, would fly off in right lines, with a uniform motion.

The endeavour is not in the opposite direction to the centripetal force, and is not 'centrifugal' in that sense, and it is also not a force, but inertial straight line motion. Even though this 'endeavour' is away from the centre of the orbit, it is not along the same straight line as the centripetal force. 'Contrary' does not mean in the opposite direction, and it does not imply a balance of forces inward and outward. The endeavour is to go in a straight line with uniform motion.

Hooke didn't really tell Newton any dynamics he didn't know, or lead Newton to reject his earlier work in orbital dynamics. What the correspondence actually did was to trigger a new insight, suggesting to Newton how he could represent it in an original and more powerful way. Hooke himself may well have derived much of his two-component view of orbital motion from Wren, as well as the analogy with the conical pendulum, which had been suggested by Horrocks as early as 1638, and which Hooke had used to illustrate his views of the 'inflection of a direct motion by a

---

[73]Nauenberg (1999, 76–77).
[74]Newton to ? Crompton, ? April 1681, *Corr.* II, 358–362.

supervening attractive principle' in a paper presented to the Royal Society on 23 May 1666.[75] Hooke never demanded to be given credit for this view of orbital motion, only for the inverse-square force.[76] It may be that, in view of his very considerable work as an architect in the rebuilding of London after the fire of 1666, he realised that Wren was the one man whom he couldn't afford to antagonise. It certainly seems that, even if Newton had derived the idea from the correspondence with Hooke, then Hooke was not the original author.

Already by 1679 Newton knew that curved paths, such as parabolas, were produced by two components of motion, one uniform, the other directed downward and accelerated. In fact, Newton was capable of evaluating approximately an orbit under a central attractive force, long before the correspondence. From early on he had a general method of dealing with dynamic orbits of any shape and involving various force laws, in which he compounded the orbital velocity along a tangent with a perpendicular change in velocity.[77] It was a *complementary* method, learned from Hooke, but already known to him for the circular case, that he applied to explain Kepler's area law. Hooke proposed a total change in velocity towards a *fixed* centre of force, compounding the tangential velocity with the radial velocity produced by the central force. Newton had used this method in his Waste Book for circular motion, but not for other kinds of motion.

Hooke, however, had been unable to make the method reveal the correct dynamics for planetary orbits because he applied the erroneous idea that the tangential velocity of the orbiting body was inversely proportional to the distance from the focus of a planetary orbit. Newton, perhaps in trying to make sense of this construction, discovered that the correct relationship was with the perpendicular distance from the focus to the *tangent* made

---

[75] Birch (1756, vol. 2, 90–92). Smeenk and Schliesser (2013, 116), point out that Hooke's formulation of inertial motion was also different to Newton's in that he didn't specify *uniform* motion in a straight line.

[76] Bennett (1982, 60–62), says that Wren 'firmly claimed priority over Hooke' in the composition of orbital motion (61). The claim is recorded in Halley to Newton, 29 June 1686, *Corr.* II, 441–442. Thomas Sprat, *History of the Royal Society*, 1667, 313–314, also attributed the conical pendulum demonstration to Wren, and Wren seems to have known about Horrocks's letter to Crabtree describing the analogy before August 1664. Bennett says that the 'cosmological significance' was entirely missing in Hooke's experiments of August 1664, 'and there is no doubt that Wren was involved in this change' (63). The fact that Wren was almost certainly the source of Hooke's view of orbital motion is strangely ignored in many popular accounts, which continue to attribute the development to Hooke.

[77] Brackenridge and Nauenberg (2002, 85–137). Brackenridge (1992, 1995). Nauenberg (1994).

by the orbiting body and that this implied that equal areas were swept out by orbiting planets in equal times. It seems that Newton discovered the significance of this Kepler area law for himself for the alternative constructions produced by Boulliau and succeeding astronomers meant that the law was not to be found in any of the astronomical handbooks with which Newton was familiar, and this may explain why Newton, who seems to have been unfamiliar with Kepler's own works, never used Kepler's name in connection with it in a published work — although it may also be because he regarded Kepler's law as a 'hypothesis' related to planetary motion rather than the general law of dynamics which he had derived himself.[78]

The change in Newton's work after the correspondence with Hooke can be seen as a change from action physics to angular momentum physics. Conservation of energy can be treated exactly by the curvature method but this is not possible for conservation of angular momentum, which only yields approximate results and only an approximate version of the area law when finite steps are used. Newton was already familiar with angular momentum in his early work, and had an equivalent of the conservation of angular momentum, but the Kepler area law, whose significance he now understood for the first time, created the opportunity to use this principle for a complete and exact solution to the orbit problem. Though Hooke had pointed the way towards shifting the entire explanation to a force towards the centre for orbits other than the circular ones which Newton had already considered, the key fact, for Newton, was that he was immediately able to derive the area law and then switch to the angular momentum mode. He had not previously seen the significance of the area law, and was the first to explain it. With area then proportional to time, he had a way of incorporating the time variable into his geometric constructions. This aspect had nothing to do with Hooke, and Newton felt therefore that he owed Hooke nothing, though he quite possibly would not have made the change without him.

---

[78]Newton was aware of Kepler's separate astronomical discoveries but neither he, nor anyone else at that time, came across them as three codified 'laws', an innovation due to Leibniz (1689). Newton credited Kepler with demonstrating the area law for the planet Mars 'by an elaborate discourse' in Phaenom. 14 in the manuscript draft *Phaenomena*, *Unp*. 385, and said that 'Astronomers' had found it to hold true 'in all the primary Planets'. See also *De motu corporum in gyrum*, Problem 3, Scholium, MP, VI, 48–49, Phaenom. 14, *Phaenomena* (MS), *Unp.*, 385. Smith (1996, 357), points out that all three 'laws' were 'in some dispute' in the 1680s.

The angular momentum method finds a single (pseudo)vector which fixes the orbital motion in a *single* centre, which is exact, unlike the noncircular method based on action. This single (pseudo)vector determines the entire orbit. Angular momentum, based on perpendicular distance, effectively represents phase, while action, based on distance along the arc, represents amplitude. The momentum mv and the total length L are now related to the radius R (with $L = 2\pi R$ for 1 cycle), but they only connect at the limit, and this information is only apparent in Newton's geometric method. In classical physics, where the quantities momentum and space are commutative and combine to produce the entire dynamic description (the 'phase space') of a system, action and angular momentum are separate quantities. Least action becomes the combination of the minimum changes in momentum and spatial position that accompany a physical transition. In quantum physics, where phase space cannot be separated below the product $\hbar/2 = h/4\pi$, action and angular momentum are effectively the same. So Planck's constant h represents both action and angular momentum, amplitude and phase, wave and particle.

Newton's continued interest in centrifugal force may make sense of his otherwise strange remark in his letter of 13 December 1679: 'And also if its gravity [central force] be supposed uniform it will not descend in a spiral to the very center but circulate with an alternate ascent & descent by it's *vis centrifuga & gravity* alternately overbalancing one another'.[79] Nauenberg believes this refers to a comparison of the magnitude of the centrifugal force with the magnitude of gravity, and has nothing to do with Borelli's idea of a balance between inward and outward forces.[80] Proposition 41 of the first book of the *Principia* shows the effect of changing from centripetal to centrifugal forces. Curvature remained an alternative method. Proposition 28 of Book III calculates the solar perturbation on the lunar motion using curvature, but this was included only in the second and third editions, giving the impression that it was a later development rather than the earliest.

Unfortunately, the inverse-square law relation led to a resurrection of the quarrel with Hooke that had begun with Newton's early optical work.

---

[79] *Corr.* II, 307–308.
[80] Nauenberg (1994, 231).

It is obvious why, having written to Newton stating his version of the relation in such explicit terms in 1680, Hooke thought that he had given his rival the initial idea. There is no way he could have known of Newton's earlier derivation, and that he had already tested the relationship in two different ways. For some reason, perhaps because he wished to terminate the correspondence, Newton did not refer to his old paper of the 1660s, when Hooke, wrote to him in 1680, but he did describe it in a letter to Halley of 20 June 1686 when Hooke demanded credit for the discovery, and then referred to his quotation of it in the letter to Huygens in a further letter to Halley on 27 July. The paper was brought out again when David Gregory visited him in 1694.[81] With the correspondence with Huygens providing the necessary independent witness, Newton would have been able to prove that it dated from at least before 1673.

Hooke, who was described by his posthumous editor and first biographer, Richard Waller, as 'melancholy, mistrustful, and jealous',[82] had plenty of form as a 'universal claimant', 'whose response to being told of some invention or discovery was to claim to have accomplished the same work himself years before'.[83] He did this on many occasions at Royal Society meetings. Sir Thomas Molyneux, a contemporary at the Royal Society, described him as 'the most ill-natured, conceited man in the world, hated and despised by most of the Royal Society, pretending to have all other inventions when once discovered by their authors to the world'.[84] Always conscious of a need to maintain his status in the scientific world, he laid claim on several occasions to discoveries that he knew he hadn't made — a fault which doesn't seem to be attributable to Newton.[85] It also seems to have been his practice to imagine that a half-formulated concept gave him an exclusive right to an 'intellectual property' claim on whatever might subsequently be produced in that area.

In addition to damaging his opponents' self-esteem, Hooke's retrospective claiming, with its implications of plagiarism on the part of his rivals, also had an insidious effect on their scientific credibility, and generated a considerable amount of ill-feeling among the members of the Royal Society. Besides attacking Newton, he had major disputes with Huygens and Oldenburg, the

---

[81] *Corr.* II, 436, 446.

[82] Waller (1705, xxvi–xxvii).

[83] Pugliese (2004).

[84] Lyons (1944, 97).

[85] Despite 'conspiracy theories' to this effect, the evidence seems to suggest that Newton knew exactly what he had a right to claim for himself and what he needed to attribute to predecessors.

Foreign Secretary of the Royal Society, whom he accused of treachery over his invention of a watch with a balance spring, in addition to a confrontation with Leibniz over his calculating machine, and there are indications that his relationship with Boyle was not exactly untroubled. His journal shows him casting aspersions on Boyle's integrity. In addition he disliked Halley and got involved in a technical dispute with the Danzig astronomer Johannes Hevelius. He was far from universally popular in his time among his peers for fairly obvious reasons, and, although some modern authors have been keen to portray him as a 'victim', few contemporaries would have seen him in this way.

Hooke was certainly a great scientist but we cannot learn much by trying to project back on him attitudes which are more intrinsically Newtonian. Though full of interesting speculations, he never has the level of insight of Newton, or the abstract ruthlessness. He challenges Newton's optical theory and explanation of colours; he finds the wrong velocity ratio for a wave theory; he disputes the finiteness of the speed of light, even though he invokes a pulse theory which surely demands it; he leads Newton astray on capillary action; and he always needs a mechanism for any theory. Every one of his physical theories, except the very simple law of elasticity, is vitiated by the fact that he is unable to separate a correct mathematical explanation from erroneous physical hypotheses. Hooke, who always conceived his work in terms of mechanistic explanations and who lacked the necessary mathematical background, never actually attained to Newton's level of abstraction, which is the real reason why Newton's final conception of the law of gravitation is regarded as a particularly significant breakthrough.

When he published *An Attempt to Prove the Motion of the Earth by Observation* (1674), Hooke said that he had not been able to establish the nature of the law of gravity, and, according to Newton, his failure to mention the law in his *Cometa* (*Lectures and Collections*) of 1678 suggested that he had not yet discovered it four years later.[86] He seems to have been looking for an experimental insight, as he had been since the 1660s. Kepler's own work on light could have led him by analogy to a law in which the Sun's influence over its satellites decreased with the square of their orbital radii, as Ismail Boulliau argued in 1645, but he refused to consider the possibility, as did Boulliau himself, who rejected all physical influences, such as Gilbert and Kepler's presumed magnetic attraction, in favour of planets moving

---

[86]Hooke (1674). Newton to Halley, 29 June 1686, *Corr.* II, 435.

by their 'proper form', that is an orbit determined by geometry alone.[87] Even so, a dynamical explanation would have required the definition of the correct dynamical or kinematical quantity to which the law applied. Hooke had, by 1680, developed an inkling that an inverse square attenuation of the Sun's influence with distance was significant for the planetary orbits, but he had in no sense formulated a 'law of gravity'. Nor had he formulated a concept of gravity that was universal, continuing, as late as 1682, to refer to gravity as the attraction of 'Bodies of a similar or homogeneous nature' in terms similar to the 'particular gravities' previously introduced by Roberval.[88]

Hooke's work, especially after 1680, was nevertheless somewhat more sophisticated than used to be thought, based on many experiments, which showed he had a physical understanding of the constant force orbit, and analogical physical reasoning, which included the conical pendulum first proposed by Horrocks in 1638, though he seemingly did not originate this line of thinking. He even developed a degree of mathematical reasoning, but only after seeing Newton's *De motu corporum in gyrum*, of November 1684.[89] The similarity of Hooke's diagram in his manuscript of September 1685, to Newton's from *De motu*, can hardly be used as evidence of independent thinking on Hooke's part, though it does provide evidence that Hooke was interested in developing a greater understanding of the dynamics of orbital motion than he had previously attained. Hooke used the diagram for a graphical (*quam proximé*) solution of the case of an orbit under a force linearly proportional to distance, but Newton had already solved the case analytically in *De motu*.

---

[87]Boulliau (1645); Wilson (2002, 204). Newton referred to the argument in a postscript to his letter to Halley of 20 June 1686, *Corr.* II, 438. In principle, Hooke's 'Keplerian' mechanism for an inverse-square 'power' could have been derived merely by reversing Boulliau's negative conclusion to an affirmation.

[88]Westfall (1971, 271), referring to Hooke (1682), in *Posthumous Works*, 176, and to Roberval (1647). Westfall (269–270) also sees the 'particular gravities' featuring in the more comprehensive *Attempt to Prove the Motion of the Earth* (1674).

[89]Trinity College, Cambridge. MS O.lla.116, September 1685. Pugliese (1989); Nauenberg (*2005*). Whiteside (1991, 56), comments that the MS shows that 'Hooke was still not au fait with the general measure of a central force which Newton had set out in his 'De motu Corporumin gyrum' the previous autumn. Instead, in the single instance of determining a non-circular orbit which he there attempted, that where the central force varies directly as the distance, he could yet only effect an approximate construction of the orbit in the manner Newton had outlined to him in December 1679'. Clearly postdating Newton's manuscript by a year, and clearly dependent on it, Hooke's work could not be regarded as an independent discovery, though there may be a degree of independence in the method.

Newton argued that, even if he had received the inverse-square relation from Hooke, that many other people had begun to perceive it by then (not a difficult calculation if one used Huygens' expression for centrifugal acceleration), and that Wren, Halley and others had considered that a demonstration was required, and only Newton had supplied it. Unlike Hooke, Wren and Halley made no attempt to claim that their hypotheses gave them the right to claim discovery, even though they had come upon them independently, and even Hooke never claimed for one moment that anyone but Newton had found the demonstration. In view of his earlier work on an inverse-square law, which, from the beginning, extended to a view of gravity beyond the laws of planetary orbits, Newton was actually relatively generous in allowing Wren, Hooke and Halley to receive a measure of credit for applying such a law to celestial motions, and in not citing his own work of c 1666 in the Scholium to *Principia*, Book I, Proposition 4. He was also relatively modest in not explicitly stating that his new universal law, now refined to the point of abstraction, completely superseded all earlier attempts at inverse-square law relationships, founded on hypotheses, however general, including his own, and those of Wren, Hooke and Halley.[90]

Newton, however, found it annoying that Hooke could make claims to big discoveries based on simplistic '*quam proximé*' reasoning and inconsistent hypotheses without having any understanding of the careful and patient reasoning that was required to generate true and exact results that would be established beyond the level of hypothesis. Many people who have taken up Hooke's cause have simply failed to understand this crucial aspect of the Newtonian procedure. To develop an approximate law '*quam proximé*' was relatively easy (although, in terms that actually make physical sense, Hooke didn't even do that — no amount of mathematical development would have made his hypothesis physically viable), but Newton went through every conceivable avenue to ensure that, when it did finally arrive, the inverse square law of gravity was a truly fundamental law of nature and not just a relatively successful hypothesis like Kepler's ellipses.[91]

---

[90]This is an absolutely crucial point. The abstract law that Newton finally created is effectively an infinite distance away from *all* the hypotheses which preceded it, including his own. In attempting to connect it with them, scholars have, seemingly inadvertently, allowed certain authors to attempt to privilege later inverse-square hypotheses over Newton's original one.

[91]Newton described Hooke's results in the correspondence of 1679–1680 as 'quam proximé', and not demonstrable as either true or based on fundamental principles, as well as frequently being inconsistent and self-contradictory. It is a common fallacy among some commentators to imagine that steps taken by, say, A that lead to a more advanced

It would seem that Newton regarded the discovery of a truly universal *law* as the important innovation, fulfilling the request that the Royal Society had relayed to him through Halley. The hypotheses that had led up to this were relatively unimportant and could be attributed to anyone who wanted to make the claim. So he referred to the contributions of Wren, Hooke and Halley, as it was their discussion of the inverse square hypothesis that had led to him establishing the law. It was only when Hooke tried to imply that his (erroneous) version of the hypothesis was equivalent to Newton's law, except for the mathematical demonstration, that Newton felt that he had to remind the members of the Royal Society that he had a prior claim on the hypothesis as well, with documentary evidence if needed. Perhaps the Royal Society members appreciated the irony of this reversal. Hooke certainly found that he had little support among them. When Newton remembered the apple story as a very old man, it had nothing then to do with priority battles he had already won. It was simply an old man remembering things he had long forgotten.

5.5. Fundamental Aspects of Gravitation

The universal law of gravitation, as it was finally presented, was a mathematical theory of an abstract action at a distance, with no need for a physical mechanism, and stripped of all such concepts as aether and *conatus*, which now appeared to be superfluous to requirements. When Newton realised that he could pare the entire theory down in this way, without any loss in power of explanation, he devised the first experiment to eliminate the aether from physics. It was one that he remembered later, having, as so many times, lost the paper on which it was written. Though the early work often contained ideas which reappeared in later contexts, it was often lacking in this quality of pure abstraction, which developed as Newton grew increasingly confident of the validity of this kind of thinking.

Though the fact that bodies with different masses fell with the same acceleration in a given gravitational field was known to Galileo and to a number of late mediaeval scientists, Newton went to great lengths to ensure

---

result by B would inevitably have led A to the same result given time. Earlier results cannot be made more significant by this kind of retrospective analysis. The Newtonian law of gravity in its final abstract form was a unique result whose meaning was never fully grasped by most of Newton's contemporaries and required a combination of mathematical and abstract reasoning that was certainly beyond Hooke's orbit. It was not just a matter of *demonstration*, as Hooke may have implied, but also one of *statement*. The Newtonian law led to a completely new way of looking at physics and not just an explanation of planetary motions.

it was true. Very early on, he wrote in his *Certain Philosophical Questions,* or *Quaestiones quaedam philosophicae,* that: 'The gravity of bodys is as their solidity, because all bodys descend equall spaces in equall times consideration being had to the Resistance of ye aire &c'.[92] He also thought of doing experiments to 'Try whither ye weight of a body may be altered by or cold, by dilatation or condensation, beating, poudering, transfering to severall places or severall heights or placing a hot or heavy body over it or under it, or by magnetisme. Whither leade or its dust spreade abroade, whither a plate flat ways or edg ways is heaviest....'[93] He would subsequently reject the possibility of any external influences on gravity. By the time he wrote Corollary 7 to *Principia,* Book II, Proposition 24, the relation of mass and weight had become encapsulated in a principle, the first statement of the so-called principle of equivalence: 'And hence appears a method both of comparing bodies one among another, as to the quantity of matter in each; and of comparing the weights of the same body in different places, to know the variation of its gravity. And by experiments made with the greatest accuracy, I have always found the quantity of matter in bodies to be proportional to their weight'. The first Corollary to Book III, Proposition 6, on the relation between weight and mass, had been that 'the weights of bodies do not depend on their forms and textures'.

Newton not only stated the principle of equivalence but regularly tested it in a series of important experiments. One of these, described in Proposition 6 of *Principia,* Book III, used two identical eleven-foot pendulums, each with a round wooden box (to ensure that they met with equal air resistance). He filled one with wood and put an equal weight of gold in the other, and repeated the experiment with silver, lead, glass, sand, common salt, water and wheat. To an accuracy of one part in a thousand, he found no difference in the oscillation period, and so in the ratio of mass to weight for any of these substances. In the same Proposition, he outlined the way in which such a variation would be detected in planets or satellites by an eccentricity of orbit. Chandrasekhar has equated this prediction of the effect due to inequalities in inertial and gravitational masses to the one 're-invented' in a later period as the 'Nordvedt effect'.[94] In fact, the concept of mass, as we use it today, was very much Newton's creation, and it was his special insight to equate the concepts which Einstein would later separate as gravitational and inertial mass, and then have to equate again through defining a

---

[92]QQP, 'Of Attomes', 121[v], NP.
[93]*ibid.*
[94]Chandrasekhar (1995, 370).

'new' principle. In Newton's world view, there never was any need for the separation; the dual nature was fundamental to Newtonian mechanics, and it was characteristic of Newton to realise this without requiring further explanation.

Newton separated gravity from magnetism, which some earlier writers, such as Gilbert and Kepler, had thought might be the force keeping the planets in their orbits. As he explained to Flamsteed, the attractions between the Sun and Earth could not be magnetic, even though Gilbert had found that the Earth was magnetic, because any magnetism of the kind observed on Earth would be destroyed by the enormous heat of the Sun.[95] In addition, the gravitational interaction was different from the magnetic one in a very significant way. A small magnet, forcibly aligned in the opposite direction to the field from a large one, will immediately turn round and align itself correctly when allowed to do so, and then remain in this alignment. So, a comet could not be at first be correctly aligned, and then force itself into the opposite alignment to be repelled from the Sun, as Flamsteed had proposed. The 'directive vertue' of a large magnet was always greater than its 'attractive vertue'. Newton here explained for the first time why gravity could not be dipolar, like the magnetic force, and why there was no gravitational repulsion, analogous to magnetic repulsion. He never deviated from this position.[96] Knowing that the Earth had a magnetic field, he considered that there could be a *slight* magnetic attraction between Earth and Moon, but left it to future observers to make the appropriate corrections if and when they found it necessary.[97]

At a later date, Newton would propose the unification of cohesive, frictional, chemical, optical and, even, magnetic forces in the form of a fundamental electrical interaction, but he always believed that gravity should remain separate. Gravity was the only force which depended on the quantity of matter; it was the only force which arose directly from the properties of mass. There appeared also to be an enormous difference in scale, which he was later able to quantify; and, though he frequently sought to make analogies between the small-scale forces within matter and the gravitational

---

[95]Newton to Crompton for Flamsteed, 28 February 1681, *Corr.* II, 340–347; Newton to ? Crompton, ? March 1681,*Corr.* II, 358–362; memoranda by David Gregory, *Corr.* III, 338.

[96]Newton to Crompton for Flamsteed, 28 February 1681, *Corr.* II, 340–347, 341–342.

[97]*Principia*, III, Proposition 37, the concluding comments after Corollary 10. As the (dipolar) magnetic force is inverse-cube, Smith (2012, 381, 393), has speculated that Newton may have considered it a candidate for providing the missing half of the precession of the lunar apse (see 6.3).

force operating on the large scale motions, he never considered, as we have seen, the possibility that a gravitational repulsion should exist by analogy with the repulsions associated with electric and magnetic forces. Believing in the ultimate simplicity of nature, he took every opportunity to reduce his active principles to a minimum; but where, as here, the evidence did not suggest the existence of a genuine unifying simplification, he was not prepared to assume that one must exist.

Newton showed that 3-dimensionality is an essential aspect of gravitation as known, when he proved that a sphere of finite radius and composed of material whose density was a function of the distance from the centre, would act gravitationally on an external object as if its entire mass were concentrated at its centre. For an inverse square law, this is true for a space of 3-D, but not for a space of any other dimensionality.[98] The philosopher Immanuel Kant would later argue formally that an inverse-square law necessarily results from a 3-D space.[99] Newton also produced a scaling law (Proposition 87, and Corollary I) which says that the force exerted externally by a distribution of mass specified by a density function, when the elements of the mass distribution attract each other according to the inverse nth power of the distance, will be as that power of the distance. According to a second corollary, the law by which the particles attract each other can only be found if that of the objects of which they are composed has an attraction 'directly or inversely in any ratio of the distances'.

Another important theorem, the proof that there is zero gravity inside a hollow spherical shell if gravity follows an inverse-square law, led to the significant inference by Joseph Priestley, in 1767, that, since electric charge is found only on the surface of a hollow conductor, then there must be an inverse-square law of force between electrical charges, the first statement of this law made on observational grounds, and an interesting example of a physical proof by analogy. *Principia*, Book I, Proposition 70 had said that 'If to every point of a sphaerical surface there tend equal centripetal forces decreasing in, the duplicate ratio of the distances from those points; I say, that a corpuscle placed within that superficies will not be attracted by those forces any way'. Priestley inferred the result for a hollow sphere based on experiments using a nonspherical hollow conductor; Cavendish, in 1773, used concentric conducting spheres to show that the inverse power of the distance

---

[98] *Principia*, I, Proposition 76.
[99] Kant (1747), §10, 11; Bochner (1966, 192–193, 248–250).

was 2 ± 0.02.[100] Newton referred to his own result in Corollary 2 to Book III, Proposition 14, where he applied it to the fixed stars.[101]

No universal law had ever been proposed for physics, and it was immediately clear that the law of gravity had a special status. All other physical laws than the universal ones are solutions of general equations which are ultimately approximate or local. Universal laws are expressed in mathematical terms by differential equations with no exact solution. Any solution that can be derived will always involve an approximation, which is not directly related to the equation. The differential equations express the relations between physical quantities, not as algebraic relations between the quantities themselves, but through expressions involving their rates of change, or rates of rates of change. 'Solving' the equations means converting from relations involving rates of change to simple and direct relations between the original quantities, and this can only be done by reducing the general equation to one or more particular cases by imposing 'boundary conditions' and this is, in principle, a process of approximation. The derivation of Galileo's or Kepler's laws from Newton's more general law of gravitation is a classic example. In effect, general and exact laws do not give us the sort of direct knowledge that we can use in observation; for this, we are obliged to reduce the infinite number of possible solutions to particular and individual cases using some kind of approximation.

Unlike the particular, local laws of Galileo and Kepler, regulating the motion of objects on the Earth and the orbits of planets around the Sun according to seemingly arbitrary rules, which Newton's work has sometimes been said to 'synthesise', the universal law applied to every particle of matter in the universe, and was a concept of an entirely new kind. Not only did it replace all local laws, showing that they were, strictly speaking, incorrect and valid only to certain degrees of approximation, but it set up a fixed standard to which all new information had to be accommodated. Unlike the laws of Kepler and Galileo, it did not apply to a particular physical system, but rather presented a totally abstract statement of a relationship between the fundamental parameters of measurement: space, time and mass.

In Newton's view, the 'planetary hypotheses' discovered by Kepler were empirically-derived phenomenological facts, not fundamental principles, which were not even strictly true within their context, and also only one set

---

[100] Priestley (1767), Cavendish (1879, 104–113).

[101] Pask (2013, 420), referring to Weinberg (1981, 45), says that: 'It is fascinating to see that a similar argument can be made in relativistic cosmology where Newton's Proposition turns into Birkhoff's theorem'.

among a group of competing approximate descriptions in use at the time (such as a number of alternative versions of the second 'law'),[102] including one by Boulliau, which he cited.[103] Newton did not accept that they were laws, and Kepler, himself, never called them such, nor did he draw attention to them in the way that later writers have done. And, of course, they are not laws, except insofar as the second law is a reflection of the fundamental law of conservation of angular momentum. Leibniz may have been the first to call Kepler's rules 'laws' in his *Tentamen de motuum coelestium causis*, of 1689, presenting the three propositions, which appeared in widely separated parts of Kepler's works, as three 'laws', in direct imitation of Newton's, effectively reversing the historical sequence in the minds of later generations.[104]

5.6. The Two-Body System

Once the universal law had been established, Newton set himself to explain the complicated motions of the members of the Solar System on the basis of the inverse-square attraction between each pair of massive bodies. The calculation for a simple two-body system is relatively easy and was solved analytically and completely by Newton himself, but the introduction of a third body leads to severe complications. No analytical solution has ever been found, and even numerical and approximate solutions are difficult. Newton, in fact, spent years trying to create a theory of the three-body system which determined the motions of the Moon as a result of the attractive forces exerted by the Sun and Earth, with a considerable degree of success in terms of fundamental principles, but seemingly without the accurate correlation with phenomenological detail that would have led to an immediate practical application such as a method of finding the longitude at sea (but see Chapter 7).

In the *Principia*, he first laid out the phenomena he wished to explain, such as the orbits of planets and comets and Kepler's empirical discoveries about them. 'The planets move in ellipses which have their common focus in the centre of the Sun...'[105] '...the greater planets, while they are carried about the Sun,... in the mean time carry other lesser planets revolving about

---

[102]Boulliau (1645), others were by Ward (1653), Streete (1661), Wing (1651, 1669), and Mercator (1676). See Wilson (1989).

[103]This used the empty focus as an equant point. Newton to Halley, 20 June 1686, *Corr.* II, 438.

[104]Wilson (2000, 225–226). Descartes, of course, had had 'Three Laws of Nature' in his *Principia* of 1644.

[105]III, Proposition 13.

them, and . . . those lesser planets . . . move in ellipses which have their foci in the centres of the greater . . . '[106] 'The motions of the comets are exceedingly regular, . . . governed by the same laws with the motions of the planets . . . '[107]

He included a lunar test, similar to the one he had first entertained in the 1660s. ' . . . the Moon gravitates towards the Earth, and by the force of gravity is continually drawn off from a rectilinear motion, and retained in its orbit. . . . the force by which the Moon is retained in its orbit becomes at the . . . surface of the Earth, equal to the force of gravity we observe in heavy bodies there'.[108] 'For gravity is one kind of centripetal force: and my calculations reveal that the centripetal force by which our Moon is held in her monthly orbit around the Earth is to the force of gravity at the surface of the Earth very nearly as the reciprocal of the square of the distance from the centre of the Earth'.[109] Revisiting his earlier calculation, he computed that the Moon or an object placed in the Moon's position with no tangential velocity would fall 15 1/2 Paris feet in one minute.[110] The same result would be found by assuming that the inverse square law applies, and calculating the acceleration at the Earth's surface.

In the first edition, he found a correlation with the Moon's gravity to 1 part in 180, but in the third edition the correlation was seemingly more accurate than would have then been possible, good to 1 part in 4,000. The initial agreement was very good, but Newton assumed that some of the Earth's gravitational effect on the Moon would be countered by a miniscule force from the Sun, so changing from his initial value to a fall of 15.093 Paris feet in a minute. This became one of the issues, along with the calculations of the velocity of sound and the precession of the equinoxes, on which he was later criticised by Richard S. Westfall.[111] As in those cases, the criticism is not really valid, though for different reasons in each case. In the case of the Moon test, the result was obviously true and any over-statement of accuracy a matter of relatively unimportant detail. Newton already had a good enough agreement from the first calculation, and so finding extra correction factors to tweak the result becomes a relatively minor issue, especially as the corrections were made quite visible for all to see.

---

[106] III, Proposition 22.

[107] General Scholium, 1713.

[108] III, Proposition 4. If it were not the same force, he argued, a moon at the level of the Earth's highest mountains and losing its centripetal acceleration would descend with twice the acceleration of any other body there (Scholium).

[109] *De Motu*, 1685; Herivel (1965, 302).

[110] III, Proposition 4.

[111] Westfall (1980, 732–734).

The procedure was unusual for Newton. More often, he makes rounding errors because he has hasn't taken the calculation far enough. However, this was a special calculation, like the calculation of the magnetic moment of the electron would be to quantum electrodynamics in the twentieth century. It was a showpiece of gravitational testing, and high-precision experiments were very much in their infancy at the time. It would only be in later times that serious attention was paid to significant figures. Even in the nineteenth century, it was quite common to push calculations to a number of figures well beyond any experimental meaning simply because the calculations led to this. Despite the extra correction factors, and the overstatement of the accuracy of the result in the third edition, the result of the first test was well within the error, as we would now express it, though that way of thinking hadn't fully developed at that time. Altogether, no major issue was involved in this procedure. No contemporary was offended by Newton's attempt to add a final flourish. For anyone other than Newton, it might have been regarded as ironic. The self-righteousness of some modern commentators is misplaced.[112]

Newton asserted that: 'Bodies attracted towards a resting centre with a force reciprocally proportional to the square of the distance from that centre move in conic sections which have a common focus in that centre and their periodical times in ellipses are in a sesquialterate proportion of their mean distances from that centre'.[113] Referring to centripetal laws in general, he said: 'Of this sort is gravity, by which bodies tend to the centre of the Earth; magnetism, by which iron tends to the load-stone; and that force, whatever it is, by which the planets are continually drawn aside from the rectilinear motions, which they would otherwise pursue, and made to revolve in curvilinear orbits'.[114] He derived the facts that an inverse-square centripetal force produces an orbit described by a conic section, while an orbit described by a conic section requires an inverse-square centripetal force. Newton's derivations required Kepler's second law, which, as we have seen, involves the conservation of angular momentum, and which he described in *The Elements of Mechanicks*: 'Bodies attracted towards a resting centre describe equal areas in equal times and their velocities in approaching that

---

[112]Nor does everyone agree with Westfall. Densmore (2003, 370), treats the calculation of the effect of the Sun, based on Book I, Proposition 45, Corollary 2, as a valid and sophisticated refinement.

[113] *The Elements of Mechanicks*; *Unp.*, 166.

[114] *Principia*, I, Definition 5.

centre increase as much as if they made the same approach by falling down directly'.[115]

The orbits of the bodies were around the centre of gravity of the system, which was either at rest or in uniform motion in a straight line,[116] which meant that Kepler's laws were not strictly true, even in a purely two-body system. Aspects of such a two-body system were outlined in *Principia*, Book I, Propositions 57–63.[117] For any system of bodies subjected to mutual attraction, 'their common centre of gravity will either be at rest, or move uniformly' in a straight line.[118] Two mutually attracting bodies will, according to Proposition 57, describe 'similar figures about their common centre of gravity, and about each other mutually'. Then, according to Proposition 58, for two bodies attracting each other 'with forces of any kind', and revolving about a common centre of gravity, 'there may be described round either body unmoved a figure similar and equal to the figures which the bodies so moving describe round each other'. Bodies which attract with forces proportional to their distance apart will describe concentric ellipses, both round each other and their common centre of gravity (Corollary 1); if they attract with forces inversely proportional to the square of their distance apart, they will describe 'conic sections having their focus in the centre about which the figures are described' (Corollary 2). Also two bodies revolving round their common centre of gravity will 'describe areas proportional to the times, by radii drawn both to that centre and to each other' (Corollary 3).

Proposition 59 modifies Kepler's third law from the $T_1^2 : T_2^2 = a_1^3 : a_1^3$, previously accepted, to the equivalent of $(M + m_1)T_1^2 : (M + m_2)T_2^2 = a_1^3 : a_2^3$, for two planets of masses m_1 and m_2 orbiting the Sun of mass M, though, to a large extent, Kepler's law will hold as a good approximation. Further modifications would follow from planetary perturbations. Following on from Proposition 59, Proposition 60 concerns the orbits of two bodies attracting each other with a force inversely proportional to the square of the distance, if each of the bodies revolves 'about the common centre of gravity'. In Book III, Proposition 15, Newton says that 'the principal diameters of the orbits of the planets' are to be found by Kepler's third law, 'and then to be severally augmented in the proportions of the sum of the masses of matter in the sun

---

[115] *Unp.*,166.

[116] *Principia*, III, Proposition 11.

[117] These propositions constitute the first part of Section XI on 'the motions of bodies tending to each other with centripetal forces'.

[118] *Principia*, I, Section XI.

and each planet to the first of two mean proportionals between that sum and the quantity of matter in the sun', as set out in Proposition 60 from Book I. Proposition 61, from Book I, shows that the motions of 'two bodies attracting each other with any kind of forces, and not otherwise agitated or obstructed' are the same as those that would result from these attractions being replaced by an attraction of the same form to 'a third body placed in their common centre of gravity'.

The centre of mass of the Earth–Moon system is used by Newton, in the third Corollary to Book III, Proposition 40, in effectively defining a constant, which Chandrasekhar says is 'commonly called the Gaussian constant of gravitation', κ, which is given by $2\pi/($sidereal year $\times$ $(1 + (M_{Earth} + M_{Moon})/M_{Sun})^{1/2})$. Newton's value is 0.01720212 compared with the modern 0.017202098, indicating that his difficulty in finding an accurate value for the mass of the Moon had a relatively negligible effect in defining the centre of gravity of the Earth–Moon system.[119]

Proposition 10, from Book I, concerning elliptical motion due to a force directed towards the centre of the ellipse, requires that 'the force is as the distance' and is totally determined by this condition.[120] Corollary 1 continues: 'And therefore the force is as the distance of the body from the centre of the ellipse; and, *vice versa*, if the force is as the distance, the body will move in an ellipse whose centre coincides with the centre of force, or perhaps in a circle into which the ellipse may degenerate'. A second Corollary equates the periodic times of all ellipses about the same centre. Chandrasekhar relates Corollary 1 to the equations of the simple pendulum

$$\frac{\partial^2 x}{\partial t^2} = -\omega^2 x \quad \text{and} \quad \frac{\partial^2 y}{\partial t^2} = -\omega^2 y \ (\omega = \text{a constant}),$$

and their solutions,

$$x = a\cos(\omega t - \delta_1) \quad \text{and} \quad y = b\sin(\omega t - \delta_2)$$

(where a, b, δ_1, and δ_2 are constants), which describe motion in an elliptic orbit.[121]

For a centre 'removed to an infinite distance', according to the Scholium which follows, the ellipse 'degenerates into a parabola', producing a constant force, and the result known as 'Galileo's theorem' describing the behaviour under constant acceleration. In effect, the part of the ellipse with the greatest

---

[119] Chandrasekhar (1995, 480).
[120] Corollary 1.
[121] Chandrasekhar (1995, 89).

curvature may be approximated by a parabola. Here, we take the ellipse

$$\frac{x^2}{a^2} + \frac{y^2}{b^2} = 1$$

to its parabolic limit. Shifting the origin to the vertex $(-a, 0)$ and then supposing that

$$a \to \infty \quad \text{and} \quad (1 - e) \to 0,$$

with the distance from the vertex measured to the nearer focus, we find that

$$a(1 - e) = \text{constant},$$

which we can define as α. The parabolic equation which emerges,

$$y^2 = 4\alpha x,$$

determines the orbit produced by a constant attractive force in the direction x, such as that acting on projectiles at a height x above the Earth's surface. Chandrasekhar expresses surprise at Newton describing this conclusion as '*Galileo's*' theorem,[122] though this is presumably because Galileo had found parabolic orbits for projectiles at the Earth's surface. The Scholium also explains that hyperbolic orbits will result when the centripetal attraction becomes centrifugal repulsion.

The famous Proposition 11 describes the problem first set by Kepler in his *Astronomia nova* of 1609, 'If a body revolves in an ellipse; it is required to find the law of the centripetal force tending to the focus of the ellipse', with Newton's celebrated answer that 'the centripetal force is ... reciprocally in the duplicate ratio of the distance'.[123] However, Newton solves the more general problem of the orbit described by a general conic by requiring the same force law, in Propositions 12 and 13, for hyperbolic and parabolic orbits. As Newton later claimed, the three Propositions combined give a complete solution to the orbits that require an inverse-square force.[124]

---

[122] Chandrasekhar (1995, 91).

[123] This proposition begins Section III on 'the motion of bodies in eccentric conic sections', which continues until Proposition 17.

[124] Newton to Cotes, 11 October 1709; and Corollary 1 to *Principia*, I, Proposition 13, added to the second edition, 61. Whiteside says that 'it is easy to show that Propositions X–XII [the draft versions of Propositions 11–13] do exhaust all the possibilities of motion under an inverse-square law central force' (*MP* VI, 148). A second corollary finds an expression for the *latus rectum* of an elliptical, hyperbolic or parabolic orbit, that Newton would need in later propositions. The *latus rectum* is the perpendicular line through the focus joining two points on the curve.

He was careful, however, to avoid inferring from Proposition 11 that *Keplerian* orbits required an inverse-square law because this is not an automatic consequence of orbits that are only elliptical *quam proximé*. An approximately elliptical orbit does not necessarily require an approximately inverse-square law. In this case, given the fact of a centripetal force from Kepler's equal area law (Propositions 1–3), where the approximate relation between areas and times does actually also hold (Proposition 3, Corollaries 2–3), he required also Kepler's 3/2 power law (Corollaries to Proposition 4) and the absence of precession in planetary orbits (Proposition 45) to fix the centripetal force law as inverse square.[125]

5.7. Applications in Celestial Mechanics

In the years immediately following the publication of the first edition of the *Principia*, Newton carried out a drastic revision of the work which involved the addition of completely new Propositions 6–10, together with many supporting Corollaries, which would have created an entirely new context for the solutions to the Kepler problem now found in Propositions 10–13.[126] In addition to the 'linear dynamics ratio' which was used, after its definition in Proposition 6, to generate the force equations in the original edition, Newton introduced methods which have been described as 'the circular dynamics ratio' and 'the comparison theorem'.[127]

Three preliminary theorems in Propositions 6–8 (the Similarity Theorem, the Proportionality Theorem, and the Comparison Theorem), the third of which (the Comparison Theorem) gave the ratios of forces directed to a point from two different force centres in the same orbit, would lead to the statement of the linear dynamics ratio in Proposition 9. The key Proposition 8 could also be used to find the centripetal force directed towards the focus of a conic section by finding that directed towards the centre, and the Propositions numbered as 11–13 in all the published editions would then follow automatically as Corollaries. In addition, Newton was immediately able to derive another new measure of force in Proposition 10, using a circular dynamics ratio, based on his early circular approximation for determining curvature, again with application to the original Propositions 11–13. Yet another solution of the Kepler problem exists in a manuscript presented to John Locke in 1690.[128] This again seemingly derives from curvature

---

[125] Smith (2012, 371–373).

[126] *MP* VI, 568–599, Brackenridge (1995, 166–210).

[127] Brackenridge (1995, 167).

[128] *A Demonstration*, March 1690.

arguments dating back to the 1660s, and includes the special case in which the proof for the force acting at the vertices of the major axis of an elliptical orbit is valid for a force centre at any point on the axis.

Newton thus created multiple solutions of the Kepler problem, using a variety of dynamical methods. Significantly, they also included a scholium with 'a general fluxional measure of the centripetal force'.[129] Unfortunately, he came to the conclusion that such a drastic revision of the *Principia* would create too much disruption of the existing text. Consequently, most of the revisions that were included in later editions were inserted as Corollaries of the existing Propositions, and the extraordinary clarity seen in the proposed new structure was lost in what Brackenridge has described as the 'labyrinth'[130] of revisions actually carried out. The result was to make the argument of the *Principia* appear much more convoluted and difficult than need have been the case, so adding to its (unwarranted) reputation for fearsome difficulty,[131] and ultimately damaging Newton's reputation with those modern critics who have emphasised the supposed superiority of algebraic methods.

Following Propositions 10–13, in the first book of the *Principia*, as it now stands, Newton proceeds in Propositions 14 and 15 to a derivation of Kepler's third law of planetary motion, that the cubes of the mean distances of the planets from the Sun are proportional to the squares of their periodic times. Proposition 14 demonstrates that, for bodies revolving about a common centre with an inverse-square force, the principal *latera recta* of their orbits are as the squares of the areas swept out by the radii of their orbits, from which Proposition 15 demonstrates that Kepler's third law follows directly, while a Corollary states that the periodic times are the same as would occur in circular orbits with diameters of the same length as the major axes of the ellipses, so being independent of the ellipticity. Proposition 16 and its nine Corollaries summarise and amplify 'the principal results of the major propositions that have been established' for orbits in conic sections. According to Chandrasekhar: 'The listing of all the special

---

[129] Brackenridge (1995, 176).

[130] Brackenridge (1995, 209–210).

[131] Brackenridge describes the unpublished revised version as an 'original and graceful development' (1995, 209). He also says that 'The problems are solved so simply that one tends to forget the magnitude of the initial challenge' (176). Whiteside comments on the *Principia*'s 'ill-deserved reputation for being impossibly difficult' and says that it does not 'set an impossibly high estimate on the mathematical competence and technical expertise of its readers' (*MP* VI, 25, 24–25). Chandrasekhar (1995) was perhaps not totally unjustified in describing his book as for the 'Common Reader'!

cases is characteristic of Newton's scientific attitude: to explore *all* matters with thoroughness'.[132]

Proposition 17 of Book I solves the problem of the motion of bodies governed by the equation

$$\frac{d^2\mathbf{r}}{dt^2} = -\frac{k\mathbf{r}}{r^3}\eta,$$

and finds solutions that satisfy given initial conditions, showing how, depending on initial conditions, the orbit can be an ellipse, a parabola or a hyperbola.[133] Newton would establish the uniqueness of the solution of the initial value problem quite generally in Book I, Proposition 41.[134]

The initial conditions, in particular the speed and angle of projection, are crucial in determining a planetary orbit. Newton later explained to Richard Bentley that, if the planets had been formed at infinity (or, in Galileo's 'Platonic' hypothesis, 'created by God in some region very remote from our Systeme') and allowed to fall under uniform gravity towards the Sun, they could not have acquired the velocities with which they now revolve in their several orbits. Each orbit must have been formed under unique circumstances.[135]

Orbits in conic sections, and, in particular, the ellipse, are quite generally observed in the Solar System, and are the products of inverse square forces. While 'The planets move in ellipses which have their common focus in the centre of the Sun',[136] '...the greater planets, while they are carried about the Sun, ...in the mean time carry other lesser planets revolving about them, and... those lesser planets... move in ellipses which have their foci in the centres of the greater...'[137] Again, '...the forces by which the primary

---

[132]Chandrasekhar (1995, 107). Pask (2013, 210–211), draws attention to Cor. 2, where the greatest speed is when the body is closest to the centre of force at the focus.

[133]Corollary 1 shows how to find the second focus given the principal vertex, *latus rectum* and first focus, while Corollary 2 shows how to find the orbit given the velocity in the principal vertex.

[134]See Arnol'd (1990, 112). Newton explicitly states the significance of these two Propositions in this context in a draft Preface to the third edition of the *Principia* dating to the late 1710s (*MP* VIII, 457–458).

[135]Newton to Bentley, 17 January 1693, *Corr.* III, 240; 25 February 1693, *Corr.* III, 255. Pask (2013, 262–263) quotes Eugene Wigner (1967) as saying that the separation of laws of nature and initial conditions is one of the 'most fruitful' ever made, one that made 'the natural sciences possible'. I don't believe, however, as Wigner does, that it is an 'artifice'. It is rather a consequence of the intrinsic properties of the parameters from which physics is constructed (Rowlands, 2007).

[136]III, Proposition 13.

[137]III, Proposition 22.

planets are continually drawn off from rectilinear motions, and retained in their proper orbits, tend to the Sun; and are reciprocally as the squares of the distances of ... those planets from the Sun's centre',[138] while ' ... the force by which the Moon is retained in its orbit tends to the Earth; and is reciprocally as the square of the distance of its place from the Earth's centre'.[139]

Corollary 1 to Book I, Proposition 13 shows the general converse theorem that an inverse square force, applied to a body with a component of linear motion, will produce a curved orbit of second degree, that is a conic section. Newton sketched out a proof in the second edition. Bruce Pourciau ended much fruitless debate, in 1991, by showing that Newton's sketch (with a minor omission rectified) 'expands into a convincing proof that inverse-square orbits must be conics'.[140]

According to V. I. Arnol'd, such proof is superfluous in any case, as there is no 'uniqueness' problem for Newton's solutions in Proposition 17, which depend smoothly on the initial conditions, so the converse is an automatic consequence.

Modern mathematicians actually distinguish between existence theorems and uniqueness theorems for differential equations and even give examples of equations for which the existence theorem is satisfied but the uniqueness theorem is not. So various troubles can arise, and if Newton's equation were troublesome, it would actually be impossible to make any deductions. A mistaken point of view arises because of the unwarranted extension of the class of functions under consideration. The fact is that in modern mathematics the concepts of function, vector field, differential equation have acquired a different meaning in comparison with classical mathematics. Speaking of a function, we can have in mind a rather nasty object – something differentiable once or even not at all — and we must think about the function class containing it, and so on. But at the time of Newton the word function meant only very good things. Sometimes they were called polynomials, sometimes rational functions, but in any case they were all analytic in their domain of definition and could be expanded in Taylor series. In this case the uniqueness theorem is no problem, and at that time nobody gave it a thought.

But in reality Newton proved everything to a higher standard.[141]

---

[138] III, Proposition 2.

[139] III, Proposition 3.

[140] Pourciau (1991).

[141] Arnol'd (1990, 30–31). See also 5.6, above, on this. Newton himself referred to I, Proposition 17 as proof of the converse theorem and of I, Proposition 45 as providing a completely general method of finding such converses in a draft for a Preface to the third edition of the *Principia*, dating to the late 1710s (*MP* VIII, 457–458).

Proposition 30 of Book I (with three corollaries) seeks 'To find at any assigned time the place of a body moving in a given parabolic trajectory', using a Cartesian method of solving a third degree algebraic equation, which is presented in the form of a geometric construction, while Proposition 31 introduces methods of approximation to solve 'Kepler's problem',[142] in which the position of an orbiting body can be found in any given time using the area law. Kepler was unable to find a 'rational' solution to the problem of finding a focal sector of an ellipse having a given simple ratio to another focal sector. He had been unable to use the area law to find exact determinations of planetary positions, so post-Kepler astronomers used approximations instead of an area law. In Proposition 31, Newton found a solution using a curve identified as a prolate cycloid.

Newton 'restates Wren's construction of the elliptical case by means of a transcendental 'stretched' cycloid'; the proof involves the derivation of what is now called Kepler's equation, $T = \theta \pm e \sin \theta$, the construction following from the geometric parametrization.[143] A Scholium investigates methods of finding approximate but accurate solutions of Kepler's equation, which a Corollary to the famous Lemma 28 had effectively shown could not be inverted to represent θ as an algebraic function of T.[144] Newton's highly accurate geometrical method of approximation, in the first edition of the *Principia*, was based on the construction he had used for the parabolic orbit in Proposition 30; but he replaced it in later editions with an analytical method (effectively, a Newton–Raphson method) which generated even more accuracy, using an iterative approach and series expansion. According to Chandrasekhar, Newton's method remained unappreciated until it was 'rescued' by the astronomer John Couch Adams, two hundred years later.[145]

Proposition 32 finds the straight-line distance through which a body will fall freely in a given time when the force is inversely proportional to the square of the distance from the centre. The three conic sections are treated as separate cases. Chandrasekhar remarks that the 'elegance and the simplicity of Newton's demonstrations are startling'.[146] A whole series

---

[142] See Guicciardini (2009, 259–266), on the hidden analysis behind the construction. Propositions 30 and 31 constitute Section VI.

[143] Whiteside, *MP* VI, 308–310, Kepler's equation, I, Proposition 31.

[144] Whiteside, *MP* VI, 309.

[145] Adams (1882), Chandrasekhar (1995, 141). See *MP* VI, 316–318. The Scholium also discusses the case of the hyperbola.

[146] Chandrasekhar, 150. Propositions 32–39 constitute Section VII on 'the rectilinear ascent and descent of bodies'.

of propositions follows in the *Principia* on bodies rising or falling in a straight line. Proposition 33 uses the limiting procedure described in Proposition 32 to compare the 'velocity of a falling body in any place C' to 'the velocity of a body describing a circle about the centre B at the distance BC', with a simultaneous treatment of both elliptic and hyperbolic cases. According to Corollary 2, 'A body revolving in any circle at a given distance from the Centre, by its motion converted upwards, will ascend to double its distance from the centre'. The parabolic case is added in Proposition 34. The cycloidal motion of a fixed *interior* point of a uniformly rolling circle in Proposition 31 becomes cycloidal motion of a fixed point on the radius in Proposition 35, with the elliptic and hyperbolic cases again treated simultaneously and separately from the parabolic case. Proposition 36 gives the times of descent of a body falling freely under an inverse-square law force, with the result then applied, in Proposition 37, to the problem of parabolic motion and the motion of 'a body projected upwards or downwards'. Chandrasekhar cites Proposition 37 as 'the very first instance of the formulation of an *initial-value problem* in physics'.[147]

Proposition 42 is a more general representation of the initial-value problem: 'The law of centripetal force being given, it is required to find the motion of a body setting out from a given place, with a given velocity, in the direction of a given right line'. Given the initial conditions regarding position and velocity (or momentum, in phase space terms), find the subsequent trajectory of the body at any given time. The solution requires the use of the key theorem of the *Principia*, Proposition 41.

Beginning a new section on moveable orbits (IX), Proposition 43 shows that a revolving orbit obeys Kepler's area law, as much as a stationary one, sweeping out equal areas in equal times. Conservation of angular momentum is preserved as Newton finds the centripetal force which will make the revolving orbit move in the same curve as the fixed orbit. 'Newton's theorem of revolving orbits', Proposition 44, then seeks to show how the centripetal force of the revolving orbit relates, mathematically, to that of the fixed orbit. According to Newton, if a body in a revolving orbit is compared with one in the same fixed orbit, the differences of the forces varies inversely as the cube ('the triplicate ratio') of their common altitudes, and this is worked out in particular for elliptical orbits directed both from the focus and from the centre.[148] As Lynden-Bell interprets it, if, in a central orbit, $r^2 d\phi/dt = h$, then $r^2 d(\alpha\phi)/dt = \alpha h$. So if we replace the angular momentum ϕ with

[147] Chandrasekhar (1995, 160).
[148] Corollaries 2–3.

$\phi^* = \alpha\phi$, then it will still be constant. To leave the motion of r unchanged we need a force to counter the extra centrifugal force $(\alpha^2 - 1)h^2 r^{-3}$. In an orbit of this kind, the apsides will rotate faster at a rate $(\alpha - 1)2\pi/P$, where P represents the mean period in the original orbit. It is equivalent to transferring the old orbit from fixed axes to ones rotating at the rate $(\alpha - 1)d\phi/dt$.[149]

The fifth Corollary to this Proposition, according to Chandrasekhar, 'establishes for the first time the notion of centrifugal potential':[150] 'Therefore the motion of a body in an immovable orbit being given, its angular motion round the centre of the forces may be increased or diminished in a given ratio; and thence new immovable orbits may be found in which bodies may revolve with new centripetal forces'. The total potential for the orbit is a combination of gravitational and centrifugal potentials, which remains fixed when angular momentum is conserved. And in the calculation in Corollary 6, Newton effectively reduces the original centripetal force to zero, changing the non-rotating orbit to an inertial straight line motion with zero angular momentum and finding that the inverse cubic 'perturbation' produces a spiral. In effect, the inverse-cube perturbation now becomes the new geodesic, replacing the Cartesian coordinates of the straight-line motion with the revolving coordinates of the perturbation. The theorem of revolving orbits is followed by Proposition 45, which sets out to relate orbits under centripetal forces of any kind that are nearly circular to ones acting under an inverse-square force that are nearly elliptical, and from this to find the precession of the line of apsides.

Proposition 46 'provides an important generalization of Proposition 42 on the initial-value problem'.[151] Given any kind of centripetal force, and the centre of force, and the plane in which the body resolves, it shows how to find 'the motion of a body going off from a given place with a given velocity' in the direction of a straight line in the plane. Here we need only the component of the force in the plane. Propositions 62 and 63 are initial-value problems to determine the motions of two bodies mutually attracting each other with an inverse-square law force, if they are 'let fall from given places' (Proposition 62), and if they go 'off from given places in given directions with given velocities' (Proposition 63).

---

[149]Lynden-Bell (2000, 131). An idea of this kind could be seen as a precursor of the modern use of geometry in physics, especially in gravitational theory, though this is now expressed in a tensor form unknown to Newton.

[150]Chandrasekhar (1995, 192).

[151]Chandrasekhar (1995, 210).

Earlier in the *Principia*, Corollary 3 to Proposition 17 had stated that, 'if a body moves in any conic section, and is forced out of its orbit by any impulse, you may discover the orbit in which it will afterwards pursue its course'. According to Chandrasekhar, Newton's method of finding this by compounding the two motions looks 'far into the future', as does Corollary 4 which states that, 'if that body is continually disturbed by the action of some foreign force, we may nearly know its course, by collecting the changes which that force introduces in some points, and estimating the continual changes it will undergo in the intermediate places, from the analogy that appears in the progress of the series'.[152] Colin Pask sees Corollary 3 as being relevant to the positioning of artificial satellites or planetary probes using booster rockets, and Corollary 4 as enabling the perturbing effect of other planets on the Earth's orbit to be calculated.[153]

A number of Propositions in the first book of the *Principia* concern the attraction of spheres on particles external to them. These are necessary for justifying the treatment of attraction by spherical or near-spherical astronomical bodies. Proposition 70 says that, if to every point of a spherical surface there tend equal centripetal forces decreasing as the square of the distances from those points, a corpuscle placed within that surface will not be attracted by those forces.[154] The 'proof is the same' as 'in any modern textbook'.[155] Essentially it relies on cancellation of the d^{-2} force relationship by the d^2 dependence of a given area on the surface, and the pairing off of equal and opposite non-distance dependent forces on opposite sides of the sphere, and, again, it is a result that is unique to 3-D space. By contrast, Proposition 71, which finds the attraction towards a sphere of a particle placed outside it (inverse-square towards the centre), solves a difficult problem of integration, effectively requiring a double integral.[156] Proposition 72 says that, if, 'to the several points of a sphere' there are equal centripetal forces which decrease as the square of the distances from those points, the force with which the particle is attracted will be proportional to

---

[152]Chandrasekhar (1995, 111).

[153]Pask (2013, 211–212).

[154]Proposition 70 begins a section (XII) on attractions to spherical bodies, which continues until Proposition 84. In a major extension of dynamics from point-particles to macroscopic objects, this includes the results famously called the 'superb theorems' by J. W. L. Glaisher, Rouse Ball (1893, 61).

[155]Chandrasekhar (1995, 268).

[156]J. E. Littlewood famously marvelled at its geometrical construction (1948), believing that it may have used some form of hidden analysis, though Chandrasekhar (1995, 270–273), dissents from this view.

the sphere's radius. The proofs of both this proposition and the preceding one use similarity arguments.[157] Corollary 3 to Proposition 72 shows the 'real power' of such arguments, when it says that, if two geometrically similar solids of the same density attract particles at proportionate distances away with inverse-square law forces, then the ratio of the attractions will be that of the diameters of the solids.[158]

A suggestion of the relevance of calculus is given in a Scholium to Proposition 73 which imagines a solid sphere as constructed from infinitesimally thin spherical surfaces whose number can be increased 'without end'. In the Proposition itself, if on the points of any given sphere, there are equal centripetal forces which decrease as the square of the distances, a 'corpuscle' placed within the sphere 'is attracted by a force proportional to its distance from the centre', and, in Proposition 74, a particle outside the sphere is 'attracted by a force reciprocally proportional to the square of its distance from the centre'. So 'attractions of homogeneous spheres at equal distances from the centres will be as the spheres themselves' (that is the masses, given by $4\pi\rho R^3/3$), and, 'at any distance whatever', the attractions are proportional to the masses ('as the spheres') and inversely proportional to the squares of the distance.[159] Corollary 3 approaches the universal law of gravitation but restricted to a 'particular context': 'If a corpuscle placed without an homogeneous sphere is attracted by a force inversely proportional to the square of its distance from the centre, and the sphere consists of attractive particles, the force of every particle will decrease in a duplicate ratio of the distance from each particle.'

It was important to establishing universal gravitation for Newton to show that spheres of matter would attract each other with inverse square laws as though all their masses acted from the point at their centres. Proposition 75 shows this for two similar homogeneous spheres; Proposition 76 extends this to spheres that are dissimilar but homogeneous. It says that, if spheres are

> however dissimilar (as to density of matter and attractive force) in the same ratio onward from the centre to the circumference; but every where similar, at every given distance from the centre on all sides round about; and the attractive force of every point decreases in the duplicate ratio of the distance of the body attracted; ... the whole force with which one of these spheres attracts the other will be reciprocally proportional to the square of the distance of the centres.

---

[157]Chandrasekhar (1995, 274–275).
[158]Chandrasekhar (1995, 275).
[159] Corollary 1, Corollary 2.

As we have said earlier, the Proposition is notable also, in a mathematical context, in using the division of ordinates now associated with the Lebesgue integral for the principal theorem.[160] The two Propositions and their Corollaries, together, finally satisfied Newton that the universal law of gravitation was exact, rather than merely approximate, as he explained in his discussion of Book III, Proposition 8, which says that two uniform spheres will attract each other with an inverse-square law force directed from their centres.

Proposition 76 is such an important result that it is supplied with nine corollaries, which say, in effect, that a collection of spheres will behave, in many respects, like a collection of point-particles. Two of them (3 and 4) combine to form the closest approach Newton gives in the *Principia* to the modern form of stating the inverse square law of gravitation, apart from that added to the second edition at the end of the General Scholium. Proposition 77 (which enumerates six separate cases) then finds that the same applies as in Proposition 76 if the force is directly proportional to the distance, and Proposition 78 shows that it is also true when there is a spherically symmetric distribution of density rather than uniform density.

Some important concepts occur effectively in Newton's work without the use of a special terminology. Examples include gravitational potential (or potential energy per unit mass) and gravitational field intensity (or force per unit mass). Since he frequently used unit masses, the quantities appear in Newton's equations, but their significance would only emerge with the work of successors such as Lagrange and Laplace. He also discusses escape velocity in connection with the thought experiments he gives on artificial satellites. In considering the special case in which the escape velocity for an astronomical object exceeds the speed of light, Michell, and, also, Laplace, would later introduce the idea of black holes.[161]

The practical consequences of Newton's work include man-made satellites, the principles of which were outlined in *Principia*, Definition 5:

> If a leaden ball, projected from the top of a mountain by force of gunpowder, with a given velocity, and in a direction parallel to the horizon, is carried in a curved line to the distance of two miles before it falls to the ground; the

---

[160] The full theory of the Lebesgue integral was introduced only in 1904, but Newton's early introduction of ordinate division is interesting as a characteristic example of his ingenuity in finding a new technique to solve a particular problem. The *Principia* has no systematic mathematical development, but uses a vast range of mathematical techniques, some only newly forged in the necessity of creating unprecedented degrees of physical explanation, to solve a variety of problems at different levels of complexity.

[161] Michell (1784); Laplace (1796, 1799).

same, if the resistance of the air were taken away, with a double or decuple velocity, would fly twice or ten times as far. And by increasing the velocity, we may at pleasure increase the distance to which it might be projected, and diminish the curvature of the line which it might describe, till at last it should fall at a distance of 10, 30, or 90 degrees, or even might go quite round the whole earth before it falls; or lastly, so that it might never fall to the earth, but go forward into the celestial spaces, and proceed in its motion *in infinitum*'.

This thought experiment, known as Newton's cannonball, has also been considered as the progenitor of the space gun approach of launching a satellite into a real space orbit. In the first edition of the *Principia* Newton also stated the principle of reaction propulsion, based on the third law of motion, which ultimately led to the first successful space rocketry.[162]

5.8. Some Consequences of Universal Gravity

All the evidence that Newton had surveyed led to the conclusion that: 'The force which retains the celestial bodies in their orbits ... can be no other than a gravitating force ... '[163], and that ' ... in general the motions of all the great bodies hitherto observed by astronomers are exactly such as ought to arise from their mutual gravities in free spaces'.[164] The large-scale uniformity of nature meant that gravity had to be universal even beyond the Solar System: 'And if the fixed stars be centres of similar systems, all these are under the same one dominion'.[165] In an unpublished Proposition 43, intended for *Principia*, Book III, Newton stated that the Earth could be at the centre of a Tychonic Solar System with the Sun and its planets in orbit round it if gravity was countered by an equal and opposite inertial force proportional to mass originating from outside the Solar System.[166] A *fictitious* force of this kind occurs if the Earth is taken as an inertial frame with respect to the rest of the universe, but a real force would only exist if the uniformity of nature was not obeyed, and the Earth was singular with respect to the other planets.

There is an interesting modern consequence of this. Attempting to apply Mach's principle to explain inertia, D. W. Sciama assumed that there must

---

[162]III, Proposition 37, first edition; Hall (1985).

[163] *Principia*, III, Proposition 5, Scholium.

[164] *The Elements of Mechanicks, Unp.*, 168.

[165]draft General Scholium (MS C); *Unp.*, 359.

[166]S. Weinberg, *To Explain the World: The Discovery of Modern Science*, HarperCollins, 2015, 251–252. See also his 1694 statement to David Gregory quoted in 6.6.

be a real long-range acceleration-dependent inertial force that was inductive or gravitomagnetic in type.[167] He derived an equation for the force on a particle of mass m distant r from the Earth of the form $Gmm_u a/c^2 r$, where m_u is the mass of the observable universe or the mass inside the Hubble radius surrounding the Earth. If we now apply the equivalence principle by equating this to the gravitational force on m within the same radius, Gmm_u/r_u^2, then we obtain $a = c^2 r/r_u^2, v = cr/r_u$, exactly like Hubble redshift with dark energy acceleration.[168]

The analysis of the effects of gravity and its origins brought out some of the most outstanding examples of Newton's mathematical skill and abstract imagination; his discussion of the subject was full of exceptionally interesting ideas, but it went beyond what many people conceive today as 'Newtonianism', and what is usually called the Newtonian system is the mechanistic conception which he explicitly rejected in all of his writings. By a process of popular vulgarisation, the Newtonian mathematical structure became associated with a universe that was materialist and determinist, a considerable deviation from Newton's own views, for a universal law of gravitation, with everything attracting everything else, requires a system that is fundamentally indeterminate. So, although Laplace considered the Newtonian gravitational system as the perfect exemplar of determinism in nature, the system actually incorporates a *fundamental indeterminacy*, a fact of which Newton was fully aware. The non-deterministic element appeared from the moment Newton began to analyse the three-body problem in gravitation. No longer was it possible, even in principle, to determine the exact positions of physical objects acting under forces. This eventually required a revision in the mathematical style, though not the substance, of Newtonian mechanics.

One of the first things that had emerged from Newton's theory was the fact that the Keplerian laws of elliptical orbits and equal areas were not exact, and that the motions of planets and comets were not those of a 'perfect' system. Even the relatively limited investigation of the many-body problem in the case of the Solar System showed that the planets and comets would lose their near-perfect Keplerian relationships with time. In addition,

---

[167] *MNRAS*, 113, 34, 1953. See Chapter 8 for a discussion of gravitomagnetic forces.

[168] Rowlands (1994, 2007). The acceleration at $c^2 r/r_u^2$ can be shown to require mass-energy density equivalent to 2/3 that of the universe. The inertial force also has the mathematical form of a (special) relativistic correction to the static gravitational force between the masses, and the fact that the universe is almost certainly not geocentric suggests that it is actually fictitious.

the Solar System would be subject to long-term instability because of the dissipative effects of friction or perturbative forces, which he attributed principally to the multitude of comets orbiting the Sun with many kinds of eccentricity. While the inverse-square law of attraction between all particles of matter in the universe replaced the perfection of the heavenly motions with another, more abstract, kind of perfection, the very universality of this law made it impossible to have an observable perfection in a system composed of such particles. Since, the motion of every particle depended on an infinite number of interactions, it was not even possible to exactly specify this motion, unless the effect of all these interactions was known; and Newton, a strong believer in the 'conformability' of nature, certainly expected this principle to apply, not only to the case of large-scale systems acting under gravity, but also to the particles of matter acted on by internal structural forces. The result was a significant break with previous ideas, which had always assumed perfection and stability for the planetary orbits; the perfection of the abstract law with which this was replaced ensured that the true motions of both planets and particles would be essentially unknowable.

Perfection, in fact, is not to be found in observed nature; it only exists in the abstract system. Planets do not move in ellipses; these are merely an approximation, they have no fundamental significance. There is no perfection in the world order, only one that could never be completely observed in the universal law. None of Newton's predecessors would have doubted that perfection of some kind was to be found in nature, but Newton's system made it impossible.[169] Kepler had been unhappy at finding that he could only explain planetary orbits using ellipses; it was the beginning of the end of the perfection represented by circular astronomy; Galileo refused to acknowledge them entirely. But, as Newton showed, the planetary motions were not even ellipses. Only coincidence made them appear so, by approximation, and Newton said that the perturbations in the many-body Solar System would eventually increase until the whole structure fell into disorder. By contrast, Cartesians such as Leibniz refused to accept this, regarding it as inconceivable that the Solar System would not preserve its perfect order in perpetuity. Newton was also aware that there was probably no such thing as an 'immovable centre' of attraction, and, in his lifetime, Halley showed

---

[169]Descartes had previously thought the planetary orbits were only approximately determined and liable to change over time (1644/1983, 98). Newton's view was that the changes were produced by perturbations from other bodies in the Solar System and would be progressive.

that even the fixed stars moved, when he found that three bright stars, Sirius, Aldebaran and Arcturus, had shifted in position by about a third to half of a degree from the positions recorded by Timocharis, Hipparchos and Ptolemey, in addition to the shift expected from precession.[170] Like the work of Copernicus in 1543, Newton's work on gravity led to a conceptual revolution in our understanding of the universe, but it also led to a much more subtle change in our understanding of what it was possible to know.

---

[170]Halley (1718).

Chapter 6

The System of the World

6.1. The Celestial Phenomena Brought to Order

As we know from over three hundred years of experience, it is one thing to find the basic laws that apply to a physical system. It is another thing to apply them so that the behaviour of that system can be explained under a variety of different conditions. The latter exercise usually involves solving the differential equations defining the basic laws, and this may require defining boundary conditions, understanding special cases and deviations from exactness in the basic laws, and developing methods of approximation. It is usually a highly nontrivial process, and often requires mathematical techniques quite unlike those used in discovering the basic laws, and it generally proceeds in a slow, iterative manner, with successive calculations approaching better agreement with observed data.

Once Newton had discovered how gravity operated at the particle level, and had developed a fundamental system of dynamics, he had to apply them to explaining the large-scale phenomena of the universe, or as he described it the 'System of the World'. His results can be regarded as the first iteration of complicated calculations. More accurate results often needed further techniques and better data. Here, Newton's methodology includes the recognition that approximation is often the appropriate way of tackling a physics problem. Commentators have divided into those stressing the amount he left for his successors to do[1] and those who point how much he actually did that is often attributed to his successors.[2] Much of this is based on the clash between supporters of analytical and of geometric methods, with the supporters of the analytical methods often assuming that Newton didn't or couldn't use them, with the innate assumption that these methods succeeded

---

[1] Wilson (2001, 2002).
[2] Nauenberg (1994, 2000, 2001a).

because they were innately superior. People have also assumed that Newton didn't know the correct method in several extensive calculations because they didn't understand *his* method.

The *Principia* included a stunning array of results, and it was mainly these results, rather than the more fundamental analytical concepts on which they were based which impressed his contemporaries. Newton had completely solved the two-body problem for the motions caused by the main interactions between the Sun and the planets when these are treated as point-centres of mass. Of course, the Sun and the planets are not point-centres of mass, but extended and complicated objects, which are not even completely solid, and, in their interactions with the Sun, the planets are not isolated from each other and from other objects in the Solar System. Book III, the System of the World, showed that mathematical treatment of even these more complicated cases based on the universal law of gravitation produced results which compared favourably to observation.

Apart from the movements of the planets, satellites and comets, there were the first valid explanations of the precession of the equinoxes and the tides. Some of the mathematical derivations involved predictions which were subsequently tested, but mostly after Newton's lifetime — an interesting corrective to the contemporary obsession with prediction as the main test of a good theory. The predictions often included subtle effects that were produced by interactions which were much more complicated than the direct one between two bodies. Two that succumbed to early testing concerned the shape of the Earth and the movement of its polar axis.

Book III, Propositions 18–19 show that a rotating planet, and, in particular, the Earth, must be an oblate spheroid, with deviations from sphericity due to the fictitious centrifugal force caused by rotation acting against gravity. (Corollary 2 to Book I, Proposition 91, which determines the attraction of an oblate spheroid at an external point on its minor axis, was a result added out of sequence when Newton was writing on the oblateness of the planets caused by their rotation in Book III, Proposition 19.) Proposition 20 then gives an explanation of variation in weight, as shown by the change in the period of a seconds pendulum (or pendulum with a time period of 1 second), as a function of latitude on such a rotating non-spherical Earth, the first such study of the variations in gravitational acceleration with latitude. Newton was aware that a shift in latitude causes a variation in rotational speed, since $v = 2r \, / \, T \times \cos \phi$, where v is the linear tangential speed at latitude ϕ; and r and T are the average values of the radius of the Earth and the period of rotation; the distance from the centre of the Earth is also affected by latitude, since the Earth is an oblate spheroid. The figure of the Earth was calculated at 1/230 on the assumption of uniform density, as against the modern value of 1/297.

Jean Richer had, in 1672, found that a seconds pendulum was shorter at Cayenne than at Paris. Edmond Halley found a similar result at St Helena in 1677, while, five years later, Varin and des Hayes observed a shortening of the pendulum on the island of Gorée. *The System of the World* had proposed that all the planets were oblate (§37).[3] Proposition 19 of *Principia*, Book III showed that Jupiter was an oblate spheroid, with a ratio of equatorial to polar diameter of 10 1/3 to 9 1/3; in the third edition it was 12/11, 13/12 or 14/13. Cassini measured the polar diameter as less by 1/15 in 1691.

The oblateness of the Earth was one of the first Newtonian predictions to be tested. The second and third editions of the *Principia* had set up this possibility by including a table showing the variation of gravity at the surface in terms of the length of a seconds pendulum against the measure of one degree of the meridian arc, both calibrated from the latitude of Paris. In 1734 Pierre de Maupertuis produced a treatise in which he asserted, following the Newtonian argument, and with some help from his teacher, Johann Bernoulli, that the Earth should be oblate, in opposition to Cassini, who saw it as prolate.[4] To make a test of the two hypotheses, the Paris Academy organised an expedition to measure the arc of the meridian, using measurements from two widely-separated places. In May 1735, Charles-Marie de la Condamine led a party of observers, including Bouguer and Godin, to Peru and Ecuador, while Maupertuis left Dunkirk on 2 May 1736, with Clairaut and Camus, on an eventful journey to Lapland. Despite the intensely cold weather and other hardships, including shipwreck in the Baltic, Maupertuis and his party returned with measurements which, allowing for a number of errors, confirmed that the Earth was oblate, as predicted by Newton. These were reported to the Academy on 20 August 1737.

Of course, Newton was aware that the density of the Earth was almost certainly not uniform, and probably increased towards the centre. So, a result deviating from the initial prediction could be used to discover how this variation actually occurred by successively refining the model. For example, as a first iteration, Newton proposed in the first edition of the *Principia*, that the increase in density could itself be assumed uniform. This process of successive refinement, which depends entirely on the validity of

---

[3] *De Systemate Mundi*, §37.

[4] Proposition 20, table. Maupertuis (1738). Despite his earlier opposition, Bernoulli did admire Newton in many ways. Thus, he used Newton's *Arithmetica Universalis* as the basis for his own course on algebra in 1738.

the universal law of gravitation, is precisely what has happened since the first measurements of de la Condamine and Maupertuis, and the process has led to significant information about the internal structure of the Earth.[5]

Proposition 21 predicts a 'nodding', or nutation, of the Earth's polar axis as one consequence of the Moon's variable perturbation on the Earth's annual orbit, manifesting itself in a variable inclination of the axis to the ecliptic and back again, arising from the variable gravitational attraction of the Moon upon a spheroidal Earth. The effect was discovered by the astronomer James Bradley in 1748.[6]

According to the great scholar, I. Bernard Cohen, Newton's theory of the tides was 'one of the great achievements of the new natural philosophy based on a force of universal gravity'.[7] Of course, it was only a first iteration, but it established that the tides were caused by the joint effect of the Sun and Moon's gravitating power. As Newton writes: '...the flux and reflux of the sea arise from the actions of the Sun and Moon'.[8] 'Those parts of the Earth and sea gravitate towards the Moon which are nearest to her and thence the sea under the Moon rises towards her and opposite to the Moon it rises from her by the defect of gravity and by rising in both places causes high tide every 12 hours. The like happens in respect of the Sun...'[9] The combined effect is responsible for spring and neap tides at different times. The basic result does not require a consideration of the rotation of the Earth, as some have maintained. '...Gravity makes the sea flow round the denser and weightier parts of the globe of the Earth...'[10] The nature of the forces of the Sun and Moon to move the Earth's seas are determined in *Principia*, Book III, Propositions 36 and 37, and a Corollary to Proposition 36 shows how to calculate the height of the solar diurnal tide at syzygies. Newton calculated the Sun's tidal force at the Earth's surface at 1/38,604,600 of the force of gravity, compared with the modern 1/39,231,000. Lacking a good value for the Moon's mass, his calculation of the lunar effect came out less well at 4.4815 that of the Sun, rather than 2.12.[11] Newton also postulated

---

[5]See Smith (2002, 158). Significantly, though the oblateness is, in principle, a test of the universality of the law of gravitation, deviations of observed results from theoretical predictions have never been seen as falsifications of this law, but as a method of exposing inaccuracies in the model of the Earth on which they were based.

[6]Bradley (1748).

[7]Cohen (1999, 238).

[8]*Principia*, III, Proposition 24.

[9]*The Elements of Mechanicks, Unp.*, 168.

[10]Q 31.

[11]The modern values are given by Cartwright (1999).

land tides in *The System of the World* but knew that they were beyond current measurement techniques.[12]

The precession of the equinoxes, as Newton established for the first time, is the result of the Moon's gravitational attraction on the rotating Earth's equatorial bulge. In *Principia*, III, Proposition 39, he shows that precession is due to the varying density of Earth, as the diameter at the equator increases with respect to the diameter at the poles. Cohen believed that: 'In retrospect, Newton's explanation of the precession of the equinoxes may be reckoned one of the greatest intellectual achievements of the *Principia*'.[13] Proposition 39 gives the mathematical explanation of the precession and computes the magnitude of the equatorial bulge in relation to the pull of the Moon and also of the Sun, so as to produce the long-known constant of precession of 50″ per year and a precession cycle of 26,000 years. This was augmented and improved for the second edition. Employing the result he had previously obtained in Propositions 20 and 21 concerning the shape of the Earth, Proposition 39 inaugurates a major aspect of celestial mechanics, the study of a three-body system, providing an example of the level which Newton was able to reach at this early date in setting out a mathematical explanation of the world.

Newton's data allowed him to calculate a value of precession of the correct order of magnitude, which is perfectly reasonable for an effect which had hitherto found no explanation of any kind. The *Principia*, however, and especially the second edition, made a much greater claim to accuracy. In effect, Newton selected values of data from a wide range of possible ones to yield a result which would give the best possible agreement between theory and observation. He was able to do this because the problem with obtaining a good value of the Moon's mass left this, in effect, as a free parameter. Newton's procedure has been heavily criticised, but perhaps for the wrong reasons. Though it is in no way a good practice, it is quite different from *fabricating* data, and it is far from unusual to see it in operation, even at the present time — we see it, for example, when a physicist invokes a particular amount of supersymmetry to explain how the weak, strong and electric forces may come together at Grand Unification. Given the opportunity to choose the best value to produce a desired outcome, researchers will frequently be tempted to do just that, which is why analyses on large experiments are often

---

[12] *De Systemate Mundi*, §54; Michelson and Gale used interferometry to observe land tides 'in the solid earth, on the grounds of Yerkes Observatory, in 1919' (Cajori, 680).
[13] Cohen (1999, 265).

done without immediate access to the theoretical or statistical predictions to which they can be compared.

However, as Newton well knew, it is not usually a very productive way of doing science, since it can give the impression that a subject is closed when it is still wide open. It consists, in effect, of taking a gambler's shot at hitting upon the right combination, rather than leaving open what cannot be satisfactorily calculated otherwise. Of course, in an investigation as complicated and unprecedented as that of the precession of the equinoxes it is most unlikely to be the correct one, and in this case it isn't, because the value selected for the Moon's mass is most definitely not a good one. Essentially, the error bars on deriving the lunar mass from tidal data are huge, and the theoretical method for doing the computation far from established, but Newton gives the idea that the value can be determined exactly because this is the one that will yield the exact constant of precession. While the basic ideas in the main calculation are right, the subsidiary components are flawed in a number of respects, and could not have yielded an accurate value without modification.

But criticism aimed at Newton's calculation is misplaced. It should be aimed, if anywhere, at the presentation. Newton sought the value of a free parameter which would give the desired outcome, and then used his calculation to attempt to fix that value, but he presented it as though he had obtained a miraculously exact result from first principles. It is clear that he felt his system had to appear strong at a critical moment, when his entire work — including his contributions to optics, mathematics and theology — was under heavy attack from Leibniz and the Cartesians. Nevertheless, though superior calculations of precession soon followed, no one in the eighteenth century thought that Newton's procedure was particularly unusual or to be criticised.[14]

One of the main criticisms of Newton's treatment of precession stems from the idea that he had no true understanding of the treatment of rigid bodies, and of the importance of conservation of angular momentum and of moment of inertia. His early work shows that this is not strictly true. Newton *did* have a basically correct treatment of the rotation of rigid bodies from early on, including an equivalent of moment of inertia. He also knew of the conservation of angular momentum, and the principle is used implicitly in his treatment of Kepler's area law. Whether this was correctly used in

---

[14]Significantly, the standards against which the calculation has been judged in more recent times were those set by Newton himself in other areas of his work, which were not those of his contemporaries.

the treatment of precession is another issue. Certainly, he understood the torque, the power 'to wheel about', exerted by an external force on a non-spherical body, which is introduced in III, Lemma 1. In this lemma and the accompanying Lemma 2, he obtains the turning moment or torque exerted on the Earth by tidal forces due to the gravitational attraction of the sun, and correctly shows that the torque, when all the matter of the exterior Earth (the bulges) is in its actual position, to the torque that would result if this matter was placed in a uniform ring about the equator, is in the ratio 2 to 5.[15]

More controversial are Chandrasekhar's claims that Lemma 2 concerns the moment of momentum or angular momentum (which measures the 'motions of the whole [body] about its axis of rotation'), while Lemma 3 incorporates the idea of *circulation* — or line integral of the velocity along a closed curve.[16] According to Lemma 3:

> The same things still supposed, I say, in the third place, that the motion of the whole Earth about the axis above-named arising from the motions of all the particles, will be to the motion of the aforesaid ring about the same axis in a ratio compounded of the proportion of the matter in the Earth to the matter in the ring; and the proportion of three squares of the quadrantal arc of any circle to two squares of its diameter, that is, in the ratio of the matter to the matter, and of the number 925275 to the number 1000000.

Chandrasekhar's view, which goes so far as to claim that Hypothesis 2 (originally Lemma 4), which treats the precession of a ring, conjectures the conservation of circulation, has been criticised as projecting the interpreter's own definitions onto the seventeenth-century context.[17] Certainly, Chandrasekhar's method was to construct his own proofs of the theorems before tackling Newton's, and many people might think this could lead to unhistorical results, but it is also possible that a creative mathematician of the highest calibre, such as Chandrasekhar undoubtedly was, might see a structure in another's work that would escape a more orthodox interpretation. But, even discounting this, and taking on board all the imperfections in the work, there is no doubt that Newton's calculation of

---

[15] Dobson (1999). Newton was certainly deficient in his knowledge of the properties of rigid bodies, and their application to this problem, but it may well be that the lack of reliable data concerning the Moon's mass constituted at least as significant a source of error.

[16] Chandrasekhar (1995, 470).

[17] Chandrasekhar (1995, 471). Smith (1996); Dobson (1999).

precession is one of his most outstanding achievements. No one previously had an inkling of such an explanation.

Another of the special celestial phenomena brought to order by Newton in Book III was the seemingly irregular motions of comets. Here he wrote that, in fact: 'The motions of the comets are exceedingly regular, ... governed by the same laws with the motions of the planets ... '[18] They 'move in some of the conic sections having their foci in the centre of the Sun'. 'Hence if comets revolve in orbits returning into themselves, those orbits will be in ellipses ....' 'But their orbits will be so near to parabolas, that parabolas may be used for them without sensible error'.[19] By 'radii drawn to the Sun', they 'describe areas proportional to the times'.[20] Book I, Proposition 5, 'There being given, in any places, the velocity with which a body describes a given figure by means of forces directed to some common centre: to find that centre', a late addition to the text of the *Principia*, proved to be particularly useful in describing the orbits of comets.

Book III, Proposition 41 showed how to fit a cometary orbit to a parabola, from three observations. This required the application of interpolation theory to the calculus of finite differences using the algorithm set out in Lemma 5, and a series of further Lemmas (6–11) to establish the necessary mathematical procedures. Of these, Lemma 6 applies the interpolation method to finding the intermediate positions of the comet at any given time from the positions known by observation. Lemma 8, using the more elementary Lemma 7, then shows how to construct the position reached by a comet in a parabolic orbit in any given time. Lemma 10, which requires the preceding geometrical theorem, Lemma 9, is 'the most significant' of this series of lemmas for 'celestial mechanics',[21] and is equivalent to a result variously known as Euler's theorem, the Euler–Lambert theorem and Lambert's theorem,

$$6\kappa t = (r_1 + r_2 + c)^{3/2} - (r_1 + r_2 - c)^{3/2} \, ,$$

which relates the time (τ) taken for a comet to traverse a parabolic arc to the lengths of the *radii vectores* drawn from the focus of the parabola to the ends of the arc (r_1, r_2), the length of the chord subtending the arc (c) and the 'Gaussian constant of gravitation' (κ).[22] A corollary to this theorem

---

[18] General Scholium; 1713.

[19] III, Proposition 40; also Corollary 1 and Corollary 2.

[20] III, Proposition 40.

[21] Chandrasekhar (1995, 508).

[22] Chandrasekhar (1995, 508, 512), says that both Lagrange and A. N. Kriloff (1926) knew of the equivalence, and quotes Kriloff as saying that, though there were many

finds the height at which the velocity would be nearly enough for the comet to traverse the given arc in the same time. The lemma following (11) says that a comet allowed to fall without initial motion towards the Sun from this height and 'impelled to the Sun by the same force uniformly continued by which it was impelled at first', would, in half the time it would take to describe the arc in its own orbit, move through a space equal to the vertical diameter of the parabola drawn from the mid-point of the chord subtending the arc. The sophistication of the preliminary apparatus provided by this set of lemmas shows how extraordinarily difficult the problem of cometary orbits actually was.

The Great Comet of 1680, which D. W. Hughes considered 'should by rights be known as Newton's Comet', was the first whose orbit was determined, with 'five of its six orbital parameters accurately calculated', the sixth being the eccentricity, which 'was set at 1.0'.[23] Proposition 42 provides a lengthy series of corrections to this procedure. Newton realised the extraordinary difficulty of establishing a comet's path posed by the many-body nature of the interactions with which it was involved. As he wrote in this Proposition:

> But, because of the great number of comets, of the great distance of their aphelions from the Sun, and of the slowness of their motions in the aphelions, they will, by their mutual gravitations, disturb each other; so that their eccentricities and the times of their revolutions will be sometimes a little increased, and sometimes diminished. Therefore, we are not to expect that the same comet will return exactly in the same orbit, and in the same periodic times: it will be sufficient if we find the changes no greater than may arise from the causes just spoken of.

Cometary motion provided what became the most famous test of the Newtonian theory, after Halley, in 1705, predicted the return in 1758 of the bright comet of September 1682 after a seventy-six year absence. Looking through previous records, Halley was able to correlate its appearances with the comets of 1456, 1531 and 1607 and assign to it an elliptical orbit. Later, he identified even earlier sightings in 1305 and 1380. He also used Newtonian gravitational theory and an observed irregularity in its period to predict that its return would be delayed by the perturbing influence of

---

proofs of the theorem in modern treatises, 'Newton's remains unsurpassed in its wonderful dynamical insight'. He also quotes Kriloff as describing Lemma 8 as a 'wonderfully exact construction' (528).

[23] Hughes (1988, 53, 61). Proposition 41, Example.

Jupiter and Saturn.[24] Though he knew he would not live to see its return, he hoped that 'candid posterity' would 'not refuse to acknowledge that this was first discovered by an Englishman'.[25] Halley's comet was first seen by Johann Georg Palitzsch on 25 December 1758, and completed its perihelion passage on 13 March 1759. Perturbations by Jupiter and Saturn did delay its return but by more than Halley had calculated. Clairaut was able to explain the discrepancy using more precise calculations and, during 1758, predicted perihelion passage on 13 April in the following year. According to Hall, these precise calculations were made possible by the geometrical techniques described in Sections IV and V of the *Principia*. 'In this respect Newton laid the foundations of specialised methods that after detecting Neptune and Pluto have gone to the direction of interplanetary probes'.[26]

6.2. Application of Various Force Laws

Newton considered many force laws besides the inverse square, including direct linear, constant, inverse linear, inverse cube, inverse fourth power, and inverse fifth power, as well as one or two more exotic ones. Examples include the inverse cube law which he found to apply to tides and the complicated laws for perturbed planetary systems, but there are also many others. Many were derivable from the more fundamental law, as in the case of perturbed planetary systems. The general strategy of the *Principia* was to derive laws for general cases, and then restricted to the inverse-square as a special case.

He had a particular interest in the inverse cube law, even before the famous correspondence with Hooke led him to Kepler's area law to derive the inverse square condition, and a group of Propositions in Book I of the *Principia* examine the case in detail. While Proposition 7 finds the centripetal force which can make a body travel in a uniform circular orbit, Proposition 8 effectively introduces the inverse-cube law for the first time. Here: 'If a body moves in the semicircumference *PQA*; it is proposed to find the law of centripetal force tending to a point *S*, so remote, that all the lines *PS*, *RS* drawn thereto, may be taken for parallels'.[27] A scholium attached to this says that 'a body will be moved in an ellipse, or even in an hyperbola, or parabola, by a centripetal force which is reciprocally as

---

[24] Halley (1705). Some of Halley's work was incorporated, with acknowledgement, into later editions of the *Principia*.
[25] Halley (1749).
[26] Clairaut (1760); Hall (1992, 219).
[27] cf Proposition 45, Example 3.

the cube of the ordinate directed to an infinitely remote centre of force'.[28] Proposition 9 then finds an inverse-cube law of force is required to make a body move around the pole of a logarithmic or equiangular spiral. This would later become one of the particular cases of the Cotesian spirals, named after Newton's follower, Roger Cotes, for such a centripetal force.[29] Later, Corollary 3 to Proposition 41, shows (though without proof) how to find the polar equation, $r = F(\theta)$, of the orbit of a point mass made to move by a centripetal or a centrifugal force of the form $f(r) = mr^{-3}$.[30]

By this time, it had become clear to Newton that there were an infinite number of possible orbits produced by an inverse-cube force, although the spiralling into (or from) the centre was a general characteristic of such orbits. Corollary 3 showed that the 'two most general species', in addition to the equiangular spiral, were when the polar equations were of the forms $r/R = \operatorname{sech} k\phi$ and $r/R = \sec k\phi$. In revising the *Principia* in October 1709, Cotes discovered two new general species: the hyperbolic cosecant spiral, with equation $r/R = \operatorname{cosech} k\phi$, and the 'reciprocal' spiral with equation $r/R = 1(\phi \cot \alpha + 1)$.[31]

Lemma 29 and Propositions 79-81 are among those theorems that use integration or 'quadratures' without explanation in the text. Here, Chandrasekhar says,

> Newton is concerned with the force exerted on an external particle by a homogeneous sphere 'to the several equal particles of which there tend equal centripetal forces' according to some specified law. The treatment is an analytical *tour de force* in which an integral over two variables is reduced to one over only one of them by inverting the order of the integrations and explicitly evaluating the integral over the other.[32]

'Newton inverts the order of the integrations by appealing to the *Book of the Sphere and the Cylinder* by Archimedes, the reference being to Archimedes' theorem establishing the area of the surface of a sphere of radius a and

---

[28] The extension to the hyperbola and parabola (not in the original MS) does not apply. *MP* VI, 137.

[29] Cotes (1714) classifies all possible inverse-cube orbits.

[30] The polar equation of the equiangular spiral is $\ln (R/r) = \phi \cot \alpha$.

[31] Whiteside, *MP* VI, 138, 353. Cotes (1714). Whiteside says that Newton 'was familiar' with the properties of hyperbolic functions in his 'geometrical model' of the hyperbola, though he had 'no explicit notation' (*MP* VI, 319). Nor had anyone else until well into the eighteenth century.

[32] Chandrasekhar (1995, 287).

that of the enveloping cylinder'.[33] In principle, the problem of evaluating the triple integral measuring the pull of a uniform sphere of radius a on an external point distant b from its centre is transformed to that of a simple integral, if we assume the attraction of a particle varies as the nth power of the distance.

Book I, Proposition 80 derives the formula for the attraction of a particle at a point on the axis of a sphere whose 'equal parts' are subject to equal centripetal forces. Corollaries 1–4 give the integrand when the centripetal force is constant, and then proportional to the inverse first, third and fourth powers of the distance.[34] Proposition 81 then makes use of the triple integral to evaluate the general 'pull' of a uniform sphere upon a point outside it. It was a theorem that Newton chose to display in a copy of the *Principia* included in one of his portraits. Examples 1–3 give the cases for $n = -1, -3$ and -4 (and may have used potential). Newton had effectively already treated the cases $n = -2$ and $n = 1$ in Propositions 71 and 77, and Maclaurin's extension to the result that the sphere attracts as if its entire mass is concentrated at its centre only for these cases was a relatively obvious consequence.[35] Proposition 82, which calculates the attraction of a corpuscle situated within the sphere by calculating that at an inverse point outside it, uses the method of images.[36] For $F \propto R^m$, where the outside and inside distances R_1 and R_2 are related to the radius of the sphere a by $R_1 R_2 = a^2$, the ratio of the forces is $(R_1/R_2)^{(m+1)^{1/2}}$. Proposition 83, finds the force with which a corpuscle placed inside a sphere may be 'attracted towards any segment of that sphere whatsoever'. Proposition 84, finds 'the force with which a corpuscle, placed without the centre of a sphere in the axis of any segment, is attracted by that segment'.

An important question in the application of Newtonian dynamics to the system of the world is, 'given a power law of centripetal attraction, is there a dual law for which a body with the same constant of areas will describe the same orbit?'[37] Newton answered this in the case of the inverse-square law, for he found that one other power law would maintain stable orbits in conic sections, one in which the force is proportional directly to distance, as in

---

[33] Chandrasekhar (1995, 289). Whiteside says that the theorem, though an immediate corollary of Archimedes' *Sphere and Cylinder*, I, 42, is not in any of his extant works (*MP* VI, 214–215).

[34] Pask (2013, 326) gives the equivalent formulas for these cases and also $n = -2$, showing that only the last is simple.

[35] Maclaurin, *A Treatise of Fluxions* (1742, 902).

[36] Chandrasekhar (1995, 294).

[37] Chandrasekhar (1995, 114).

Hooke's law for the vibrating spring (to which Hooke had drawn his attention in 1679).[38] A law in which, in Newton's words, forces 'increase in a simple ratio of their [that is, the bodies'] distances from the centres' is the only law other than the inverse-square law in which bodies move through equal areas in equal times and conserve angular momentum.[39] With this law, also, unlike the inverse-square law, there is no decay of orbits due to perturbation, and Newton was able to solve the n-body problem completely and uniquely, from any initial conditions, for this particular case (Proposition 64), and for spheres of symmetrically distributed density as well as point-particles (Corollary to Proposition 78). The n-body case is worked out by induction after the 2-body case has been extended to 3 bodies. Newton's method has the advantage of showing, physically, what is exactly happening.

Newton recognised that the duality meant that planetary orbits were dynamically similar to the regularly periodic simple harmonic motion exhibited by the spiral spring, which was governed by this direct distance relation, though the infinite forces, accelerations and velocities at infinite distances suggested that it could not be the force law governing the system of the world. As Newton knew, the inverse-square force and harmonic oscillator orbits are the only stable classical ones that can be formed under arbitrary initial conditions. Bertrand formalised this in a well-known mathematical theorem in 1875.[40]

Newton used a simple geometric construction for the mapping that he had discovered (the projective transformation between planes and lines) rather than complex analysis, but, in the complex plane, trajectories for Hooke's law become conic sections, and a Newton's law orbit can be mapped onto a Hooke's law orbit using the transformation $z \to z^2$, and vice versa.[41] This conformal mapping, which was introduced by Böhlin and Sundman for gravitational orbits,[42] exactly parallels the case in quantum mechanics, where the Schrödinger equation has exact solutions for the equivalent harmonic oscillator and Coulomb potentials ($V \propto r^2$ and $V \propto r^{-1}$), which

---

[38] Hooke's law, *'ut tensio sic vis'* ('as the extension, so the force'), was first published in *De potentia restituta*, 1678.

[39] I, Proposition 64.

[40] Bertrand (1875). The two force laws also lead to linear differential equations in the single- and two-body cases. The problem of infinite force will not arise if the linear force is a cosmological constant arising from a finite Hubble radius (see 5.8).

[41] Hall and Josić (2000).

[42] Böhlin (1911), Sundman (1907). See Arnol'd (1990), Saari (1990). The idea of the complex plane was, at this time, just beginning to emerge in Wallis's work (1685), but would not be fully established till the time of Argand (1800).

can be exchanged by substituting the oscillator variable r_{ocs} for the squared Coulomb variable r_C^2, in effect also exchanging the energies and coupling constants (or charges) and rescaling the orbital angular momenta.[43]

Some propositions in the *Principia* were specifically worked out for the linear force law. Proposition 38 works out the times of descent, velocities and distances traversed for a body falling freely under a centripetal force which is proportional to the distance, finding them respectively proportional to the arcs, the sines, and the versed sines of the arcs. Proposition 47 says that, under such a force law, 'all bodies revolving in any planes whatsoever will describe ellipses, and complete their revolutions in equal times; and those which move in right lines, running backwards and forwards alternately, will complete their several periods of going and returning in the same times'. Chandrasekhar calls this 'a beautiful theorem which is at the heart of another beautiful Proposition'.[44] The Scholium which follows says that the same principles will apply when we replace the plane of motion by a curved surface.

An attractive force directly proportional to distance is also produced within the body of a uniform sphere made up of particles attracting each other with an inverse-square law force, according to *Principia*, Book I, Proposition 73,[45] because the mass within the distance r from the centre becomes $4\pi\rho r^3/3$, and so the force on a mass m within the body of a planet $(4\pi G\rho r^3 m/3r^2 = 4\pi Gmr/3)$ can be supposed to be very nearly of this nature.[46] This additionally connects with the fact that a spherically symmetric distribution of particles subject to a linear distance law again acts as though all the mass were placed at the centre, as Newton recognises in the Scholium to Proposition 78. Of course, the Earth is not a uniform sphere, nor did Newton think it was — he saw violent activity going on under its crust and at its centre — but the idealisation to a linear law is the closest that can be achieved without the extensive experimental investigations subsequently made possible by seismology.

---

[43] Grant and Rosner (1993).

[44] Chandrasekhar (1995, 201).

[45] In 1686 Newton, who seemingly had no access to the original correspondence and had to rely on memory, pointed to the fact that Hooke had asserted that the inverse square relationship must go right to the Earth's centre, 'which I do not' (Newton to Halley, 20 June 1686, *Corr.* II, 435–437, 435–436, and 437–440, 439), although he was actually inconsistent in intimating elsewhere that the 'power' of gravity below the Earth's surface decreased with depth (Hooke to Newton, 6 January 1680, *Corr.* II: 309). In fact, as explained in 6.2, the analogies that Hooke used meant that the force had to be constant and only the velocity would decrease.

[46] III, Proposition 9.

The linear law in this case also allowed Newton to create a 'limit-case idealization' from uniform Galilean gravity on an idealised plane surface to universal gravity. In Corollary 2 to Proposition 52, Newton took the limit of the centripetal acceleration of a hypocycloidal pendulum under gravity at the surface of a perfect sphere as the radius is increased to infinity and the curvature to zero and stated that it would act equally in lines perpendicular to the surface and parallel to each other, in the same way as the centripetal acceleration on a pendulum oscillating according to the common cycloid in Huygens' measurements. In this case, the pendulum is isochronous only for gravity proportional to the distance from the centre, and not for gravity acting under an inverse-square law. So, by using the law as relating to an idealised uniform sphere, Newton was able to effect a smooth transition between the respective conditions required for Galilean and for universal gravity, while preserving the condition of isochronicity required for Huygens' pendulum measurements.[47] He also specifically related this finding to the cycloidal motion of 'pendulums, in mines and deep caverns of the Earth', where experiments could be carried out.

Corollary 3 to Book I, Proposition 91, another late addition, showed that the total gravitational force on a particle placed in a shell bounded by two concentric, similar ellipses (i.e. with the same ellipticity and similarly situated) is equal to zero. The problem of the attraction of a particle internal to an ellipsoid could in this way be reduced to the problem of a particle placed on the surface, with the further consequence that in the interior of the ellipsoid the attraction varies directly as the distance from the centre.

Section IX of the *Principia*, Book I, examines the effect of a number of different central force laws on the stability of dynamic orbits. The Scholium to Proposition 78 explains the two principal cases of attractions, inverse-square and direct distance relationships, and, for a large part of Book I, Newton treats the two cases as parallel:

I have now explained the two principal cases of attractions: to wit, when the centripetal forces decrease in a duplicate ratio of the distances, or increase in a simple ratio of the distances, causing the bodies in both cases to revolve in conic sections, and composing sphaerical bodies whose centripetal forces observe the same law of increase or decrease in the recess from the centre as the forces of the particles themselves do; which is very remarkable.

---

[47] The case is discussed and explained in Smith (2012, 387–389).

(Chandrasekhar remarks that this is the one place in the *Principia* where Newton expresses surprise at any of his results.[48]) It is only when he considers the phenomenology of the Solar System (Kepler's third law) that he opts for the inverse-square law relation. The two laws are combined in the case of the force for the lunar apogee, where a repulsive direct distance force is subtracted from the attractive inverse-square law force to find the overall centripetal force.[49] It was always possible, he considered, that the direct distance relationship should hold, and, of course, it does in the case of the cosmological constant. In the sense that he considered the direct distance relationship as possible, Newton is the originator of this concept.[50]

Newton also established a number of other dual force laws. In addition to the attractive inverse-square law and attractive force proportional to distance, for elliptical orbits, there is also a duality for the attractive inverse-square law and repulsive force proportional to distance, for hyperbolic orbits,[51] while Corollary 1 to Proposition 7 of Book I shows that an inverse-fifth power of attraction is self-dual for motion in a circle 'with the centre of attraction on any point of the circumference'.[52] The 'conjugate branches of a hyperbola', with a repulsive force directly proportional to distance, are also dual to each other; while dual repulsive and attractive inverse-square laws apply to 'hyperbolic orbits with the centre of force at the locus of the conjugate hyperbola'.[53]

In more general terms, classical or quantum mechanical systems governed by potentials proportional to r^n can be related to those with potentials proportional to r^m, where $(n + 2)(m + 2) = 4$. In terms of force laws proportional to r^a and r^b, this becomes $(a + 3)(b + 3) = 4$. Newton actually cites a relation which is equivalent to this in 'discussing the rate of rotation of precession of orbits which deviate slightly from circular ones in arbitrary power-law potentials'.[54] Of the integer pairs, a and b, Newton considers 1 and -2, and -5 and -5, but not -4 and -7 or the -1 and -1, which is 'generally excluded on physical grounds'.[55]

---

[48]Chandrasekhar (1995, 287).

[49]*Principia*, I, Proposition 45, Corollary 2.

[50]See, for example, Calder and Lahav (2008). It is interesting that Newton's later aethers, like dark energy, effectively act in opposition to gravity.

[51]*Principia*, I, Proposition 12.

[52]Chandrasekhar (1995, 118).

[53]I, Proposition 12; Chandrasekhar (1995, 118).

[54]Grant and Rosner (1993, 4).

[55]Chandrasekhar (1995, 125). The pairing of -4 and -7 brings to mind the connection between Casimir and Van der Waals forces, which have exactly these powers of distance.

Book I, Proposition 45 finds the precession of the line of apsides for an arbitrarily defined centripetal force in orbits that are very nearly circular. For a uniform centripetal force, the angle of rotation will be $180°/\sqrt{3}$ (Example 1) and if the centripetal force is proportional to the distance to the power $n - 3$, the angle will be $180°/\sqrt{n}$, the body returning 'from the lower to the upper apsis, and so on *in infinitum*'. Particular cases considered include the force laws proportional to r^{-3}, $r^{-11/4}$, r^{-2}, r^{-1}, and r, where the respective angles are ∞, $360°$, $180°$, $180°/\sqrt{2}$, and $90°$ (Example 2). For the inverse linear case ($F \propto r^{-1}$, equivalent to constant energy at all distances), the body in orbit 'will by a perpetual repetition of this angle, move alternately from the upper to the lower and from the lower to the upper apsis for ever'. For $F \propto r^{-11/4}$, 'the body parting from the upper apsis, and from thence perpetually descending, will arrive at the lower apsis when it has completed one entire revolution; and thence ascending perpetually, when it has completed another entire revolution, it will arrive again at the upper apsis; and so alternately for ever'. For $F \propto r$, as already established in Proposition 10, the body will 'revolve in an immovable ellipsis, whose centre is the centre of force'. After a quarter of a revolution it will arrive at the lower apsis, and then after another quarter at the upper apsis, 'and so on by turns *in infinitum*'. The power law for the central force can thus be found from the motion of the apsides and *vice versa* (Corollary 1). The inverse-cube force law (dealt with in Corollary 3 to Proposition 41) presents us with a singularity, long known to Newton, an orbit with 'an infinite number of spiral revolutions' into the centre.[56]

Book I, Proposition 7, Corollary 2 sets out, given a body in circular orbit, with the same period of revolution under centripetal attraction from two different centres, to find the ratio of the centripetal attractions towards the two centres.[57] Corollary 3, added in 1713, gives a more general result, in which an orbit (not necessarily circular) is described under a known law of centripetal attraction of two different laws, provided the constant of areas is the same. Given one orbit, with known period and known law of centripetal attraction, it is always possible to find the other. The same result, applied to conic sections, also appears in a Scholium following Proposition 17, which closes Section III, and which was also added in 1713. Chandrasekhar: considers the Corollaries to be of 'profound theoretical significance' and to 'display Newton's deep insight'.[58] They enable him to switch from one

---

[56] Newton to Hooke, 13 December 1679, *Corr.* II, 307–308.
[57] Chandrasekhar (1995, 80).
[58] Chandrasekhar (1995, 113).

case to another without doing a new calculation. Lynden-Bell describes the Proposition as a 'lovely theorem', and refers to 'Newton's wonderful reciprocal corollary on how the same orbit can be swept out with the same angular momentum when the force is directed either to one centre or to another. This enables him', he says, 'to show that the centred elliptical orbit swept out by a simple harmonic oscillator is directly related to the Keplerian ellipse swept out when an inverse square force is directed towards a focus'.[59]

Book I, Propositions 90 and 91, together with the third Corollary to Proposition 91, were used subsequently in Propositions 18–20 of Book III. In Proposition 90 (which again employed quadratures without explanation) Newton found the attraction of a circular lamina or disc on a point situated on the straight line orthogonal to the plane of the lamina and passing through the centre of the lamina. Reconstructed in modern notation,[60] if $F(s)$ is the force due to an element of the lamina at a distance s from the point, the total force becomes the equivalent of $2\pi h \times$ the integral between limits h and S of $F(s)ds$, where h is the perpendicular height of the point from the centre of the lamina, R is the radius of the lamina, and $S = (h^2 + R^2)^{1/2}$. Corollaries 1 and 2 give the result when $F(s)$ is proportional to s^{-n} and s^{-2}, while a third corollary finds the limiting value when the disc becomes infinitely large for $F(s) \propto s^{-n}$. In modern notation, if $F(s) = gs^{-n}$, for some constant g and for $n > 1$, then the total force becomes $2\pi g/((n-1)h^{n-2})$. As with the hollow spherical shell, the disc could be assumed in the limiting case to be infinitesimally thin and used as a component in determining the forces on solid bodies constructed from an infinite number of such discs. In Proposition 91, which follows, Newton showed how one could obtain the attraction of a corpuscle in the axis of a solid of revolution with circular cross-sections of variable radii, 'to whose several points there tend equal centripetal forces decreasing in any ratio of the distances whatsoever'. The three corollaries apply this to a cylinder, an oblate spheroid and a homoeoid (a shell between two concentric similar ellipsoids or spheroids).

Another kind of force law appears in Proposition 93, where Newton gives a solution in the form of an infinite series, for the attraction of a body towards an infinite plane. Here, the 'attractive force of the whole solid, in the recess from its plane surface, will decrease in the ratio of a power whose side is the

---

[59]Lynden-Bell (1996, 254). This is in a part of Proposition 17 under the heading 'The same otherwise'.

[60]Pask (2013, 158–163) uses this as a simple example of Newton's geometrical integration.

distance of the corpuscle from the plane, and its index less by 3 than the index of the power of the distances'. Newton considers two cases. In the first the attraction is found at a point at a normal distance from the bounding plane. In this case the index of the inverse power of attraction must be greater than 3, as the integral diverges logarithmically at $n = 3$, and 'the attraction of the remoter part of the infinite body is always ... infinitely greater than the attraction of the nearest part' (Corollary 3). In the second case, for the attraction at a point z_0 below the bounding plane, the attractions of the slab z_0 above and below the point will cancel, leaving only the contribution below $2z_0$. 'Hence the attraction of the slab at image points, equidistant from the bounding plane, are equal'.[61] Corollary 1 finds the attraction for a slab of finite thickness. For a slab of thickness tending to infinity, according to Corollary 2, the force dependency will tend asymptotically to proportionality with $z_0^{3-n}/[(n-1)(n-3)]$, while the same will apply, according to Corollary 3, for a body of arbitrary shape with a plane boundary so long as z_0 is much smaller than any of the linear dimensions of the body. The Scholium which follows shows how to calculate the motion of a body to a plane under a given attraction or the attraction under a given equation of motion 'by resolving the ordinates into converging series' (in fact, a Laurent series with odd and even solutions), including the cases where the motion is parabolic or hyperbolic.[62]

Book III of the *Principia* introduces perturbation theory into physics, particularly for interplanetary perturbations. This is now a standard technique, used in calculations involving all the known physical interactions. Of I, Proposition 44, Donald Lynden-Bell (referring to Chandrasekhar) says that: 'while it is clear that the addition of an inverse cube force is equivalent to a change of angular momentum, Newton proves the lovely theorem that the path traced out relative to rotating axes is precisely the same as that traced in fixed axes without the additional force'.[63]

The Scholium to I, Propositions 1–2 states that a force perpendicular to a described surface (along the direction of angular momentum $\mathbf{h}$) makes the orbital plane change direction but not its surface area. It doesn't change motion in that plane; $r^2\phi = h$ in the moving plane, but with gyrations of h. However, motion in the plane is unchanged. Lynden-Bell says that if we add an additional acceleration $= \mu r^{-3}$ when h and r sweep around cones, a new force of form $\mathbf{v} \times \mathbf{B_1}$ results, with $\mathbf{B_1} = \mu\hat{\mathbf{r}}/r^2$; this is the gravomagnetic

---

[61] Chandrasekhar (1995, 318).
[62] See Chandrasekhar (1995, 320–321).
[63] Chandrasekhar (1995, 187), Lynden-Bell (1996, 255).

monopole known to relativists as NUT space. The result, however, is not in the *Principia*.[64]

Another important question relates to general relativity, the idea that has had most influence in gravitational theory since 1915.[65] While it is quite normal in texts on general relativity to invoke the Newtonian equation for a perfect unperturbed elliptical orbit and imply that the Einsteinian calculation from general relativity leads to a precessing orbit of a different kind, this can lead to a false impression of the relation between the two theories, for the true Newtonian equation also leads to exactly the same kind of precessing ellipse. In fact, the calculation of the precessing orbit due to the additional non-inverse-square terms caused by planetary perturbations is due to Newton himself, for Newton showed that the effect of such extra perturbative terms would be a rotation of the apsides, leading to a precessing ellipse. In *Principia*, Book I, Proposition 45, he writes: 'if the centripetal force be as

$$\frac{bA^m - cA^n}{A^3},$$

the angle between the apsides will be found equal to

$$180° \sqrt{\frac{b - c}{mb - nc}}.$$

Here, A represents the altitude or distance, b, c, m and n are constants, and the angle is that between successive apses. Where the force law includes only a single term of the form bA^{m-3}, the angle separating successive perihelia will be $2 \times 180° \times \sqrt{b/mb} = 2\pi/\sqrt{m}$. Newton also has a numerical calculation for the effect on Earth, Venus and Mercury in the Scholium to Proposition 14 in Book III of the *Principia*.

The general effect of perturbation theory is to add extra terms to the expression for centripetal force, which are in powers of the distance, depending on the geometrical configuration of the system. In first order, as in Newton's theory of the lunar apogee, this can emerge as a repulsive force directly proportional to distance. The perturbations on a planet orbiting the Sun due to the other planets are of this kind. In the Solar System, the planet most affected is Mercury, and in the nineteenth century, Leverrier and then Newcomb found that the perihelion of Mercury would

---

[64]Lynden-Bell (1996, 255).

[65]The relation between Newtonian theory and general relativity is such an important subject in historical, philosophical and physical terms that more extensive discussions will be given in Section 6.6 and Chapter 8.

advance by 531 seconds of arc per century due to such perturbations. Observations showed an advance of 574 seconds, leaving 43 seconds of the advance unexplained.[66] The effect of the general relativistic calculation which explains this is to introduce an inverse-fourth power term to account for this additional advance. Ultimately, the origin of the term is special, rather than general, relativistic, and can be done in a simple way using a relativistic expression for the potential energy. It is then relatively straightforward to substitute this directly into Newton's equation to produce the correct result.

There are many ways of doing this using the more restricted theory of special relativity.[67] For example, we can define a potential with 'relativistic' mass and 'contracted' distance:

$$V = -\frac{GM/\gamma}{r\gamma} = -\frac{GM}{r\left(1 - v/c\right)^2},$$

recognising that any contraction due to gravity will be radial, in rods placed in the direction of the field, and not tangential, in the direction of the planet's motion. Assuming the angular momentum per unit mass for the orbit is a constant $(vr = \alpha)$, a first-order binomial series expansion produces an 'inverse-fourth' law force correction. Substituting α^2/r^2 for v^2, we now obtain

$$V = -\frac{GM}{r} - \frac{GM\alpha^2}{r^3 c^2}.$$

Differentiating this expression to find the gravitational field strength or force per unit mass, $-dV/dr$, we find

$$F = -\frac{GM}{r^2} - \frac{3GM\alpha^2}{r^4 c^2},$$

which is the precise equation for force which emerges from the Schwarzschild solution of Einstein's general relativistic field equations.

---

[66] Le Verrier (1859), Newcomb (1882).

[67] For example, Strandberg (1986). Papers are regularly published claiming that the arguments that the light deflection and perihelion precession can be obtained without the full field equations must necessarily be invalid. A recent example is Kassner (2015), which quotes 'refutations' by Sacks and Ball (1968) and Rindler (1968) as authoritative, but these early 'refutations' have themselves long been refuted (for example, in Rowlands, 1994). It would, in fact, be (to say the least) quite implausible if equations that are identical to those that can be generated by special relativity should be inaccessible to a theory that is accepted as a special case of the general one. Such 'coincidences' do not occur in physics, and the limiting transitions to the Lorentzian metric and Newtonian gravity are fundamental and *non-provable* assumptions of general relativity, a purely mathematical theory which provides *no direct access* to the physical effects which it sets out to describe.

In principle, the term which produces the additional perihelion precession is a special relativistic correction of the gravitational potential by a factor $(1 + v^2/c^2)$, or, in a relatively low speed orbit, by a factor $1/(1 - v^2/c^2)$. A potential of the form

$$V = -\frac{2GM}{r}\left(1 + \frac{v^2}{c^2}\right)$$

was derived by Harvey in 1978 in a discussion of the 'Newtonian limit for a geodesic in a Schwarzschild field'.[68] We can now use Newton's own Proposition 45 to complete the solution numerically. Leaving out the common term $-GM$, we have $b = 1$, $c = -3\alpha^2/c^2r^2$, $m = 1$, and $n = -1$; so the precession per orbit becomes

$$2\pi\sqrt{\frac{1 + 3\alpha^2/c^2r^2}{1 - 3\alpha^2/c^2r^2}} \approx 2\pi\sqrt{1 + \frac{6\alpha^2}{c^2r^2}},$$

which, again using a binomial approximation, becomes an advance of

$$\frac{6\pi\alpha^2}{c^2r^2} = \frac{6\pi v^2}{c^2} = \frac{6\pi GM}{rc^2}$$

radians for a near circular orbit. If we take into account the ellipticity of Mercury's orbit, r will be replaced by $a(1 - e^2)$, in which the a is the length of the semi-major axis $= 5.79 \times 10^3$ m, and e is the eccentricity $= 0.2056$. With the values 1.5475×10^3 m for GM/c^2, and 0.24084 years for the sidereal period, we obtain the additional 43 arcseconds per century.

There is also another way to approach the question, which brings it closer to Newtonian thinking. Newton was fully aware that results in celestial mechanics could be obtained only at the best order of approximation and that new factors determining the motions of bodies in such a complex system as that of the Sun and planets might be discovered and lead to significant changes in numerical predictions. He was also fully aware that the theory he was presenting depended on defining an inertial frame, even though there is no such thing in nature. Sometimes, this problem can be overcome by using 'fictitious' inertial forces, as in his explanations of the shape of the Earth, the tides and the precession of the equinoxes. However, a major source of future difficulty, which appeared in his time with Rømer's discovery of the finite speed of light in 1676, could not be addressed until improvements in

---

[68] Harvey (1978).

the accuracy of observational techniques made it necessary to confront the problem.[69]

Essentially, a finite speed of light means that, in a pure Newtonian theory, with gravity assumed (though by default, rather than by necessity) to act instantaneously at a distance, there will be a mismatch of positions of astronomical bodies as calculated from their assumed mutual interactions, and as measured optically using the finite speed of light. It was exactly this mismatch, in the case of Jupiter's satellites, that led to Rømer's discovery of a finite light speed, and Newton would use it himself in 1691 to ask the Astronomer Royal, John Flamsteed, to perform a test of one of his optical theories using the same satellites.[70] In a more general context, after it was realised that *all* observation was limited by the speed of light, and in some cases by a speed of light that was further slowed down by the effect of gravity on the signal, then it was no longer possible to present Newtonian theory by equations which used the distances to measured positions as though they were the same as the 'absolute' ones used in the calculation.

To overcome this, relativity introduced an observer-centred or epistemological physics, but it is still possible to construct a physics that is ontological (or 'God-centred'), like that of Newton, if we have a systematic way of incorporating the changes that will result from the mismatch between theory and observation. And, of course, there is such a possibility through the use of inertial forces. Essentially, the Newtonian equations demand a Newtonian space-time for *gravity*, that is one in which there is no relativistic connection between space and time and an instantaneous interaction. If we carry this logic through, we find that the space-time of relativity's Lorentzian metric (used for all forces of *observation*), which is itself distorted by the presence of a gravitational field, will experience an 'aberration' caused by the rotation of any local coordinate system with respect to the absolute one, assumed by Newton to apply to gravity. In the vicinity of the Sun's gravitational field, taking the Sun as a spherically-symmetric point-source,

---

[69]Rømer (1676, 1677).

[70]Newton to Flamsteed, 10 August 1691, *Corr.* III, 164; Flamsteed to Newton, 27 August 1691, *Corr.* III, 165. It is important to realize that the principle of relativity, in both Newtonian and Einsteinian forms, requires that a body, such as a planet, moving with near-inertial motion under the action of a central source of gravitation such as the Sun, is subjected to an acceleration from the *instantaneous position of the source*, and not from the point from which the source's light would appear to come. Relativity, and indeed the inertial principle of Newtonian physics, demand that the static gravitational field *does not propagate*. Only noninertial changes, equivalent to gravitational waves, are transmitted at the speed of light.

the aberration will be calculable from the geodesic of a light ray and will produce exactly the same effect as is seen in the Schwarzschild solution.[71] The aberration will also cause the deflection of a ray of light in a gravitational field, and a slowing down of electromagnetic signals, as well as appearing as an analogue to the magnetic field of electromagnetic theory, so producing analogous 'gravitomagnetic' effects, including a gravitational analogue to electromagnetic waves.[72]

The idea that modifications have to be brought into Newton's law of gravity to explain apparently anomalous astronomical observations has quite a long history. Clairaut, for instance, in the mid-eighteenth century, was on the point of adding an inverse fourth power term to the law of gravity to explain the motion of the apsides of the Moon, having reached the same initial conclusion as Newton that the result was only half the value expected (as also did d'Alembert and Euler in independent investigations) when he took his calculations to a further degree of approximation, and found that they agreed with the Newtonian law after all.[73] James Challis, in 1859, sought in a similar way to introduce an inverse-fourth power law to explain the anomaly that Leverrier had uncovered in the perihelion precession of Mercury, which, as we have seen, is exactly the effect produced by the relativistic correction which subsequently explained it. Then, Hugo von Seeliger, in 1895, sought to multiply Newton's law by an exponential term $(e^{-\lambda r})$, which would be effective in the case of great distances, to avoid the infinite gravitational field which he predicted to exist everywhere in space, though this was soon discounted because the coefficient calculated from the precession of Mercury was inconsistent with the results from the other planets.[74] At the present time, there are people who have considered explaining 'dark matter' and 'dark energy' through fundamental modifications to Newtonian dynamics or Newtonian gravity. Ultimately, we don't know where such ideas will eventually take us, but the historical record seems to suggest that the Newtonian inverse-square law of gravity, stemming ultimately from the 3-dimensionality of space, has the fundamental and ultimate significance of a principle that will survive many such challenges.

---

[71] *Sitzungsberichte, Preussische Akademie der Wissenschaften*, 1916, 189–196, 424–434.

[72] As explained above, the instantaneity of gravity in systems at rest or under inertial motion is *as essential to relativity as it is to Newtonian physics*. The choice between God-centred and observer-centred coordinate systems is then a philosophical, rather than a physical, one. In either case, we would expect the limit on the velocity of observation to apply in exactly the same way with the same mathematical results.

[73] Clairaut (1745).

[74] Challis (1859); Seeliger (1895, 1896).

6.3.   The Three-Body Problem: The Lunar Motion

Though Newton's laws are strictly deterministic, the systems they describe are nevertheless potentially chaotic, so that very slight changes in initial conditions can affect long-term behaviour in almost arbitrary and unpredictable ways. Newton already knew that orbits were never repeated because of the perturbing influence of other bodies. In *De Motu Sphaericorum Corporum in fluidis*, one of the antecedent works leading to the *Principia*, he says:

> By reason of this deviation of the Sun from the centre of gravity the centripetal force does not always tend to that immobile centre, and hence the planets neither move exactly in ellipse nor revolve twice in the same orbit. So that there are as many orbits to a planet as it has revolutions, as in the motion of the Moon, and the orbit of any one planet depends on the combined motion of all the planets, not to mention the action of all these on each other. But to consider simultaneously all these causes of motion and to define these motions by exact laws allowing of convenient calculation exceeds, unless I am mistaken, the force of the entire human intellect.[75]

In *Principia*, I, Proposition 65 he showed that bodies attracting each other according to an inverse-square law would not move exactly in ellipses, because of perturbations, but he did consider the possibility that certain special configurations of the bodies (such as were later found by Euler and Lagrange) might lead to stable elliptical orbits. He was mostly concerned with those cases where the orbits would 'not much differ from ellipses' with areas 'very nearly' proportional to the times. These included a number of lesser bodies revolving about a very great one at different distances; then the case in which the whole system was attracted by a 'vastly greater body' at such a large distance that the attractions of the body on the different parts of the system were parallel. In the first instance, the orbits would be very nearly ellipses. In the second the motions of the bodies amongst themselves would not be affected by the attractive force affecting the system as a whole. This is, of course, the case with our own Solar System, which is attracted to the centre of mass of the Milky Way, but retains its local motions relatively undisturbed. In such cases, the motion of the centre of mass of the local system could be separated from that of the motions of the components within the system. Of course, if the 'very great body' should approach much more closely to the system, the internal motions of the parts of the system would become perturbed, as the 'great body' attracted the parts of the system unequally and no longer in parallel directions. The 'system' and its

---

[75] *De Motu Sphaericorum Corporum in fluidis*, Herivel (1965, 301).

centre of mass would then need to be redefined to include the larger body as an intrinsic component. The perturbations would be even greater if the attractions were no longer governed by an inverse-square force due to already being perturbed. As a result of this, if a system should be observed with its parts in elliptical or circular motion without any significant perturbation, any disturbing force must be very weak or 'impressed very near equally' on the parts of the system 'and in parallel directions upon all of them'.[76]

As well as completely solving the 2-body problem, and a special case of the many body problem, Newton also initiated the study of the crucial 3-body problem, both in the restricted case of Sun, planet and satellite, and in the more general case of Sun and two planets. In Proposition 5, Corollary 3, of Book III, he showed that, 'near their conjunction', Jupiter and Saturn, since their masses are so great, 'sensibly disturb each other's motions'. He knew that Jupiter caused Saturn to slow down while approaching conjunction and then to speed up subsequently, while Saturn had the opposite effect on Jupiter.[77] The value, which is estimated in Proposition 13, is relatively small (Jupiter's effect on Saturn being 1/211 of that of the Sun, causing a perturbation with amplitude of about $7''$ and period 19.86 years), and Newton is able to say in Proposition 14 that the aphelions, the planes and the nodes of the orbits of the planets are effectively fixed. He says: 'It is true, that some inequalities may arise from the mutual actions of the planets and comets in their revolutions; but these will be so small, that they may be here passed by'. Anomalies in the motions of Jupiter and Saturn had been known since at least the time of Horrocks, but a theory of such anomalies which matched the observations remained out of reach until 1785 when Laplace discovered the Great Inequality in the planets' motions, with a period of 890 years. Newton, however, had anticipated the probability of a significant effect, and identified it as a problem to be solved using his gravitational theory.

In the Scholium following proposition 14, he includes numerical estimates of the effects on other planets:

> Since the planets near the Sun (viz., Mercury, Venus, the Earth, and Mars) are so small that they can act with but little force upon one another, therefore their aphelions and nodes must be fixed, excepting in so far as they are disturbed by the actions of Jupiter and Saturn, and other higher bodies. And hence we may find, by the theory of gravity, that their aphelions move forwards a little *in consequentia*, in respect of the fixed stars, and that in the sesquiplicate [3/2 th] power of their several distances from the Sun. So

---

[76] I, Proposition 65, including Cases 1–2 and Corollaries 1–3.

[77] Gregory (1702), Proposition 51, reprint English translation (1972, 488).

that if the aphelion of Mars, in the space of a hundred years, is carried forward 33′ 20″ *in consequentia* in respect of the fixed stars, the aphelions of the Earth, of Venus, and of Mercury, will in a hundred years be carried forwards 17′ 40″, 10′ 53″, and 4′ 16″, respectively. But these motions are so inconsiderable, that we have neglected them in the Proposition.

Newton notably finds 256 seconds of arc per century for the precession of Mercury's orbit due to the larger planets, which is about half the total calculated from every single possible source of perturbation (531 seconds of arc per century, excluding the 43 seconds due to the relativistic correction). His aim isn't to do such a calculation, which would require knowledge of bodies in the Solar System not discovered until long after his time, as well as the masses of the inner planets, Mercury, Venus and Mars, of which he had no direct knowledge, but to show that the main sources indicate that it is inconsiderable within the accuracy of measurements available to astronomical observers in his own day.[78]

Newton investigated the coordinate frame in which the perturbations of a three-body problem could be most conveniently represented. In Proposition 67, he considered an 'interior' system of two bodies, with an 'exterior' perturbing body. The exterior body performs an orbit closer to an ellipse with the centre of gravity of the interior system as focus than to one with the focus at the position of the larger body in the interior system, and, according to Proposition 68, it is closer to such an orbit if the larger interior body is agitated by the attractions rather than at rest. He concludes: 'Therefore the perturbation is least when the common centre of the three bodies is at rest; that is, when the innermost and greatest body T is attracted according to the same law as the rest are; and is always greatest when the common centre of the three, by the diminution of the motion of the body T, begins to be moved, and is more and more agitated'. Chandrasekhar comments that this 'is precisely the coordinate frame in which the variational equations' involved in the three-body problem of Earth, Moon and Sun may be derived.[79]

For the kind of many body problem represented by such systems as Jupiter and its satellites, Newton says that the orbits of the many small bodies round the great one

will approach nearer to ellipses; and the descriptions of areas will be more nearly uniform, if all the bodies attract and agitate each other with

---

[78] His value of 2000 arc-seconds for Mars may be compared with the current observational value of 1600 arc-seconds.

[79] Chandrasekhar (1995, 265).

accelerative forces that are as their absolute forces directly, and the squares of the distances inversely; and if the focus of each orbit be placed in the common centre of gravity of all the interior bodies (that is, if the focus of the first and innermost orbit be placed in the centre of gravity of the greatest and innermost body; the focus of the second orbit in the common centre of gravity of the two innermost bodies; the focus of the third orbit in the common centre of gravity of the three innermost; and so on), than if the innermost body were at rest, and was made the common focus of all the orbits.[80]

Chandrasekhar has pointed out that these are what have been called 'Jacobi coordinates' in a modern textbook.[81]

In Proposition 69, Newton considers a system of bodies, in which two, A and B, attract all the others with inverse-square forces. In this case, he says, 'the absolute forces of the attracting bodies A and B will be to each other as those very bodies A and B to which those forces belong'. In principle, the 'absolute attractive force' will be as the masses of the bodies. This can be generalised to the attractions of all the bodies to each other, and to 'any common' law of attraction, that is according to any power or other function of the distance.[82] Newton again emphasises, as he had in the third corollary to Proposition 65, that

> if the lesser revolve about one very great one in ellipses, having their common focus in the centre of that great body, and of a figure exceedingly accurate; and moreover by radii drawn to that great body describe areas proportional to the times exactly; the absolute forces of those bodies to each other will be either accurately or very nearly in the ratio of the bodies. And so the contrary [i.e. conversely].[83]

The lunar motion became a celebrated instance of the restricted 3-body problem, involving the effect of both Earth and Sun. In the case of the lunar orbit, the Sun is the perturbing body and has a considerable effect. Cohen says that: 'In some ways … one of the most revolutionary parts of the *Principia* … set the study of the moon in a wholly new direction which astronomers have largely been following ever since'.[84] The Moon's motion is actually very complicated. The late nineteenth century theory by G. W. Hill and E. W. Brown required about 1,500 terms, and there are

---

[80]I, Proposition 68, Corollary.

[81]Chandrasekhar (1995, 265), citing Brouwer and Clemence (1961, 588).

[82]I, Proposition 69, Corollaries 1 and 2.

[83]I, Proposition 69, Corollary 3.

[84]Cohen (1999, 246).

many irregularities in its motion.[85] The 'great' Proposition 66, Book I,[86] along with 22 corollaries, gives an extended account of a 3-body system of the same structure as the Sun, Earth and Moon, with particular attention to the effects on the moonlike body. Book III, Proposition 22, enumerates the 'inequalities' established in the various corollaries of Proposition 66, as applied to the motion of the Moon, stating that the various planetary satellites will have similar inequalities, due to the additional disturbing action of the Sun. Proposition 23 points out how we can infer the inequalities to be expected in other planet-satellite systems, in particular those of the satellites of Jupiter and Saturn, from what we know of one of them.

In Proposition 25 Newton finds the 'forces with which the Sun disturbs the motions of the Moon' and, in Proposition 26, shows how these forces produce an inequality ('horary increment') of the area described by the Moon, assuming, for simplicity, a circular orbit. Then, from the hourly motion of the Moon, Proposition 27 demonstrates how to find its distance from the Earth. Corollary 1 states: 'Hence the apparent diameter of the Moon is given; for it is reciprocally as the distance of the Moon from the Earth. Let astronomers try how accurately this rule agrees with the phaenomena'. And Corollary 2: 'Hence also the orbit of the Moon may be more exactly defined from the phaenomena than hitherto could be done'. Proposition 28 finds 'the diameters of the orbit in which, without eccentricity, the Moon would move', showing that it is a prolate ellipse.[87]

Proposition 38 explains why the Moon is spheroidal (with the equatorial exceeding the polar diameter by 186 feet) and why, apart from the 'exceedingly slow' librations, it always shows the same face to the Earth.[88] According to Proposition 17, the diurnal motions of the planets are uniform, and the libration of the Moon arises from its diurnal motion (here Newton refers to the letters he wrote to Mercator on the theory of the libration in longitude of the Moon). Newton must have traced the longitudinal libration of the Moon to the difference, $u - \phi$, between the mean and the true anomalies. In similar fashion, he showed that the utmost satellite of Saturn must show the same face to Saturn; and 'so also the utmost satellite of Jupiter'.

Of the four irregularities in longitude known to Newton when he began his theory, one, evection, he explained in terms of Jeremiah Horrocks's

---

[85] Hill (1905–1907), Brown (1896, 1919).
[86] Guicciardini's epithet (1999, 65).
[87] See Chandrasekhar (1995, 425–426).
[88] Proposition 38 and Corollary.

kinematic theory before proposing a solar force that might produce it (I, 66, Corollary 9). The other three explanations are qualitatively the same as in modern theories. Newton also identified a number of new lesser inequalities in longitude. He gave quantitative results on the variation,[89] the motion of the nodes (or points where the orbit intersects the ecliptic)[90] and the varying inclination.[91] In addition, he was able to explain correctly the main inequalities in latitude, and, using Corollary 6 from Proposition 66, to derive the annual equation.[92]

The first inequality or anomaly in the Moon's motion, observed by Hipparchus in the second century BC, is the rotation of the line of apsides in an eccentric orbit. The second inequality, later known as evection, and also pointed out by Hipparchus, is a combination of a monthly variation in eccentricity and a twice monthly fluctuation in motion of the line of apsides. It resulted in a deviation of more than 2 1/2 degrees in the longitude from the positions calculated by the epicyclic motions which had successfully accounted for the positions from the Sun at conjunction and opposition. A series of astronomers, beginning with Ptolemy, proposed mechanisms which would produce this anomaly; one, which included his newly-found area law of planetary motion, was proposed by Kepler and then modified in 1640 by Horrocks. Horrocks attached an epicycle to the centre of the moon's elliptical orbit, using Kepler's choice of rotation angle. Many centuries after Hipparchus, Tycho found unequal motions of the nodes and inclination, two smaller variations in longitude, and a twice monthly fluctuation in speed — *variation*; the annual fluctuation in speed is known as the *annual equation*.[93] In addition to these were the parallactic inequality, and the secular acceleration of the Moon's mean motion, discovered by Halley in 1693.[94]

Horrocks's kinematic theory was explained by Newton in terms of his gravitational dynamic theory based on perturbations of the Moon's orbit due to the gravitational force of the Sun. Though eighteenth and nineteenth

---

[89] III, Proposition 29, achieved to within 7/8 of the observed value.

[90] III, Propositions 30–33 and Corollaries, achieved to within 0.3% of the observed value. An alternative derivation of the motion of the nodes, by John Machin, was added in the third edition in a Scholium to Proposition 33.

[91] III, Propositions 34 and Corollaries, and Proposition 35.

[92] III, Proposition 35, Scholium; I, Proposition 66, Corollary 6.

[93] Smith (1999).

[94] Halley (1693).

century mathematicians understood this and praised the lunar sections of the *Principia* as 'one of the most profound parts of this admirable work' (Laplace)[95] and 'the most valuable chapter that has ever been written on physical science' (Airy),[96] a number of twentieth century science historians — notably Whiteside, Westfall and Kollerstrom[97] — missed the connection, and referred to the Newtonian theory as if it had remained kinematic and failed to advance significantly beyond that of Horrocks. Michael Nauenberg has shown that this was not at all the case.[98] In addition, the detailed account of Newton's first-order perturbation theory in a manuscript in the Portsmouth collection[99] 'shows that Newton's method anticipated the perturbation methods of Euler and Laplace'. Whiteside was seemingly wrong in his assessment, as he was in some other areas, but he was followed by other writers because of the great authority he acquired as Newton's mathematical editor.

According to Nauenberg, Newton's method 'corresponds to the modern method of variation of orbital parameters attributed to Euler, Lagrange and Laplace, although this connection has not been generally appreciated in the past'.[100] He says that 'Newton's physical and geometrical approach leads directly to differential equations [...] for the parameters of the revolving

---

[95] *Theory of the Moon's Motion* (1702, reprinted 1975), 41. Laplace (1825, 409): 'Parmi les inégalités da mouvement de la Lune en longitude, Newton n'a développé que la *variation*. La méthode qu' il suivie me parait être une des choses le plus remarquables de l'Ouvrage des Principes'. Nauenberg (2011b).

[96] G. B. Airy, *Gravitation*, London, 1834, vii.

[97] Whiteside (1976); Westfall (1980, 540–548); Kollerstrom (2000).

[98] Nauenberg (2000, 2001a); Wilson (2001). Incorrect views of Newton's lunar theory according to Nauenberg (2001b) include the assertion that it 'was a retrogresive step back to an earlier kinematic tradition which he had hoped once to transcend, and to a limited Horrocksian model which was not even his own invention' (Whiteside); that the most important of the corrections which Newton made in the second edition of the *Principia* was 'a kinematic theory of the motion of the center of the moon's orbit which had no foundation in gravitational dynamics' (Westfall); that 'Newton could not get his gravity theory to work and had to reach to an earlier kinematical theory' (Kollerstrom).

[99] *MP* VI, 508–537. The Portsmouth collection refers to the main collection of Newton's manuscripts which passed through his collateral descendants to the Earls of Portsmouth. The 'scientific' ones were presented to the University of Cambridge in 1876, the others were sold by public auction in 1936.

[100] Nauenberg (2000, 168).

ellipse [...] in a method that 'corresponds precisely to the modern' one.'[101]
Using this, Newton was able to calculate the periods of the lunar inequalities,
though not all of the magnitudes.

Of Propositions 65–69, Félix Tisserand, who had seen the Portsmouth
papers, had written: 'I am inclined to think that he [Newton] knew all the
formulae [the equations generally known as *Lagrange's planetary equations*]
but that, instead of publishing them, he preferred deducing a large number
of geometric propositions from them which he obtained by considering in
each case the effect of just one component [element]'.[102] Tisserand was thus
of the opinion that Newton had derived for himself the equations expressing
the variations with time of the elements of the Kepler orbit under the action
of an external force F — elements that would otherwise have been constants;
and that Proposition 66 and its 22 corollaries provide the principal basis for
this inference. Laplace wrote that 'The method [Newton's development of
the method of the variation of the elements] appears to me as one of the
most remarkable things in the *Principia*'.[103]

According to Nauenberg, Newton's solution of the lunar variations in
Book III, Proposition 28 'corresponds to the periodic solution obtained later
by L. Euler and in full detail by G. W. Hill'.[104] Curtis 'Wilson agrees with
Nauenberg in recognizing that Newton in Propositions 26 and 28 employed
algebraic techniques, most notably binomial expansions, trigonometrical
relations, integrations of sinusoidal functions, and calculations of curvatures
of algebraic curves, but he did so in a completely different way than Euler or
Hill'.[105] Wilson seems to imply that Newton's methods incorporate *ad hoc*
elements derived from physical arguments and don't have the same degree
of 'control' as those of the later mathematicians, but this is what one
would expect in a pioneering mathematical analysis. The results of Euler
and his successors were an astonishing achievement and an essentially new
development, based on sixty years' development of algebraic calculus and
mathematical dynamics, but it is equally remarkable that Newton achieved
results with any 'correspondence' to the later work at the very beginning of
such developments.

---

[101]Nauenberg (2001, 191). Palmieri's book review (2004) objected to this idea of 'correspondence', claiming that Galileo's verbal statements of kinematics did not 'correspond'
exactly to the equations now used for them. However, as far as Newton was concerned,
they did, and they still do for most commentators.

[102]Tisserand (1889), Tome III, Chaptire 3, p. 33; Chandrasekhar (1995, 206).

[103]Quoted in Chandrasekhar (1995, 206).

[104]Nauenberg (2001, 198).

[105]Guicciardini (2003, 422–423).

Newton made no secret of the difference between the theory and the semi-empirical model needed for practical purposes. To make his work more useful for obtaining tables of lunar positions, he adjusted the values using Flamsteed's high-quality data, but the work was by no means a failure, and did have practical consequences (see Chapter 7). The adjustments were made to make the theory useful on a semi-empirical basis, and were not disguised within the data, although they were not always explicit. Such a lack of explicitness is typical of many aspects of the *Principia*.

The motion of the lunar apsis or apogee is worked out only in a manuscript calculation. In the *Principia*, Newton shows how to account for half of it, and he showed in the first edition how to account for the rest, but leaving out the actual calculation because it wasn't precise enough and involved arbitrary assumptions. Clairaut was later able to obtain precise agreement by extending Newton's solution. The first half is calculated as due to the radial component of a perturbing force from the Sun; if the line of apsides advances by 3° per revolution, then the centripetal force will vary as the inverse of r to the power 24/243 (the value depending on the relative size of the contributions of attractive inverse square force and repulsive force proportional to the distance). Newton is quite explicit in saying that, from the values used in the semi-empirical calculation that follows, the advance is only 1°31′28″, while 'The apsis of the moon is about twice as swift'. He correctly assigned the remaining half to the transradial component of the solar perturbation, with the tidal effect of the Sun creating the required eccentricity in the lunar orbit, but was clearly quite prepared to expose discrepancies between calculation and observation where a resolution might be achieved by a successor.[106] As we have seen, the calculation proved to be difficult, with Clairaut, d'Alembert and Euler each obtaining the same solution as Newton. Clairaut finally succeeded in obtaining exact agreement in 1749 by extending the calculation to 'next to next to leading order' when he included higher-order terms involving the squared and cubed values of the eccentricity.[107]

Ultimately, the Newtonian method proved to be more important than any of the specific results achieved in *Principia*, and George E. Smith has outlined nine different ways in which the method permanently changed

---

[106] *Principia*, I, Proposition 45, Corollary 2; III Proposition 35, Scholium. Chandrasekhar (1995, 453), says 'his solution of the variational equation is entirely correct'.

[107] Clairaut (1752). The fact that Clairaut, d'Alembert and Euler all failed using algebraic methods suggests that Newton's 'geometric' style was not the main source of the problem. That Newton had, in fact, realised the importance of second-order effects is evident from Corollary 12 to I, Proposition 66.

physics. In principle, Newton had defined abstract relationships between abstract parameters, only some of which had observable 'measures', within a generic theory, in which forces were assumed to act without requiring a mechanistic explanation according to exact laws which could be inferred from phenomena in which they were only approximately true. By defining what Smith has called 'Newtonian idealizations', presumed to hold exactly in some specified circumstances, he was then able to change the focus of future research into a search for deviations from them. The success of the explanations found for such 'second-order phenomena' could consequently be used in support of the original theory, or, in the case of theory change, to define 'limit-case idealizations' to effect a smooth transition from the old theory to the new.[108] The ultimate breakthrough was that a physics defined purely by abstractions could actually be used to create an explanation of a vast array of physical phenomena with quantitative and qualitative precision, even to the extent of including phenomena which had yet to be discovered.

6.4. Newton and Flamsteed

Newton's difficulties with the lunar theory became the cause of the second of his three famous quarrels with contemporaries, after the first with Hooke and before the third with Leibniz. Flamsteed, the Astronomer Royal, was a great observer who certainly never achieved sufficient recognition for his outstanding contributions to observational astronomy. Throughout his career, he faced constant tribulations and minimal financial reward, for, although paid a basic salary as Astronomer Royal, he had to provide his own instruments and pay for his own assistants. He had every reason to feel disgruntled, but he was in addition a very difficult man, a seriously disturbed individual who would 'have tried the patience of an archangel'[109] — which Newton certainly was not — and he was even more inclined to paranoia than his great adversary. He was also a perfectionist who regarded his data as his personal possession and was unwilling to make compromises on its presentation for the good of science.

Already in 1678 he was in disagreement with his patron Sir Jonas Moore and the Royal Society about publishing data that he considered still only preliminary. Then, at some early stage, he fell out with Halley (which Newton somehow never quite managed to do). Halley had very quickly,

---

[108]Smith (2012). Markus Fierz is quoted as saying that Smith 'makes very clear that Newton's celestial mechanics was something truly novel, namely that it displays the currently used method of doing mathematical physics' (*op. cit.*, 395).

[109]Hall (1992, 272).

in the same year as Flamsteed had refused to divulge his data on stellar positions, published the catalogue of southern stars he had made on a visit to St. Helena. As a free-thinker, he was, according to the deeply religious and impossibly self-righteous Flamsteed, an infidel, as well as a slapdash astronomer. Further incidents led to considerable bitterness. Flamsteed had disputes also with another astronomer, Johannes Hevelius of Danzig, though these stopped short of a quarrel, while Hooke annoyed him by publicly challenging his knowledge of optics in 1681. Clearly he had form, even before he became associated with Newton.

Two such prickly and uncompromising personalities as Newton and Flamsteed seemed destined to clash, though at first relationships between them were surprisingly good. Flamsteed supplied Newton with data which was used, with acknowledgement, in the *Principia*, though Flamsteed thought Newton should have given him proper credit for identifying the comets of 1680 and 1681 as a single body, going round the Sun. Newton, on his part, thought that Flamsteed's theory was defective in that he had said the comet had been deflected from one straight line path to another by the Sun's magnetic field before reaching the Sun when he had found that the comet's path was a *gravitational* orbit *round* the Sun. He had corrected Flamsteed on this point even before accepting that this particular 'pair' of comets was a single one.[110] He also thought that Flamsteed was unwilling to provide data for theorists whom he thought were superfluous to the real business of providing observations of the heavenly bodies for practical purposes such as navigation; theories were ephemeral, but data were permanent.

Flamsteed, like many others, simply didn't understand that theory, as Newton understood it, involved grand conceptualisations, and the most general equations possible before what we would now call 'phenomenology', or fitting in the particular details, was attempted. Newton wrote to him in 1694 in some exasperation a letter which, better than any other, explains why his method has stood the test of time beyond any immediate application:

> I believe you have a wrong notion of my method in determining the Moons motions. For I have not been about making such corrections as you seem to

---

[110]Newton to Flamsteed, draft letter, ? April 1681, *Corr.* II, 358–362, 360–361. This is an important corrective to the suggestion sometimes made that Flamsteed 'put Newton right' on comets. He was certainly correct in showing that the 'two' comets of 1680 and 1681 must be the same one, but Newton was correct in defining the only possible orbit in which this would be the case, and in assigning the cause to gravity rather than a magnetic attraction.

suppose, but about getting a general notion of all the equations on which her motions depend and considering how afterwards I shall go to work with least labour and most exactness to determine them. For the vulgar way of approaching by degrees is bungling and tedious. The method which I propose to my self is first to get a general notion of the equations to be determined and then by accurate observations to determine them. If I can compass the first part of my designe I do not doubt but to compass the second... And to go about the second work till I am master of the first would be injudicious...[111]

The quarrel with Flamsteed stemmed from the latter's determination to publish his observations in the perfect form that he himself desired. Flamsteed was slow in providing Newton with data for his work in celestial mechanics, always seeming to hold something back. Newton, who required the measurements to complete his difficult and unrewarding work on lunar theory, became increasingly impatient for immediate information. Newton was urging Flamsteed to publish a star catalogue as an accurate reference for the lunar measurements as early as 1691, but the latter prevaricated, putting the perfection of his observations above any utility they might have had.[112] Newton needed the data right away while he was working at the immensely difficult lunar theory, and felt that, as Flamsteed was a public servant, the data should be regarded as public property. The simmering discontent gradually developed into a long and bitter dispute.

Newton came to regard Flamsteed, a highly skilled observer and more than capable theoretician, as a mere lackey, whose obdurate attitude was holding up the progress of research. Flamsteed certainly *was* difficult and frequently impossible; but, in principle, he was partially right — the observations had been obtained with *his* labour with instruments paid for out of his own money. Newton, who felt that Flamsteed's position as Astronomer Royal made his observations public property, was also partially right and took his opportunity to impress his views after his election as President of the Royal Society. On hearing that the Lord High Admiral, Prince George of Denmark had expressed interest in supporting the publication of Flamsteed's observations, he visited Greenwich on 12 April 1704, ostensibly to give his support. Flamsteed's estimate of the costs was read out at a Royal Society meeting on 15 November, and the Society resolved to approach the Prince. Articles of agreement were signed on 17 November 1705, and printing

---

[111] 16 November 1694, *Corr.* IV, 167.
[112] 10 August 1691, *Corr.* III, 164.

commenced. A partial star catalogue, missing six Northern constellations, was supplied by Flamsteed in 1706 in a sealed package.

By April 1708, Flamsteed had received £ 125 from Prince George, and the printing of the first volume of his proposed *Historia Coelestis Britannia* was complete. This consisted of observations taken before he acquired the Mural Arc quadrant in 1689 and was able to take more accurate data. Flamsteed, according to the signed agreement, was to deliver the second volume (of post-1689 observations) along with the star catalogue without delay. This never happened. After various efforts at sweet reasoning had failed, the project came to an abrupt halt when the Prince died in October. Flamsteed used the break to complete his catalogue, but no further printing was possible without funds. Newton, however, came up with a new plan. On 14 December 1710, he instituted a Board of Visitors to Greenwich Observatory, again ostensibly as a means of support for the Astronomer Royal, in preparing his data for publication but, in fact, with the express object of forcing Flamsteed's observations from him. Typical of the durability of British institutions, the Board of Visitors survived its immediate purposes down to the ending of Greenwich Observatory in the late twentieth century.

After it became clear that Flamsteed, a man incapable of understanding the importance of keeping to business schedules, was only prepared to work at his own pace, Halley was drafted in to use the observations supplied in 1706 as public property and to work them up in a catalogue, edited by himself. Flamsteed heard early in 1711 that an order from the Queen had instructed him to complete the *Historia Coelestis* with funds supplied by the Treasury, but he immediately came up with a new reason why the project would have to be delayed: he needed help in constructing new planetary tables. Then, when he heard that his long-standing enemy was the editor, Flamsteed refused to supply the remaining parts of the catalogue. Halley was instructed to fill in the missing six constellations using Flamsteed's observations with the Mural Arc quadrant, which would not now be printed separately, either to save time or money. Flamsteed considered this a travesty of his original intentions, and he furiously rejected Halley's offer to correct the proofs when they became available in the summer of 1711.[113] Following a particularly unpleasant scene between Newton and Flamsteed, recorded by the latter with a considerable show of self-justification,[114] the volume, with all its imperfections, was published early in 1712.

---

[113] Flamsteed (1712).

[114] Flamsteed (1835). Newton's known lack of verbal eloquence certainly contributed on this occasion.

The situation was clearly one in which no party could escape with a hundred per cent credit. Flamsteed was impossibly difficult, Newton impatient and aggressive, desperate for the data which Flamsteed regarded as his hard-won personal possession, and ready to use his position as President of the Royal Society to force Flamsteed's formerly independent institution to come under his control. Flamsteed's account of the case was used by the later President of the Royal Astronomical Society, Francis Baily, in his support, and many subsequent authorities have quoted it, almost as if it were the work of an independent witness.[115] Many writers have also stressed how, after a change in government in 1714, Flamsteed managed to acquire the unsold copies of the 'defective' catalogue, from which he removed the offending pages, burning them in a public ceremony. The part retained, the pre-1689 observations, would become the first volume of a three-volume 'corrected' edition. This has been presented as a 'victory', but, if so, it was a hollow one, for the data now were publicly available, and Newton now had no interest in any further action that Flamsteed might take.

In addition, though the saga has often been presented as a personal battle between two egotistical and unbending personalities, something else was always at stake: all the while that the data remained unpublished and inaccessible to theoretical astronomers, ships were being lost at sea. The search for a method of finding the longitude had, after all, been the motivation for the original founding of the Royal Observatory at Greenwich, and the problem had become even more acute after the wreck of Admiral Sir Cloudesley Shovell's fleet off the Scilly Islands in 1707. This had been the motivation for Prince George's involvement in the project, and it was certainly one of which both Newton and Halley were acutely aware. The lunar method relied on the occultation of known stars by the Moon and so an accurate star catalogue was absolutely crucial. By no means everyone believes that the 'moral' rightness was on the side of Flamsteed, however ruthless and personably disagreeable Newton may have become in pursuit of his aims.[116] As in the case of the invention of the calculus, 'moral' lessons

---

[115] Baily (1835).

[116] Silverman (2003), for example, considers that 'The conflict with Flamsteed is not simply the vengeful act of a self-centered Newton trying to cheat a poor astronomer out of his data. In large measure, Newton and Flamsteed differed widely in their perceptions of Flamsteed's official position — and it seems to me, by the standards of today at least, that Newton's viewpoint is the more sustainable. To Newton, the Astronomer Royal was a civil servant, a scientist paid by the State to obtain results of use to the State. Is it not, then, almost a malfeasance of duty for a government scientist to refuse to relinquish data until such time as he can complete some personally conceived magnum opus?' Newton, as

have also been drawn too readily from Newton's supposed failure with the lunar theory. It has been assumed that Newton's aggression may have originated in a sense of his own failure, that it led to no useful result, and that it may have hindered the project rather than enhanced it. However, in Nauenberg's estimate, as we have seen, the lunar theory was anything but a failure, and the practical version published in 1702 produced genuinely useful results and served as a very important basis for the lunar longitude method developed later in the century by Tobias Mayer (see Chapter 7).[117]

Flamsteed might have felt that Newton had been dictatorial and that Halley's editing was defective, but he couldn't argue that the data had been 'stolen'. The original decision to publish had been made by agreement between all the parties, and the data supplied by Flamsteed. The catalogue was issued in his name and he was given credit for the observations, and, as the paid Astronomer Royal, he couldn't deny that the material should be made available for public use. He could justifiably argue that he had never received sufficient recognition or financial recompense for the hardship he had had to endure in collecting it, and that the Royal Observatory had been created under his initiative and that he was entitled to a say in how the data he had collected should be published. In fact, a catalogue conforming to his own perfectionist standards was only issued in 1725, six years after his death, by his assistants, and far too late to be of the slightest use to Newton.[118] Ironically, though he had resolved to eliminate all the features that Halley had introduced in 1712, including the numbers assigned to the stars in the 'defective' catalogue, these survived Flamsteed's purges and are now known as 'Flamsteed numbers'.

6.5. The Causes of Gravity

The Newtonian system attributes all events in the universe to action at a distance between point particles of matter. 'Particles', in this sense, are not mechanistic atoms, but infinitesimally small mathematical constructs, and the action between them can be conceived in a purely abstract way, irrespective of any mechanism devised to explain it. Gravitation is the result of an active power, the spirit acting directly: ' ... in general the motions of

---

a public servant, always put the 'King's business' ahead of all other interests, and he no doubt expected Flamsteed to do the same.

[117]Nauenberg (2000, 2001a, b); Wepster (2010).

[118]Flamsteed (1725).

all the great bodies hitherto observed by astronomers are exactly such as ought to arise from their mutual gravities in free spaces'.[119]

From an early period, Newton's calculations of the dynamical effects of gravity outpaced any of his attempts at 'physical' or metaphysical explanations of the phenomenon, though there would be many of these, including the impact forces produced by particles within a Cartesian aether; the pressure of Cartesian aethereal vortices; various vital agents including the incorporeal one responsible for 'vegetation', the process responsible for changes in materials beyond the mechanical rearrangements of material particles; an aethereal density gradient; an 'active principle' of the kind imagined by the Stoic philosophers; the direct action of God; and the action produced by an 'elastic and electric spirit'. In his mature work, besides excluding some of these hypotheses as contradictory to experimental evidence, he reached a position where the mathematical laws of gravity could be separated from such discussion.

At the time of the *Principia*, in 1687, Newton had abandoned his earlier belief in a Cartesian aether on the basis of experiments with pendulums, though he later returned to it in a modified form. The motions of the astronomical bodies seemed to rule out any material plenum of the Cartesian kind, which would require component particles with resistive properties.

> The heavens and universal space some persons fill with a most subtle fluid matter, but the existence of this is neither evident to the senses nor is proved by any arguments, but is precariously assumed for the sake of an hypothesis. Moreover, if we trust reason and the senses, that matter will be in exile from the nature of things. For one cannot understand how movement can be accomplished in a plenum.[120]

'This subtle matter in which the planets float, and in which bodies move without resistance is not a phenomenon. And what are not phenomena, and subject to none of the senses, have no place in experimental philosophy'.[121]

Starting with the atmosphere, there is no limit to the rarefaction that can occur as we go further away from the Earth's surface:

> Now if the Resistance in a Vessel well emptied of Air, was but an hundred times less than in the open Air, it would be about a million times less than in Quick-silver. But it seems to be much less in such a Vessel, and still

---

[119] *The Elements of Mechanicks, Unp.*, 168.

[120] Manuscript associated with revision of *Principia*, early 1690s; McGuire (1968, 162–163).

[121] Draft for the *Principia*, c 1716; McGuire (1968, 180).

much less in the Heavens, at the height of three or four hundred Miles from the Earth, or above. For Mr. *Boyle* has shew'd that Air may be rarified above ten thousand times in Vessels of Glass; and the Heavens are much emptier of Air than any *Vacuum* we can make below. For since the Air is compress'd by the Weight of the incumbent Atmosphere, and the Density of Air is proportional to the Force compressing it, it follows by Computation, that at the height of about seven and a half *English* Miles from the Earth, the Air is four times rarer than at the Surface of the Earth; and at the height of 15 miles it is sixteen times rarer than at the surface of the Earth; and at the height of 22 1/2, 30, or 38 Miles, it is respectively 64, 256, or 1024 times rarer, or thereabouts; and at the height of 76, 152, 228 Miles, it is about 1,000,000, 1,000,000,000,000, 1,000,000,000,000,000,000 times rarer; and so on.[122]

Corollary 3 to *Principia*, Book III, Lemma 4 asserts that the motion of comets shows that 'the celestial spaces are void of resistance'. And in Query 28 of the *Opticks*, Newton says: 'And against filling the heavens with fluid mediums, unless they be exceeding rare, a great objection arises from the regular and very lasting motions of the Planets and Comets in all manner of courses through the heavens. For thence it is manifest, that the heavens are void of all sensible resistance, and by consequence of all sensible matter'.

6.6. Newtonian Gravity and General Relativity

Newtonian theory proved surprisingly robust to early challenges. It was set up to provide a fixed standard which would eventually incorporate results that initially seemed anomalous, and it managed to do this for more than two centuries. Only in the twentieth century was a challenge mounted, and then only by a single theory armed with the explanation of a single anomalous result, and a few predictions relating to light, rather than the system of the world. The way this happened has more to do with cultural history than with the fundamental meaning of physics. Despite the journalistic appeal of a tiny anomaly leading to a revolution, while everyone is complacently imagining that there is nothing new to be discovered, this isn't really how science normally operates and most of the other examples that are usually cited

---

[122]Q 28. The argument is broadly correct. The density of the Earth's atmosphere decreases exponentially up to about the final two layers, the thermosphere and the exosphere, and at about 100 km (the boundary where, by convention, 'outer space' is said to begin) it is reduced to 10^{-6} of that at the Earth's surface, comparable with Newton's figure at 76 miles. After that it decreases more slowly, but in interplanetary space it is reduced to between about 5×10^6 and 10^8 particles/m^3, which is comparable to Newton's final figure (for the exosphere), which is equivalent to 5×10^7 molecules/m^3.

can be seen to be based on exaggeration, mythology or misinterpretation. However, the early twentieth century was a time of cultural challenges in virtually every field — for example, in art, music, literature and politics, as well as science — not always with happy results. Monoliths had to be overthrown and Newtonian physics was an obvious target. Einstein's general theory of relativity of 1915, together with Eddington's 'confirmation' of 1919, provided the opportunity, generating headlines such as 'Revolution in Science. New Theory of the Universe. Newtonian Ideas Overthrown'.[123]

So, the creation of a new explanation of the anomaly of Mercury's perihelion precession had to be seen as a *revolution*, excluding Newtonian science. The change had to be dramatic and not an evolution in the long process of explaining the detailed motions of celestial bodies in which the Newtonian theory still had the most significant place. This rhetoric is still used and still widely believed, even though the implication that Einstein's general relativity is a theory *sui generis* is enormously damaging to our understanding of the place of the theory within fundamental physics, to the extent of making it seem like an anomaly itself. Einstein's theory had to be seen as a *rival* to Newton's, creating an entirely new world-view.

The position of general relativity with respect to Newtonian science is, in this sense, unique. No other theory has actually deviated from the Newtonian world-view, not even special relativity or quantum theory; statements to this effect can easily be shown to be premature or misconceived. So all discussions of the Newtonian system of the world have to come up eventually against the question of how it stands in relation to the theory which, according to the dominant rhetoric of most of the previous century, finally 'replaced' it. Not only does Einstein's theory give more accurate predictions, we are told, but the 'force' of gravity which was the basis of Newton's theory simply does not exist. What we imagine to be due to the actions of a force of gravity is simply the effect of curvature in regions of a combined 4-dimensional space-time. Alongside this approach to the technical issues involved with the Newtonian theory, we have been informed that the revolution in science brought about by Einstein's theory generated even more fundamental issues about how science relates to fundamental truth.

According to this, Einstein's general theory has brought with it an entirely new philosophy of nature. We also have a new attitude to science. The old certainties are gone. A theory may last as a solid foundation for centuries but, ultimately, an effect is discovered or a prediction is made

---

[123] *The Times*, 7 November 1919.

which requires an entirely different physical 'paradigm'. Newton's theory served its purpose for a long time but its failure to explain the perihelion shifts of the planets showed that it was fundamentally faulty. Then general relativity came along, explained this effect and predicted a few more which were eventually discovered. General relativity, however, was not just a modification of Newton's theory; it required a completely new conception of physics, and so Newton's theory was not only inaccurate but fundamentally 'wrong'. In addition, physical space could no longer be considered as purely Euclidean — it had intrinsic Riemannian curvature, which manifested itself as gravity, and affected the construction of the entire universe. Gravitational 'force' had no existence as a physical entity or source of physical action; 'gravitating' objects simply moved along lines of space-time curvature. The whole episode shows that Newton's theory was merely a 'hypothesis'. If a theory of this magnitude can be shown to be 'wrong', then there is no such thing as certainty in science. All scientific theories are simply hypotheses which serve their time until facts arrive which they cannot explain; they are best fits to data, useful for a time but ultimately to be discarded.

Such statements are made frequently with the confidence that the whole scientific world accepts them, but they are nonetheless philosophically naïve, and far from robust in either scientific or historical terms. The classic logical error involved is the fact that like is not being compared with like. First of all, the geometrical representation of general relativity does not mean that the idea of gravitational force has been eliminated from physics. The fact that the curvature term $(G_{\mu\nu})$ is equated to an *energy-momentum* tensor $((8\pi G/c^4)T_{\mu\nu})$ suggests otherwise. It is, in fact, possible to represent *any* field theory by a covariant geometrical construction of the same type, *even Newtonian gravity*. General relativity, also, is *not* an explanation of gravity, but a way of mathematically representing how gravity affects a system. The theory, in fact, is a mathematical construction devoid of primary physical content. The field equations of general relativity are purely a description of space-time curvature, which is only *assumed* to be intrinsic. They give no indication of the physical meaning of the curvature or of how curvature is manifested in any particular locality. To do this, as Einstein himself proclaimed, we have to reintroduce the Newtonian potential GM/R (through the energy-momentum tensor), not as a limit in weak fields, but as the theory's only physical content.[124]

---

[124]We may note here that general relativity is so intrinsically complicated, with equations that can be solved only in the simplest of cases, where the extreme symmetry drastically reduces the complexity, that there is no serious possibility of using the field equations to

In addition, the 'curvature' of general relativity is not of the familiar physical space but of a mathematical construct called space-time, which relativistic quantum mechanics doesn't recognise as a simple 4-D or 4-vector quantity as it requires an extra level of structure in the gamma matrices which link the units of space and time. It is also purely local, as global curvature of the kind widely assumed by the early pioneers of general relativistic cosmology simply doesn't exist.

Even more significantly, Newton's own analysis *is also presented as pure geometry*. When he set out to calculate the perturbations by a third body on a two-body system in *Principia*, Book I, Proposition 44, Newton took this to be equivalent to finding a new coordinate system varying in space and time in which the perfect ellipse of the two-body system would be preserved. This is exactly what general relativity does, except that general relativity's null-geodesic is a straight line. Newton takes the ellipse and the inverse-square law of gravity as his starting-point, and then finds the new coordinate system in which this is preserved under the action of the additional force produced by the perturbing body. Einstein takes the straight line and zero gravity as his starting-point and finds the new coordinate system in which the straight line structure is preserved under the action of the total force involved. The principle is identical — only the starting-point is different — and Élie Cartan, by writing Newtonian theory using the same mathematical structure as Einstein's, effectively showed how Newton could also have started from the straight-line position if he had thought it appropriate.[125] In fact, in the special case of Corollary 6 to Proposition 44 (as we saw in 5.7), he did just that, and derived an inverse-cube orbit from straight-line motion using a rotating coordinate system rather than a centripetal force. Chandrasekhar's description of this as 'centrifugal potential' implies that, in the new coordinate system, it can be regarded as a purely fictitious inertial effect. Physical effect becomes defined by pure coordinate geometry. In fact, defining a new coordinate system as a new 'geometry' means little more than defining a non-Euclidean geodesic.[126]

---

reproduce any of the more advanced classical analyses of the Solar System, such as the Hill-Brown theory of the Moon's motion.

[125] *Ann. Ecole Norm.*, 40, 325, 1923; 41, 1, 1924. We can write Newtonian gravitation as $R_{ab} = 8\pi T_{ab}$ by comparison with general relativity's $R_{ab} = 8\pi(T_{ab} - g_{ab}T/2)$.

[126] We should note here that the Newtonian definition of gravitational force assumes no physical cause — it is only an effect given certain conditions. It would be consistent with Newton's own statements to make space-time curvature as much a possible cause as any 'physical' action.

In the Einstein system, of course, there is an extra degree of perturbation due to the finite speed of light, which is itself affected by the gravitational force acting on it. In relation to this additional perturbation, however, it is clear that the theories are not actually discussing the same thing. Newtonian theory is designed to be applicable as an ontological description of what is actually happening in a gravitating system. General relativity, on the other hand, is epistemological and observer-centred. Newton is fully aware that the velocity of light is finite, and that an assumption that a finite time for light to reach the Earth from the Sun implies a discrepancy between the observed positions of gravitating objects and the gravitational effects they produce. The experiment that Rømer would do to measure the speed of light in 1676 was founded on exactly that principle, in which the moons of Jupiter did not occupy the positions that the existing laws of planetary motions (that is, Kepler's) required. Newton told David Gregory on 4 May 1694 that 'In the system of the Earth moving round its axis the same results occur whether light comes from a star instantaneously or in observable time; in the Ptolemaic system, the opposite is true'.[127] He is perfectly aware of the effect of a finite velocity of light on the results of observations. He also believes that light rays will be deflected by gravitational fields, further complicating the picture. And if we do take account of this extra source of perturbation, the full precession effect will come within the compass of Newtonian gravitational theory.[128]

Of course, Newton had enough difficulty dealing with the Solar System at the limit of accuracy then known. Adding the finite time of observation, which might be further affected by gravitational attraction of the light signal carrying the information, would have been taking calculations well beyond the practical level of observation for the period. However, doing so would have been well within the Newtonian paradigm, and, to compare Newtonian and general relativistic results *on the same footing*, it is an absolute necessity to do this. In addition, when we investigate gravitational effects acting upon light itself, it is also necessary to take into account the fact that Newton believed that it had many of the properties of a

---

[127] *Corr.* III, 317; cf the idea that an inertial effect due to the finite speed of light is fictitious (5.8).

[128] It is worth observing here that, even if relativistic theory assumes that the innate speed of gravity is 'really' the same as that of light, the anomalies will always cancel in such a way that the speed of gravity in an inertial gravitating system or one at rest will necessarily *appear* to be infinite. In an observer-centred physics, the distinction between 'real' and 'apparent' observations has no intrinsic meaning. It becomes an arbitrary choice with the coordinate system.

material particle, including what we now call 'relativistic mass'. When we do this, it quickly becomes apparent that the three effects predicted by general relativity on the behaviour of light in a gravitational field — gravitational redshift, gravitational light deflection and the gravitational delay of an electromagnetic radiation signal — are well within the possibilities of 'Newtonian' explanation (see Chapter 8). The reason for this, of course, is that, since a light photon without rest mass has no kinematics, its trajectory is determined only by its energy relations, and these are structured in relativity theory to coincide with those of a purely classical dynamics. Interestingly, this becomes apparent through Newton's work on light, though he didn't make numerical predictions for any of these effects.

There is, in fact, a curious historical circumstance relating to the establishment of the General Theory, which casts an interesting light on its validity as evidence for the hypothetico-deductive model. General relativity owed its great public success to Einstein's prediction of the gravitational bending of light as tested by Eddington at the total solar eclipse of 1919. According to Einstein's calculations, the general relativistic bending would be twice the 'classical' value; when Eddington's measurements showed that the bending was as predicted by general relativity, Einstein's theory was established. As soon as the measurements reached Europe, the mathematician John E. Littlewood sent a telegram to Bertrand Russell with the news that 'Einstein's theory is completely confirmed', and so it has remained in the historical record.[129] However, the 'Newtonian' calculation presented for comparison had taken account of time dilation due to gravity but not of the contraction of measuring rods, which is necessary to complete the match between relativistic and classical energy equations; the full classical calculation should have predicted exactly the same result as general relativity. The same result could also have been obtained directly from classical dynamics using the *kinetic energy equation* from Newton's *Principia*. There is a detailed discussion of this in *Newton and Modern Physics*, where we show that the classical equation subtly includes special relativity by making the photon's dynamic velocity emerge from the gravitational attraction, rather than from the 'velocity of light' associated with a free massless particle. While this has no direct bearing on the validity of general relativity as a theory, it does suggest that the change of physical 'paradigm' which

---

[129] 'Dear Russell, Einstein's theory is completely confirmed. The predicted displacement was $1''.72$ and the observed value $1''.75 \pm 0.6$. Yours. J. E. L'. Clark (1973, 227).

occurred as a result of the 1919 episode was not a response to overwhelming evidence.

The light-related effects have been generated in many ways by a number of authors without using the full field equations of general relativity. A popular method has been to use special relativity, which was early accommodated to Newtonian mechanics by redefining Newtonian vector terms as 4-vectors. By 'special relativity', of course, we mean anything which supplies a 4-vector relation between space and time or energy and momentum in a Lorentzian metric, or a system in which a gamma factor, dependent on the squared speed of a moving object (*not* its velocity), simultaneously affects length, time and mass measurements. It has nothing to do with whether or not the motion is inertial. Though Einstein showed how the principle of relativity led to the construction of gamma factors *even for* inertial systems, involving only uniform motion in a straight line, no physically real or measurable system can be constructed in this way, and all the well-known, applications, including clocks taken round the Earth, the GPS, muons arriving at the Earth's surface, and virtually all aspects of relativistic quantum mechanics, are non-inertial in this way. Many, in addition, have radial symmetry in which the squared speed term originates in a radially-directed acceleration, and in which the speed is not directed along the distances to be measured. In principle, the 'relativistic' aspects are determined by energy considerations and not, in the first instance, from kinematics.

The Schwarzschild solution of the general relativistic field equations for a point source, which predicts the gravitational light deflection, has exactly this symmetry. The radial symmetry eliminates all 'curvature' except in the radial direction, and effectively produces a purely Lorentzian metric with a gamma factor on the radius and on the time component. Although this was noticed by several people very early on, a considerable number of authors have tried to deny its validity right up to the present day, usually on the completely erroneous idea that 'special relativity' excludes non-inertial motion. However, general relativity is specifically *designed* to reduce in the limit to the Lorentzian metric. If our solution is identical in structure to such a metric, then this is *exactly what it is*.

Now, since light is electromagnetic, the 'relativistic' aspects come from the nature of light, and its origin in local interactions between discrete charged particles, and not from anything to do with the gravity; and, again, the energy relations are designed to be compatible with classical ones. So purely 'classical' Newtonian calculations, based on energy, also yield correct results. However, even the comparisons relating to material objects, such as the planet Mercury, rather than light signals, are not at all that they seem.

Eddington and those who followed him presented the 'alternative' Newtonian theory as though it could be used in its purest form. But, as Newton knew well, there is no such thing as an inertial frame, so we can never use a pure Newtonian theory. We must always incorporate inertial forces. Fictitious forces are never brought into the so-called 'Newtonian' calculations of relativistic effects, even though they are an essential part of any classical theory. In the relativistic interpretations of Newton, all forces become instantaneous; observation equals structure of the world; measured quantities are set up as absolutes; time equals measurement by clocks. So it is impossible to have observation at c with the instantaneous speed of gravity. If Newton is correct in assuming that gravity is transmitted instantaneously in an inertial system, the mismatch between theory and observation will produce aberration effects which will have the same effect as inertial forces, rotating the coordinate system, and these will also produce gravitating effects, including gravitational waves.

In fact, gravity will, in this sense, have the effect of curving the space-time of measurement or observation — which always takes place using local nongravitational forces. In this reading, Einstein's theory becomes the most general way known of incorporating all such effects, and, far from 'replacing' Newtonian theory, it requires Newtonian theory to provide its gravitational content. Though general relativity started by assuming the validity of three fundamental principles — Mach's principle, the principle of general covariance and the equivalence principle — it eventually abandoned all physical content in favour of a mathematical expression, which, though described as 'Einstein's field equations', is simply a mathematical description of curvature, with physical meaning to be supplied by the user in the energy-momentum tensor. It is remarkable that a theory 'of gravity' actually says nothing about gravity at all, simply borrowing the Newtonian potential for all known solutions of the field equations. In addition, the equations that determine the behaviour of an expanding universe, the most popular theory of modern cosmology, are purely Newtonian, like the Schwarzschild radius for black holes, in accord with the idea that Newtonian theory provides the principal gravitational content of general relativity. *Tests of general relativity are also tests of the theories that it incorporates*, specifically Newtonian gravity and the Lorentzian metric.

Though a great deal of physical content has been supplied by many in subsequent years — intrinsic transmission of gravity at the speed of light, infinite gravitational collapse, wormholes, quantum gravity, and in earlier times a finite but unbounded universe with non-Euclidean geometry — none of it is required by the theory and, in the absence of an independent source of information, much of it could be completely spurious. This is now believed to be the case for light-speed transmission of the static force, for this seems to be incompatible with Lorentz invariance. In addition, the once widely-held assumption that the actual space-time of the observable universe is curved seems to be incompatible with observations that indicate that the universe is actually flat and its space is totally Euclidean and probably infinite in extent. Real space is not curved. Quantum gravity, again, which seems to be required by intrinsic transmission at the speed of light in a *local* interaction, has been unable to overcome the problem of unrenormalizable infinities. It also predicts a value for the cosmological constant that is wrong by more than 120 orders of magnitude. The reason for the infinities appears to be that quantum field theory requires a spin 2, rather than a spin 1 boson, for a local *attractive* interaction between (like) discrete sources, and this extra amount of spin introduces exponentially increasing complications. Newtonian gravity, however, was always nonlocal, with only the repulsive (and, therefore, spin 1) inertial reaction a local force. The evidence from both Newtonian theory and general relativity is that gravity is something other than a 'real' local quantised force, like electromagnetism or the weak and strong interactions. Precisely this role, however, is available to the inertial reaction and it may be that we can discover the link to this force through the equations of general relativity and the effects it predicts.

It is frequently said that general relativity (unlike Newtonian theory) is in conflict with quantum mechanics, but the conflict only occurs with respect to certain assumptions about the physical meaning of the field equations.[130] If we were able to able to accept fully that the theory is not an 'alternative' to Newtonian gravity, but a significant addition to it, providing a way of reconciling gravity with inertia, we might see the Newtonian (and

---

[130] Quantum mechanics has resisted the Eisnteinian attempt to extend the concept of localization with time-delayed interaction to the whole of physics, by requiring an additional concept of *vacuum* which has properties akin to Newtonian gravity — nonlocality, instantaneity and negative energy.

relativistic) instantaneous transmission of the static gravitational force as a route to understanding the instantaneous correlation of quantum mechanics, and overcoming the problems of quantum gravity and the cosmological constant. Ultimately, we might find that the reason why general relativity currently appears to be anomalous with respect to the rest of physics is because Eddington and his followers created a mythological account of its meaning to produce a cultural revolution at a time in the early twentieth century when such things were considered desirable.

Chapter 7

Astrophysics and Cosmology

7.1. The Constitution of Planets and Stars

Newton found, not unexpectedly, that the discovery of universal gravity came with a bonus: the discovery of many new results, often with a significance very different from that of the original research programme, and emerging purely as deductive consequences. Newton's theory immediately led him to uncover many significant facts about the main astronomical bodies: stars, planets and comets. For example, the respective densities of the Sun, Jupiter, Saturn and Earth in the first edition of the *Principia* were 100, 90 (misprinted as 60), 76, 386, and, in the third edition, 100, 94.5, 67, 400, which may be compared with the present day densities of 100, 94.4, 50.4, 390.1.[1] The relative densities of these bodies could be calculated accurately because of the motions of their satellites under the inverse square law of gravity; and since he had already guessed (correctly) that the density of the Earth was between 5 and 6 times that of water, the relative values also gave *qualitative* information about the composition of these bodies.[2] The Earth being about 4 times denser than the Sun made the latter not much denser than water, and so 'in a sort of rarefied state'. 'The Sun, therefore, is a little denser than Jupiter, and Jupiter than Saturn, and the Earth four times denser than the Sun; for the Sun, by its great heat, is kept in a sort of rarefied state'.[3] The large planets, Jupiter and Saturn, the ones we now know as gas giants, were even less dense than the Sun and so again different in kind from Earth. The high degree of oblateness of Jupiter might seem to imply a high degree of fluidity in its state, and more so at its equator than

---

[1] III, Proposition 8 and Corollaries 1–3; Cohen (1999, 220–227).

[2] *Principia*, III, Proposition 10.

[3] Q 11. Despite being in a 'rarefied' (gaseous) state, the Sun has a relatively high density because the gas is ionized.

at the poles due to the greater heating effect of the Sun. It certainly had an atmosphere, obscuring its solid body. Newton also calculated the masses. In the first edition, the relative masses of the Sun, Jupiter, Saturn and Earth were, respectively, 1, 1/1,100, 1/2,360, 1/28,700. By the third edition, these had become 1, 1/1,067, 1/3,021, 1/1,69,282, compared to the present day 1, 1/1,047, 1/3,498, 1/3,32,946.[4] Again, since he had a figure for the radius of the Earth, these could also be translated into absolute values. His value for the Earth's mass would not be significantly different from the modern 6×10^{24} kg, and his value for the Sun's mass would be of order 1 or 2×10^{30} kg, which is of the correct order to indicate its enormous value. A body of such massiveness, a million times that of the Earth, would clearly be different in nature to the much smaller planets, and the same would be true of other fixed stars, which were clearly Sun-like objects.

In paragraph 17 of *The System of the World*, Newton uses information from that other inverse square law, for light intensity, to speculate further on the composition of the planets, with special reference to a comparison with the metallic-rich nature of the Earth, which had been so important in *Of Natures Obvious Laws*:

> It may be that the remoter planets, for want of heat, have not those metallic substances with which our earth abounds; and that the bodies of Venus and Mars, as they are exposed to the Sun's heat, are also harder baked and more compact. For from the experiment with the burning-glass, we see that the heat increases with the density of light; and this density increases inversely as the square of the distance from the Sun; when the Sun's heat in Mercury is proved to be seven-fold its heat in our summer seasons. But with this heat our water boils; and those heavy fluids, quicksilver and the spirit of vitriol, gently evaporate, as I have tried by the thermometer; and therefore there can be no fluids in Mercury but what are heavy, and able to bear a great heat, and from which substances of great density may be formed.

Here, and in Corollary 4 to Proposition 8, he speculates that the Sun's heat will make the planets closer to the Sun more dense than those further away, and there is, in fact, a systematic decrease in density in the planets from Mercury to Saturn, except that the Earth is slightly denser than Venus.

Unfortunately, the direct gravitational calculation could not be used for the Moon as it has no satellite of its own, and Newton had to estimate its mass indirectly using data from tidal observations, a much less satisfactory process. Consequently, his value at 1/40th of the mass of the Earth is only

---

[4]Cohen (1999, 221). Densmore (2003, 466) corrects the 1/1,69,282 to 1/1,96,282 on the basis of the third edition containing a typographic or copying error.

order of magnitude, approximately 2 times the actual mass; and led to the conclusion that the Moon was slightly denser than the Earth, when it is, in fact, 1.6 times less dense.[5] However, it was very likely of a roughly similar nature in composition.

The stars are clearly bodies of a very different kind from the planets, even those like Jupiter and Saturn that are even more rarefied than the Sun. In Phaen. 5. of *Phaenomena* Newton writes:

> The Sun & fixt stars are lucid bodies shining with a very strong light: but the earth & Planets are dark bodies shining only by the reflected light of the Sun. ... The fixt stars therefore are exceedingly more lucid than the Planets & so are bodies of another kind & by the great strength of their light resemble the Sun. ... And since in removing from a lucid body the light thereof decreases in a duplicate proportion of this distance, if we were 30,000 ... times remoter from the Sun then we are at present, his light would appear equal to that of Saturn without his ring or to that of a star of the first magnitude. ...[6]

The Sun and fixed stars are certainly extremely hot. The huge gravity of the Sun and of its atmosphere has the effect of strong compression, condensing the gases which would otherwise emerge, and conserving the heat by the great mass.

> And are not the Sun and fix'd Stars great Earths vehemently hot, whose heat is conserved by the greatness of the Bodies, and the mutual Action and Reaction between them, and the Light which they emit, and whose parts are kept from fuming away, not only by their fixity, but also by the vast weight and density of the Atmospheres incumbent upon them; and very strongly compressing them, and condensing the Vapours and Exhalations which arise from them?[7]

Newton even *calculated* the temperature of the Sun using the inverse square law of light intensity and his conclusion that 'the heat increases with the density of light'. The heat of the Sun, he said, must be 50,000 times that of a hot summer's day.

> Were we ten times nearer him no doubt we should feel him an hundred times hotter for his light would be there an hundred times more constipated and & ye expt of ye burning glass shews that his heat is answerable to ye constipation of his light. So then were a body hard by the sun, his light there being about 50000 times more constipated, his heat would be 50000

---

[5] *Principia*, III, Proposition 37, Corollary 3.
[6] *Phaenomena* (MS), *Unp.*, 380–381.
[7] Q 11.

times greater than we feel it in a hot summers day, wch is vastly greater than any heat we know on Earth.[8]

In our terms, this is $50,000 \times 300$ K or 15 million K, which is, in fact, a good estimate for the temperature at the centre of the Sun. Interestingly, John James Waterston, in 1861, used Newton's law of cooling to deduce the first modern figure, of 13 million degrees, for the temperature of the Sun.[9]

In the *Opticks* Newton emphasises the effect of the Sun's massive gravity on its retaining its atmosphere, despite the enormous heat:

> ...the great weight of the Atmosphere which lies upon the Globe of the Sun may hinder Bodies there from rising up and going away from the Sun in the form of Vapours and Fumes, unless by means of a far greater heat than that which on the surface of our Earth would very easily turn them into vapours and fumes. And the same great weight may condense those Vapours and Exhalations as soon as they shall at any time begin to ascend from the Sun, and make them presently fall back again into him, and by that action increase his Heat much after the manner that in our Earth the Air increases the heat of a culinary Fire. And the same weight may hinder the Globe of the Sun from being diminish'd, unless by the Emission of Light, and a very small quantity of Vapours and Exhalations.[10]

A description of the solar corona in a manuscript called *Atmosphaera solis* comes from observations probably made at the total eclipse of 22 April 1715, but ends with a projection that looks forward to the much later concept of the solar wind, and the idea that the Sun extends far beyond its visible presence and well into the Solar System:

> That the Sun is indeed surrounded by a huge Atmosphere appears, from eclipses of the Sun, in which the Moon where it covers the whole Sun appears as a black circle, surrounded by a shining corona like a halo. The interior limb of the corona where it touches the circle of the Moon is most brilliant. It shines less than the Sun itself, but its light exceeds that of the brightest clouds. And the more distant this light is from the circle of the Moon, the less brilliant it is, reaching a distance of more than two or three degrees from the centre of the Moon before it stops. It stops gradually so that its exterior limit cannot be defined and around the whole circle of the Moon it appears of the same breadth, colour and brightness. Indeed, it no less illuminates the whole sky by its light than does the dawn a little before the rising of the Sun when fixed stars of the second magnitude are just becoming invisible. Imagine that the atmosphere of the Sun does not

---

[8]Letter to Crompton for Flamsteed, ? April 1681; *Corr.* II, 358–362, 359.
[9]Waterston (1862).
[10]Q 11.

end where it ceases to be visible but that it extends as far as the orb of Mercury and far beyond as a more tenuous medium. It is also conducive to the ascent of vapours...[11]

The stars, though 'fixed' in apparent position,

are subject to various changes. For there are seen frequently spots upon the Sun some of wch are darker some brighter then the rest of his body & these spots are generated & corrupted like scum upon a pot & seldome last above a month & while they last they move round with him once in about 26 days. And to the like mutations the fixt stars are subject: For some of them have grown brighter others darker, some have vanished & others appeared anew & some have appeared & disappeared & appeared again by many viciccitudes.[12]

The Sun and the stars are in today's understanding the most visible and observable sign of the action of nuclear forces in nature. In an alchemical context, the Sun was the 'type' of the inner forces within material particles. According to Betty Jo Teeter Dobbs: 'The 'fire at the heart of the world' was also to Newton the creative fire at the heart of matter, the informing form of all ordinary substances, hidden but active in the elemental or terrestial realm'.[13] Unlike much of the chemistry that Newton included in the *Opticks*, this seems never to have been translated from the alchemical context into a scientific statement. It is notable in this connection that Newton's comments on the constitutions of the planets in *The System of the World*, written during the period when he was still active in 'alchemy', make no concession to the alchemists' conflation of the planetary bodies with specific metals, an idea that he would have regarded as a corruption of the truth occurring in early historical times. The planet Mercury certainly has no connection with metallic mercury or 'quicksilver' and Jupiter and Saturn, supposedly representing tin and lead, appear to be devoid of those metals in which the Earth is rich. Though Newton necessarily studied texts in which such connections were made, 'alchemy' was, to him, an experimental science, not a mystical experience or a source of religious allegory.

---

[11]*Unp.*, 319.

[12]*Cosmography*, *Unp.*, 374–377.

[13]Dobbs (1991), 164. Although Newton may have had no real idea that the fire at the heart of the Sun was fundamentally different from that of an ordinary 'culinary fire', he was not incorrect in *Principia III*, Rule II in equating the source of the *light* from each, since these (at respective temperatures of $\sim 1400\,\mathrm{K}$ and $\sim 5800\,\mathrm{K}$) are both examples of black body radiation, a process he would effectively describe on a number of occasions, along with its dynamical origin, including in Query 8 (see 17.1).

7.2. Stellar Distances and Distribution

Newton was the first to provide a correct estimate of the distance of the
nearest stars, but this time he was fortunate in having a ready-made method
created by a colleague who was not in a position to use it satisfactorily
himself, and in possessing a copy of the rare book in which it was described.
The method he used was entirely due to James Gregory, but Gregory, writing
in 1668, did not have a good value of the solar parallax, for determining the
Earth-Sun distance or astronomical unit (a.u.).[14] By the time that good
values for the parallax were established, in 1672, Gregory only had a few
years to live and he did not return to the problem.

Gregory had proposed, in his book, *Geometriae pars universalis*, that, to
a first approximation, the stars can be assumed to be identical in absolute
magnitude or brightness. The apparent magnitude, as observed, would then
give a measure of a star's relative distance if we apply the law that this
decreases with the square of the distance. If we observe a planet to be equal in
brightness or apparent magnitude, at some moment, to Sirius, the brightest
star in the northern hemisphere, and the nearest star that can be seen with
the naked eye, then find the ratio of the amount of light that comes directly
from the Sun and that coming from the Sun but reflected off the planet, then
using the known astronomical distances, we can calculate the ratio between
the apparent brightness of the Sun and that of Sirius, and, if Sirius is like
the Sun, we can find a direct estimate of its distance. Using this technique,
Newton reported, in *De Systemate Mundi*, §57, that the nearest stars were
(in modern units) 15 light years away, which is about a million times further
away than the Sun, or 1,000,000 a.u. In unpublished work, he reduced this
to 5,00,000 a.u., which is accurate to within a factor 2 for the nearby stars
– and actually true for Sirius.[15]

Other stars, he thought, were effectively suns with planets of their own.
'Sun and stars shine by their own light and are bodies of the same kind
scattered throughout all the heavens'.[16] They were not necessarily fixed, and
they were not necessarily all of the same composition, as their differences

---

[14]Gregory (1668).

[15]Harrison (1987, 84–85).

[16]*Machinae mundanae descriptio brevis*, Add. 4005, f. 542, early 1690s; McGuire (1978);
Hoskin (1977), 90. The General Scholium (544) implies that they may be centres of other
planetary systems, which are prevented from falling into one another by their immense
separations: 'if the fixed stars are centres of other like systems, these being formed by the
like wise counsel, must all be subject to the dominion of One; especially since the light of
the fixed stars is of the same nature with the light of the sun, and from every system light
passes into all the other systems: and lest the systems of the fixed stars should, by their

in colour would indicate.[17] He also allowed that: 'The fixt Starrs may move inter se by their mutual actions'.[18] In 1717, Halley showed that they did.[19] Newton asked (according to David Gregory's annotations): 'If all bodies truly have a gravitational attraction for each other, why do the fixed stars not move towards each other by gravity and come together? Is there need for a continual miracle to prevent this outcome? Or is gravity enfeebled [*languescit*] in the immense distance that separates them? Or do they revolve around various centres, turning after the manner of the planets?'[20]

Far-reaching conclusions emerged from Newton's investigation of atmospheric refraction. The blue colour of the atmosphere suggested that blue-making rays from the Sun were disturbed (or, in modern language, scattered) by the atmosphere more than those producing the other colours (because, as Lord Rayleigh showed in the nineteenth century, it is proportional to the fourth power of the frequency),

> and hence of the direct solar rays the yellow-making ones usually predominate and cause the sun, which would otherwise appear white, to become yellow. The sun's atmosphere, which perhaps forms a sphere around it, can also contribute to this effect. You should not, however, on that account deny that some kind of ray in the original light is frequently in excess, since the colours of flames and stars are diverse.[21]

Apart from distinguishing photosphere and chromosphere, and implying an influence of the latter on the light reaching the Earth from the former, this is an astonishing prevision of the spectroscopy that was used in the nineteenth century to begin to understand stellar constitutions; and it is significant that Newton had not only begun this science with his experiments on the dispersion of white light with a prism, but had also taken the first spectrum of light from an astronomical source, observing the spectrum of Venus, effectively a point source, as a white line, in the process showing that

gravity, fall on each other mutually, he hath placed those systems at immense distances from one another'.

[17] *Optica*, II, Lecture, 8, 68, *OP*, I, 504, translated 505.

[18] Recorded by David Gregory as 'Mr Newton's exceptions ag[ains]t my book', RS Greg MS, f. 176ʳ; Hoskin (1985), 92. *Principia*, III, Proposition 14, Corollary 1, assumed the stars were immovable, but this was with respect to the more obvious motions of the components of the Solar System.

[19] Halley (1718).

[20] David Gregory, 'Notae' on the *Principia*, Royal Society MS 201, f. 47. Hoskin (1985), 92. *Principia*, III, Proposition 14, Corollary 2, claimed that the fixed stars were too far away 'to produce any effect in our system', and, again, that their distribution might be arranged to cancel their 'mutual attractions'.

[21] *Optica*, II, Lecture, 8, 68, *OP*, I, 504, translated 505.

the elongation in the solar spectrum was not a consequence of the shape of the Sun's image; he also proposed, in his optical lectures, to take the first spectrum of a star other than the Sun.[22] It is equally significant that he suggested doing astronomical spectroscopy because it might lead to new results. Newton, as we have already seen, produced many new leads to be followed up by his successors even when he had neither time nor inclination to follow them up himself, and this one, in its consequences, turned out to be as important as any of them.[23]

The immense distances of the stars, now revealed for the first time, had an immediate consequence: 'The fixed stars being, therefore, at such vast distances from one another, can neither attract each other perceptibly, nor be attracted by our Sun'.[24] Stars should have a roughly uniform distribution in space. However: 'we are not to account all the fixt starres exactly equal to one another, nor placed at distances exactly equal nor all regions of the heavens equally replenished with them'. The passage continues:

> For some parts of the heavens are more replenished with fixed stars then as the constellation of Orion with greater or nearer stars & the milky way wth smaller or remoter ones. For ye milky way being viewed through a good Telescope appears very full of very small fixt stars & is nothing else then ye confused light of these stars. And so ye fixed clouds & cloudy stars are nothing else then heaps of stars so small & close together that without a Telescope they are not seen apart, but appear blended together like a cloud.[25]

Nebulae, which could be identified with star clusters, and the Milky Way are visible proof of how far the real universe departs from the perfectly regular model.[26]

In a 'pioneer investigation of stellar statistics', in the 1690s,[27] Newton attempted to arrange the stars in concentric spheres accommodating stars of equal magnitude, beginning with 12 or 13, and then increasing this by 4

---

[22] *Lectiones opticae*, Lecture 2, 23, *OP*, I, 77; Lecture 2, 21, *OP*, I, 72–73, also, in published form, in '*The Observations, made upon these proposals*', *Phil. Trans.*, no. *83*, *4060–4062, 10 May 1672*. The elongation meant that 'all the rays' from Venus 'were not equally refracted'.

[23] According to Shapiro (*OP*, I, 70–71), although Newton had an arrangement with narrow aperture, prism and lens which was essentially a spectroscope (*Lectiones opticae*, Lecture 2, §18, *OP*, I, 69–71), he did not report observing spectral lines, yet the fourth Lord Rayleigh (1948) saw them with the same arrangement without requiring a lens.

[24] *De Systemate Mundi*, §58.

[25] *Cosmography*, *Unp.*, 374–377.

[26] *Cosmography*, *Unp.*, 374–377; Hoskin (1977, 90–91).

[27] Hoskin (1989).

and then by 9, 16, and so on. He found the Sun was surrounded by 15 or 16 first magnitude stars, rather than the predicted 12 or 13. He therefore supposed that the Sun was about 20% larger than the mean.[28] In fact, the Sun is in the top 10% of stars by magnitude. Many stars in the Milky Way have only half the mass of the Sun, and some only one tenth.

To avoid the problem, Newton decided that he didn't need to work from the conventional magnitudes; by redefining the problem he observed that there were four times as many stars at two units of distance as at one, and that each was a quarter of the brightness; then, there were nine times as many at three units of distance, and so on. The implication of these results was that the sky must eventually assume a uniform brightness in an infinite universe. Thomas Digges, who had first proposed an infinite Copernican universe, had come upon this problem in 1576,[29] but there is no evidence that Newton considered it at this stage.

He also made sixth-magnitude stars correspond to 8 or 9 units of distance compared to those of the first magnitude, and so, though rather fortuitously, was 'very close to the correct relationship between first and sixth magnitudes'. The modern definition, 'brilliantly anticipated' by Halley in 1721, makes this factor 10, which is equivalent to 100 in distance. So, an equal factor between successive magnitudes would be $\sqrt[5]{100} = 2.512$, which is equivalent to a distance factor of $(\sqrt{2.52})^3 = 3.98$.[30] Newton says, in one document connected with the *Principia*:

> There are about 12 or 14 stars of the first magnitude, about 50 of the second magnitude about 160 of the third about 4 or 5 hundred of the fourth & so on. For the stars of the smaller magnitudes being distinguished by their light decreasing in a geometric proportion, there are recconed about three times as many stars of the third magnitude as of the second & about three times as many of the fourth magnitude as of the third, & so on.

M.A. Hoskin has commented:

> To modern eyes this passage is truly remarkable. When he says of successive magnitudes that the brightnesses decrease 'in a geometric proportion', Newton is anticipating the nineteenth-century application of Fechner's Law which... 'implies that between successive magnitudes there is a constant ratio, which must be derived from photometric measurements'. Indeed, if

---

[28] ULC, Add 3965, f. 74^r; Hoskin (1977, 81–90).
[29] Digges (1576); reprinted in Harrison (1987, 211–217).
[30] Hoskin (1977, 88–89, 99–100); Halley (1721a).

we replace Newton's 'about three times' by an idealised proportion of 3.98, we have exactly the modern definition of magnitudes![31]

In another manuscript of this time Newton discussed the Galilean 'double star' method of detecting annual parallax, and so directly measuring stellar distance, which would become particularly significant in the nineteenth century.[32]

Two decades later, the young physician and antiquarian William Stukeley, of Lincolnshire, and friend of Newton in his declining years, suggested that if the universe of stars was infinite the entire night sky would be covered with the 'luminous gloom' of the Milky Way. 'We see every night, the inconvenience of it', he said. 'The whole hemisphere w[oul]d have had the appearance of that luminous gloom of the milky way. We sh[oul]d have lost the present sight of the beauty and the glory of the starry firmament'.[33] The idea was discussed over breakfast with Newton and Halley on 23 February 1721.[34] On 9 March, two weeks after this, Halley presented to the Royal Society a paper 'On the infinity of the sphere of fix'd stars' where he said: 'Another Argument I have heard urged, that if the number of Fixt Stars were more than finite, the whole superficies of their apparent Sphere would be luminous'.[35]

Halley gave two arguments against this view, one of which was based on the corpuscular nature of light, which meant that, 'at very remote distances', the usual inverse square law of attenuation would not apply. In a second presentation, a week later, Halley for the first time presented what we now know was Newton's model. The argument for uniform brightness was resurrected numerous times in the subsequent centuries, becoming known as Olbers' paradox after Olbers' version appeared in 1823.[36] By contrast with Halley, Stukeley tackled the paradox by assuming that the stars were not uniformly distributed but that the brightest stars formed a sphere about the Solar System, with the Milky Way forming a ring, separated by empty

---

[31]Phaenom. 6, *Phaenomena, Unp.*, 382. Hoskin (1977, 91). Hoskin quotes Pannekoek (1961, 446).

[32]Add 3965, f. 175^r, related to drafts of Theorem XV, c 1693; Turnbull (1939, 306–307), n. 10.

[33]Stukeley, 'Memoirs of Sr Isaac Newton's Life', RS MS App XXXVI, f. 69^v. Hoskin (1985), 94.

[34]Stukeley (1752, 14–15). Stukeley to Richard Mead, King's College Cambridge, Keynes MS, 136, f. 10. Hoskin (1985).

[35]Halley (1720a,b); Harrison (1987, 218–219 and 219–220).

[36]Chéseaux (1744), reprinted Harrison (1987, 221–222); Olbers (1823), reprinted Harrison (1987, 223–226); Harrison (1987).

space from this sphere — a significant conception in its own right.[37] Many suggestions have been made to answer the problem, but the most likely solution appears to be the one proposed by Lord Kelvin in 1901, which showed that there are insufficient stars in the *observable* universe, limited by an event horizon determined by the velocity of light, to occupy every conceivable sight-line.[38] Even if Newton's gravitational universe is infinite, the light-emitting matter in the observable universe is not.

7.3. The Constitution of the Earth

A draft related to Query 23 of the Latin *Optice*, which became Query 31 of the second English edition of the *Opticks*, postulates an internal heat of the Earth, responsible for geological phenomena, in 'active principles' similar to those involved in the Sun's shining. This may be compared with the lack of such a source in such nineteenth century thinkers as Lord Kelvin.

> Seeing therefore the variety of motion (wch we see) in the world is always decreasing, there is a necessity of conserving & recruiting it by active principles; such as are (the power of life & Will by which animals move their bodies with great & lasting force;) [*bracketed and cancelled*] the cause of gravity, by which Planets & Comets keep their motions in their Orbs & bodies acquire great motion in falling; & the cause of fermentation, by wch the heart and blood of animals are kept in perpetual motion & heat, the inward parts of the earth are constantly warmed, bodies burn and shine, mountains take fire, the caverns of the earth are blown up, & the Sun continues violently hot & lucid, & warms all things by his light (For we meet with very little motion in the world besides what is owing to these active principles & therefore we oght to enquire diligently into the general Rules or Laws observed by nature in the preservation or production of motion by these principles as the Laws of motion on which the frame of Nature depends & the genuine Principles of the Mechanical Philosophy & the inward parts of the earth are constantly warmed & generate hot sulphureous unhealthful exhalations wch breaking forth with violence cause earthquakes, tempests, & hurricanes, raise or subvert Islands & Mountains, sink Lakes & carry up the sea (partly) in columns, (partly) in drops & thick mists wch convening above fall down in spouts, & sulphureous steams, set mountains on fire & the inward parts of the earth are constantly warmed [*all material in brackets is cancelled*] For we meet with very little motion on the world

---

[37]Stukeley, 'Memoirs of Sr Isaac Newton's Life', RS MS App XXXVI, ff. 69^r–71^r, Hoskin (1985, 81–84).
[38]Kelvin (1901), in Harrison (1987, 227–228).

besides what is (visibly) owing to these active principles & the power of the will.[39]

In the published *Opticks* this becomes:

And even the gross Body of Sulphur powder'd, and with an equal weight of Iron Filings and a little Water made into Paste, acts upon the Iron, and in five or six hours grows too hot to be touch'd, and emits a Flame. And by these Experiments compared with the great quantity of Sulphur with which the Earth abounds, and the warmth of the interior Parts of the Earth, and hot springs, and burning Mountains, and with Damps, mineral Corruscations, Earthquakes, hot suffocating Exhalations, Hurricanes, and Spouts; we may learn that sulphureous Steams abound in the Bowels of the Earth and ferment with Minerals, and Explosion; and if pent up in subterraneous Caverns, burst the Caverns with a great shaking of the Earth, as in springing of a Mine. And then the Vapour generated by the Explosion, expiring through the Pores of the Earth, feels hot and suffocates, and makes Tempests and Hurricanes, and sometimes causes the Land to slide, or the Sea to boil, and carries up the Water thereof in Drops, which by their weight fall down again in Spouts. Also some sulphureous Steams, at all times when the Earth is dry, ascending into the Air, ferment there with nitrous Acids, and sometimes taking fire cause Lightning and Thunder, and fiery Meteors. For the Air abounds with acid Vapours fit to promote Fermentations, as appears by the rusting of Iron and Copper in it, the kindling of Fire by blowing, and the beating of the Heart by means of Respiration.[40]

Newton also says, in a letter to Richard Bentley, that the internal heat is more significant than the Sun in heating the Earth: 'our Earth is much more heated in its bowells below the upper crust by subterraneous fermentations of mineral bodies then by the Sun'. [41]

Also in *Opticks*, he writes, in an argument which has connections with philosophical views expressed earlier by the Cambridge Platonists, Ralph Cudworth and Henry More: 'And if it were not for these Principles, the Bodies of the Earth, Planets, Comets, Sun, and all things in them, would grow cold and freeze, and become inactive Masses; and all Putrefaction, Generation, Vegetation and Life would cease, and the Planets and Comets would not remain in their Orbs'.[42] Objects such as white dwarfs, neutron stars and black holes are, indeed, such 'inactive Masses', in which the 'active Principle' of degeneracy pressure has ceased to operate.

---

[39] draft related to Q 23 of the Latin *Optice*, Add 3970.3, ff. 255^r–256^r, NP.

[40] Q 31.

[41] Newton to Bentley, 10 December 1692, Trinity College Library, Cambridge, 189R447, f. 5^r, NP; a version also in *Corr.* III, 233–238.

[42] Q 31.

In one of his more alchemically-inspired early works, *Of Natures Obvious Laws & Processes in Vegetation*, Newton had put forward a view of the Earth as animate, like that of the Stoics (and not massively different from the modern theory of 'Gaia'), a view which fed into his idea that all natural systems must eventually run down.

> And this is very agreeable to nature's proceedings, to make a circulation of all things. Thus this earth resembles a great animal or rather inanimate vegetable, draws in aethereal breath for its daily refreshment and vital ferment and transpires again with gross exhalations. And, according to the conditions of all things living, ought to have its time of beginning, youth, old age, and perishing.[43]

In correspondence on Thomas Burnet's *Sacred History of the Earth*, Newton countered Burnet's argument that an Earth emerging from a uniform initial chaos would be smooth, by showing that it could equally congeal to the kind of uneven shape that it had at present. He cited three experiments in which 'uniform liquids' congealed 'into solids with rough surfaces' to explain how the Earth could have acquired its present unevenness.[44] The idea that the landscape formed through 'ye breaking out of ye vapours from below before the earth was well hardened, the setting and shrinking of ye whole globe after ye upper regions or surface began to be hard' was an early suggestion of mountains as the consequence of a shrinking Earth, an idea long held in geological theory, though Newton's later views on the absorption of cometary matter seemed to suggest that the earth, after formation, might be expanding.[45]

The Earth certainly had to have structure. Newton had estimated that the average density of the Earth was between 5 and 6 times that of water. However, since the surface matter was only about twice as dense as water, while the matter below the surface in mines was about 3, 4 or 5 times as dense, the matter in the centre had to be even denser, perhaps approaching that of metals such as iron (7.8 times that of water), in line with Newton's view that metals were created inside the Earth. Newton presented his calculations without discussion, but a centre containing iron would additionally enable the Earth to develop a magnetic field.[46]

---

[43] *Of Natures Obvious Laws*, NPA.

[44] Newton to Burnet, *Corr.* II, 329–334.

[45] *Principia*, III, Proposition 42: 'and hence it is that the bulk of the solid earth is continually increased'.

[46] *Principia*, III, Proposition 10.

A fascinating connection between Newton and Hooke (probably not intended by either party) produced some interesting possibilities regarding the migration of the Earth's poles. Newton realised that the Earth's rotation meant that it must be an oblate spheroid, and inserted two Corollaries after *Principia*, Book I, Proposition 91, on the attraction of a particle outside and inside a spheroidal body, as well as discussing the oblateness and many related effects in section 37 of *The System of the World*.[47] He had also, in Corollary 22 to Proposition 66 of Book I, which had been at the Royal Society since April 1686, shown how the 'constitution of the globe' (the Earth) could be known 'from the motion of the nodes', and that its shape could be derived from the constant of precession, where the precession resulted from the Moon's gravitational pull on the Earth's equatorial bulge. According to Birch's *History of the Royal Society*, the Corollary had discussed 'the direction and position of the axis of a globe turning about itself, ... shewing, that by the addition of some new matter on one side of a globe so turning, it shall make the axis of the globe change its position, and revolve about the point of the surface where the new matter is added'.[48]

Then, on 26 January 1687, Hooke, following up earlier geological work, read a lecture to the Society,

> giving his hypothesis how shells and such like substances come to be found deep in the earth, and far above the surface of the sea, as it is at present. He supposed, that the diurnal rotation of the earth by its *vis centrifuga* taking off part of the gravity, formed the surface of the sea into a compressed spheroid; [i.e. an oblate spheroid] ... which some experiments of the shortning of the pendulum near the equator seem to make out. Then, if it may be supposed, that the poles and axis are moveable, the equinoctial and greatest diameter will be likewise altered, and by consequence the parts of the land towards which the poles approach, will be raised, and the sea retire; but, on the contrary, those parts, from which the poles recede, will sink, and the water rise upon them: and that the poles may be altered, he endeavoured to prove by alledging the latitudes of several places considerably different from those assigned by PTOLOMEY and the old geographers.[49]

When Hooke added further details of his 'hypothesis of the mutability of the poles of the earth' at a meeting of the Society on 2 February 1687, Halley,

---

[47] §37.

[48] Birch (1756–1757, IV, 528).

[49] Birch (1756–1757, IV, 521–522). Considering the strained relations between the two men at the time, Hooke was conceivably trying to pre-empt Newton's publication of the *Principia*.

the Secretary, wrote a letter to John Wallis, dated 15 February, with a full account of Hooke's hypothesis, which was read on 9 March, together with the relevant paragraph from the *Principia*. Birch records that: 'It was thought, that the same translation of the axis might be occasioned in the globe of the earth by the blowing up of mountains by subterraneous fire'.[50]

The Earth has an atmosphere and Newton imagined that, viewed from space, it would shine with the light from its clouds, and its solid body would be largely hidden by them — as we now know is true for Venus.[51] The lack of a dense Cartesian aether, however, meant that both interplanetary and interstellar space must be almost empty. Projecting from his results in *Principia* II on real fluids to ideal ones, Newton writes:

And though air, water, quicksilver and the like fluids, by the division of their parts *in infinitum*, should be subtilised, and become mediums infinitely fluid, nevertheless, the resistance they would make to projected globes would be the same. For the resistance considered in the preceding Propositions arises from the inactivity of the matter; and the inactivity of matter is essential to bodies, and always proportional to the quantity of matter. By the division of the parts of the fluid the resistance arising from the tenacity and friction of the parts may be indeed diminished; but the quantity of matter will not be at all diminished by this division; and if the quantity of matter be the same, its force of inactivity will be the same; and therefore the resistance here spoken of will be the same, as being always proportional to that force. To diminish this resistance, the quantity of matter in the spaces through which the bodies move must be diminished; and therefore the celestial spaces, through which the globes of the planets and comets are continually passing towards all parts, with the utmost freedom, and without the least sensible diminution of their motion, must be utterly void of any corporeal fluid, excepting, perhaps, some extremely rare vapors and the rays of light.[52]

In draft A for the General Scholium of 1713, he writes:

Projectiles suffer little resistance in our air. In the Boylian vacuum there is none, as may be gathered from bodies falling [in it]. The resistance of the air ceases in the heavens above the Earth's atmosphere and there bodies should move very freely and conserve their motion for a very long time. And thus they will obey the laws of gravity and perpetually describe the orbits in which they begin to move. For from the markedly eccentric motions of Comets it is demonstrated that celestial bodies can move in all directions with complete freedom. And all celestial bodies whether Planets or Moons

---

[50]Birch (1756–1757, IV, 528).
[51]*Principia*, III, Proposition 41.
[52]*Principia* II, Proposition 40, Scholium.

or Comets conserve for a very long time the motions that they have once begun. But these motions cannot at their beginning arise from mechanical causes.[53]

7.4. Comets and Observational Astronomy

Newton considered comets to be closely related to planets, shining with the Sun's light, a view now generally accepted, though, like others of his day, he also thought they were comparable in size.[54] He established by calculation that they were members of the solar system orbiting the Sun in conic sections, though, by their highly eccentric orbits, they showed that gravity acted in regions well beyond the planets and the visible parts of the solar system. It was in this connection that Newton probably did his best work as an observational astronomer, making direct measurements of the positions of several comets, and expertly using those of the comet of 1680, together with the positional measurements of several other observers, in deriving with precision the elements of its orbit.[55] This was an extraordinarily difficult calculation, the most difficult, in Newton's opinion in the whole of the *Principia*.

The calculation also led to a number of physical interpretations concerning the nature and structure of comets. As the first to calculate a comet's distance from the Sun at perihelion, Newton was also able for the first time to indicate how close comets came to this enormous source of heat. For the comet of 1680, the perihelion distance was only 0.006 A.U. This meant that the heat it received was about 28,000 times the solar heat received by the Earth, and 2000 times that of red hot iron. The 1680 comet emitted 'a much longer tail, and much more splendid', after absorbing the Sun's heat, from which Newton concluded that 'the greatest and most refulgent tails always arise from comets immediately after their passing by the neighbourhood of the sun'. From this, he correctly inferred that 'the tail is nothing else but a very fine vapour, which the head or nucleus of the comet emits by its heat'.[56]

Concerning cometary tails, Newton noted that, where there was curvature of the path, bright tails were 'more resplendent and more exactly defined on the convex than on the concave side', but the form did not depend

---

[53]Draft for General Scholium (MS A), Add 3965, ff. 357–358; *Unp.*, 349–352, translation, 352–355, 352.

[54]*Principia*, III, Lemma 4, Corollaries 1–3. Comets at 100 m to 40 km are considerably smaller than regular planets, but are comparable in size to the minor planets or asteroids, which range from tens of metres to 1000 km. Minor planets were unknown until 1801.

[55]*Principia*, III, Propositions 41 and 42.

[56]*Principia*, III, Proposition 41. See Hughes (1988, 66).

on whereabouts a comet was in the sky. He estimated the velocity of the components by finding that the end of the tail of the 1680–1681 comet, as seen on 25 January 1681, 'had spent in its whole ascent 45 days'. For the cometary atmospheres or comae, he quoted observations by Hevelius which said that 'they are seen least of all just after they have been most heated by the Sun', when the tails are 'longest and most resplendent'. The intense heat, however, must surround the cometary nuclei, 'environing' them 'with a denser and blacker smoke in the lowermost parts of their atmosphere'.

Concerning the material composition of comets, Newton says in the *Principia*:

> Now if one reflects upon the orbit described, and duly considers the other appearances of this comet, he will be easily satisfied that the bodies of comets are solid, compact, fixed, and durable, like the bodies of the planets; for if they were nothing else but the vapours and exhalations of the Earth, of the Sun, and other planets, this comet, in its passage by the neighbourhood of the Sun, would have been immediately dissipated; for the heat of the Sun is as the density of its rays, that is, reciprocally as the square of the distance of the places from the Sun.[57]

It was important for Newton to show that comets actually travel through the solar system in orbits higher than the Moon, as the extraordinary variety of these orbits (some of them retrograde) in itself demonstrates that Descartes' system of vortices is impossible. The proof is in *Principia*, Book III, Lemma 4. The closeness of cometary orbits is demonstrated from their luminosity which decreases as the fourth power of their distance, r, from the Sun, with one factor of $1/r^2$ produced by the decrease in solar flux, and another $1/r^2$ due to the reduction in apparent diameter.

In *De Systemate Mundi*, paragraph 68, he says that the tail of a comet is extremely tenuous despite its brilliance, for one can see stars shining through it. If we suppose 'that the air and vapours are extremely rare in celestial spaces', then 'a very small amount of vapour may be sufficient to explain all the phenomena of the tails of comets'. In *Principia*, Book III, Proposition 41, he says that the vapour from comets' tails is dispersed throughout the heavens but 'by little and little' attracted to planets by their gravity and becomes mingled with their atmospheres. In the same way as the sea exhales the vapours which water the earth, so comets are needed 'that from their exhalations and vapours condensed, the wastes of the planetary fluids spent upon vegetation and putrefaction, and converted into dry earth,

---

[57]*Principia*, III, Proposition 41.

may be continually supplied and made up'. There is an implication here that cometary vapour has a watery component, in line with the modern belief that a substantial part of their composition (about 50%) is ice.

In the same passage he considers cometary matter to play a part in increasing the bulk of the earth: 'and hence it is that the bulk of the solid earth is continually increased; and the fluids, if they are not supplied from without, must be in continual decrease, and fail at last. I suspect, moreover, that it is chiefly from the comets that spirit comes, which is indeed the smallest but most subtle and useful part of our air, and so much required to sustain the life of all things with us'.[58] The Earth is increasing in mass by about 40,000 tons a year mainly through the impact of cometary and meteoric material. Also, some part of the water now in the Earth's oceans is thought to have come from comets and similar objects, particularly asteroids, during the Late Heavy Bombardment, which occurred between about 4.1 and 3.8 billion years ago, due to violent resonances between Jupiter and Saturn, which were then closer to the Sun, though another part may have been produced during the planet's formation from ice crystals trapped within dust grains, and then forced out by pressure from below the Earth's mantle. Perhaps significantly, the Late Heavy Bombardment seems to have coincided with the time when bacteria first appeared on the Earth's surface.

Comets' tails, which reach their greatest length when nearest the sun, always point away from the sun; so Newton thought that a nongravitational force must overcome the gravitational attractions of the sun and other stars for the cometary matter. In *An Hypothesis of Light*, he had proposed that cometary vapours replenished the sun and stars; in the first edition of the *Principia*, however, he intimated that the force of repulsion would prevent this.[59] According to Curtis Wilson: 'Nongravitational forces are now accorded a role in cometary motion. ... Current opinion assigns these nongravitational accelerations to outgassing in the comet's near approach to the Sun...'[60]

Paragraph 69 of *De Systemate Mundi* describes the formation of comet's tails in terms of convection currents of aether particles brought about by radiation pressure:

---

[58]Gjertsen (1986, 127) compares this idea with the views of Hoyle and Wickramasinghe (1978, 1979) on the origin of life (and various diseases) on Earth. He also points out that Newton had no interest in the superstitious aspects of cometary lore.

[59]Kubrin (1967), reprinted in Cohen and Westfall (1995, 289, 291).

[60]Marsden *et al.* (1973); Wilson (2002, 222).

Kepler ascribes the ascent of the tails of comets to the atmospheres of their heads, and their direction towards the parts opposite to the sun, to the action of the rays of light carrying along with them the matter of the comets' tails; and without any great incongruity we may suppose that, in so free spaces, so fine a matter as that of the ether may yield to the action of the rays of the Sun's light, though those rays are not able sensibly to move the gross substances in our parts, which are clogged with so palpable a resistance. Another author thinks that there may be a sort of particles of matter endowed with a principle of levity, just as others are with a power of gravity; that the matter of the tails of comets may be of the former sort, and that its ascent from the sun may be due to its levity; but, considering that the gravity of terrestial bodies varies as the matter of the bodies, and therefore can be neither more nor less in the same quantity of matter, I am inclined to believe that this ascent may rather proceed from the rarefaction of the matter of the comets' tails. The ascent of smoke in a chimney is due to the impulse of the air with which it is entangled. The air rarefied by heat ascends, because its specific gravity is diminished, and in its ascent carries along with it the smoke with which it is engaged. And why may not the tail of a comet rise from the sun after the same manner? For the Sun's rays do not act any way upon the mediums which they pervade but by reflection and refraction; and those reflecting particles heated by this action, heat the matter of the aether which is involved with them. That matter is rarefied by the heat which it acquires, and because by this rarefaction the specific gravity, with which it tended towards the sun before is diminished, it will ascend therefrom like a stream, and carry along with it the reflecting particles of which the tail of the comet is composed; the impulse of the Sun's light, as we have said, promoting the ascent.

The early twentieth century editor and translator of the *Principia*, Florian Cajori, citing J. Ph. Wolfers' German edition of the *Principia*, of 1872, drew attention to a passage in *De Systemate Mundi* as leading towards a possible prediction of Uranus or other planets beyond Saturn:

But astronomical observations seem to confirm a very slow progress of the aphelions, and a regress of the nodes in respect of the fixed stars. And hence it is probable that there are comets in the regions beyond the planets, which, revolving in very eccentric orbits, quickly fly through their perihelion parts, and, by an exceedingly slow motion in their aphelions, spend almost their whole time in the regions beyond the planets; as we shall afterwards explain more fully.[61]

Though at least one planet (Neptune) would be discovered in this way in 1846 after Uranus had been found by Herschel's large telescope in 1781, it

---

[61] *De Systemate Mundi*, §30, 576; Cajori (1934, 680), quoting Wolfers (1872, 659).

looks as much like a prefiguring of the Oort cloud of debris left over from the formation of the solar system, which is hypothesised as the source for long-period comets. Neither Wolfers, writing in 1872, nor Cajori, writing in 1930, would have known about this mass of planetesimal objects, proposed in 1932 and 1950, and thought to be about a light year away or a quarter of the distance to the nearest star, Proxima Centauri, and, though it has yet to be observed directly, it is clearly of great significance for the understanding of the Solar System. Long-period comets, as described by Newton in this passage, are thought to originate in this very distant region, while most short-period comets originate in the much closer region beyond Neptune now described as the Kuiper belt. Some short period comets, however, like Comet Halley, were originally long-period comets from the Oort cloud, but were captured by the strong gravitational fields of the giant planets Jupiter and Saturn, and sent into new short-period orbits.

The heat that the 1680 comet received from the Sun, even led Newton to a conclusion in *Principia* III, Proposition 41, which was seemingly at odds with a recent Earth, less than 6,000 years old:

> This comet, therefore, must have conceived an immense heat from the Sun, and retained that heat for an exceeding long time; for a globe of iron of an inch in diameter, exposed red-hot to the open air, will scarcely lose all its heat in an hour's time; but a greater globe would retain its heat longer in the proportion of its diameter, because the surface (in proportion to which it is cooled by the contact of the ambient air) is in that proportion less in respect of the quantity of the included hot matter; and therefore a globe of red hot iron equal to our earth, that is, about 40000000 feet in diameter, would scarcely cool in an equal number of days, or in above 50000 years.

Of course, he added by way of possible explanation: 'But I suspect that the duration of heat may, on account of some latent causes, increase in a yet less proportion than that of the diameter; and I should be glad that the true proportion was investigated by experiments'. Though clearly uncomfortable with his conclusion, it is interesting that Newton chose to record it, and the doubt concerning the Biblical chronology had already begun. The Comte du Buffon took up the challenge, and similar experiments, recorded in his *Les époques de la nature* (1778), gave the Earth a history of '74,047 years approximately'.[62] James Hutton soon found, from geological considerations, that it was unimaginably older. Newton, despite his religious scruples, had

---

[62] Buffon (1778).

made the first quantitative move in this direction.[63] D. L. Simms considered that Newton's attempts 'to apply the laws of local heat transfer to terrestrial and celestial questions' were 'astoundingly original', though defective, for 'Newton, the fanatical Biblical scholar, was the first to try to determine the age of the Earth by experiment'.[64]

7.5. Practical Astronomy

Newton's work as a practical astronomer also included several contributions to astronomical instrumentation, including the first construction of a reflecting telescope (to overcome both spherical and chromatic aberration) and the first of Newtonian design,[65] a proposal to shorten telescopes, including the use of a right-angled prism as reflector,[66] and a suggestion that astronomy would be most successful in places where there was 'a most serene and quiet Air, such as may perhaps be found on the tops of the highest Mountains above the grosser Clouds'[67] — a procedure that became standard after the mid-nineteenth century, when it was first investigated by Charles Piazzi Smyth.[68] His tables for atmospheric refraction, published in 1721, were useful irrespective of their theoretical background, as the results of such calculations are not particularly sensitive to the theory used.[69]

Newton's construction of a working reflecting telescope (in fact, at least three) led to major developments in optical astronomy. Large telescopes (including radio ones) are nearly all reflectors and the Newtonian design is still used widely today. He gave good instructions for grinding metallic mirrors, and was apparently the first to use pitch for polishing them. He also experimented with producing a reflector using a silvered objective lens.

---

[63]Hutton (1785, 1788, 1795). III, Proposition 10 (419) had asserted that the interplanetary medium encountered by Jupiter would be so rare that it would not lose a millionth of its motion in a million years. While Newton did not imagine that the age of the Solar System could be anything like this order, the calculation is interesting in showing that he could contemplate calculating regular and continuing physical processes for times on this scale.

[64]Simms (2004, 72). Pask (2013, 457), notes that Lord Kelvin later followed this procedure in 'calculating an age for the Earth by modelling its cooling processes'.

[65]'A New Theory'; *Opticks*, Book I, Part 1, Proposition 8, 102–107. A letter of 4 May 1672, published in the *Philosophical Transactions* on 20 May, indicates that Newton had considered alternative designs, on the basis of Gregory's *Optica Promota* (1663), and that he had opted for the one which he considered most practicable.

[66]*Opticks*, Book I, Part 1, Proposition 8.

[67]*Opticks*, Book I, Part 1, Proposition 8.

[68]Smyth (1858).

[69]*Tabula refractionum*, 1695; Halley (1721b).

In the case of refractors, he estimated with quantitative precision the lack of focus produced by chromatic and spherical aberration in lenses,[70] including a description of the circle of least confusion (here one 55th part of the whole aperture).[71] He recognised that chromatic aberration was the more serious problem, with errors increasing with the cube of the aperture, whereas errors from spherical aberration increased only with the aperture itself.[72]

Though Newton seemingly ultimately despaired of refracting telescopes as being unable to overcome the problem of chromatic aberration, he had, in fact, at one time considered that an achromatic lens might be composed of two media of different densities.[73] A section 'Of Refractions' in Newton's Mathematical Notebook includes a description of a compound achromatic lens of this kind, in two folios that were, for some reason, torn out of the manuscript and found their way to Shirburn Castle, the home of the Earl of Macclesfield, pupil and then patron of Newton's protégé, William Jones.[74] The suggestion later appeared in the published lectures of David Gregory, which modelled the idea of the two materials on the crystalline and vitreous humours of the eye, quite possibly at Newton's instigation, despite the fact that Newton knew and had demonstrated that the eye, though compound, was still subject to chromatic aberration.[75] Even as late as the *Principia*, Newton had not given up on the possibility. He certainly suggested a compound lens could correct spherical aberration.[76] He also attempted to construct an achromatic combination, and a prism that is nearly achromatic can be made using water and the kind of Venetian glass that Newton used in some of his experiments, because their dispersive powers, as he knew, were nearly the same. An experiment in a draft reply to Hooke from 1672 suggests that for such an achromatic prism, colours would be seen in beams that emerged parallel to the incident ones, but not in those that did not.[77] It is not certain whether the experiment was ever performed, though it could

---

[70] *Opticks*, Book I, Part 1, Proposition 7.

[71] *Opticks*, Book I, Part 1, Proposition 7, with additional reference to *Lect. Optic.*, Part I, Sect. IV, Prop. 37.

[72] *Opticks*, Book I, Part 1, Proposition 7.

[73] draft reply to Hooke, 1672, MS Add. 3970, ff. 433^r, 445^r, 447^r. Shapiro (2005, 99–125).

[74] 'Of Refractions', ULC Add. MS 4000, 26^r–33^v; Whiteside, *MP* I, 559–576, 572–576, especially 575–576; Bechler (1975).

[75] Gregory (1695); Gregory's *Tres lectiones cursoriae* (Aberdeen University Library, MS 2206/8, fol. 1) had used the description of the optic nerve which Newton had explained to Pitcairne. *Optica*, II, Lecture 14, 129–131, *OP*, I, 578–582 (on chromatic aberration in the eye).

[76] *Opticks*, Book I, Part 1, Proposition 7; also *Principia*, I, Proposition 98, Scholium.

[77] draft reply to Hooke, 1672, MS Add. 3970, ff. 433^r, 445^r, 447^r. Shapiro (2005, 109–111).

have yielded the result that Newton projected. Later, using separate glass and water-filled prisms, Newton found that, when the opposing refractions of the two media cancelled, so did the deviations, leaving the emergent beam parallel to the incident one.[78]

Newton's subsequent negative opinion on the possibility of constructing achromatic lenses, which informs Observation 8 of Book I, Part 2 of the *Opticks*, seems to have been based on his desire to mathematise optics using his erroneous linear law of dispersion.[79] In this observation, he found that if light emerged from a combination of media in rays parallel to those in which it was incident, it would remain white, the dispersion being corrected by 'contrary Refractions', but, if it emerged at an angle, it would be coloured at the edges, a conclusion which seems opposite to the result he found for his achromatic prism in 1672. Shapiro says that 'While the observation is in general invalid, it is true or very nearly true for the observation he actually claims to have made'.[80] So that, while the experiment might at first seem a candidate for the use of spurious data, it seems to have been rather a case of the use of an invalid deduction from a particular case when the general case is untrue.

Since Newton's *Opticks* did not make significant inroads on the Continent until after the second French edition of 1722, it is not really possible to argue that his authority prevented research on achromatic lenses for decades *there* even if it could be argued for Britain. In fact, such lenses were made in Britain as early as 1729 by the barrister Chester Moor Hall, and used by him to construct a number of refracting telescopes by 1733.[81] His successful method involved a combination of two types of glass (crown and flint) with different dispersive powers, and Newton had found that, with two different prisms, the dispersive powers could vary by as much as 13%.[82] Hall attempted to keep his work secret by contracting out the manufacture of the crown and flint glass lenses to different opticians, but they both subcontracted the work to George Bass, who, realising Hall's subterfuge, put the two parts together and noticed the achromatic properties of the combined lens. Bass was making and selling such lenses as early as 1733, but they were not generally known until he told John Dollond of the invention in the 1750s. Dollond had, in 1753, criticised Euler's 1747 suggestion that achromatic lenses could

---

[78]*Opticks*, Book I, Part 1, Proposition 2. Shapiro (2005, 118).
[79]Shapiro (2005, 118).
[80]Shapiro (2005, 118).
[81]Dollond (1789).
[82]Shapiro (2005, 117).

be made out of a water-glass combination. Euler's article had stimulated Samuel Klingenstierna, professor of mathematics at Uppsala University, into a mathematical investigation of Newton's experiments. The inconsistencies that he found led to his conclusion that Newton's linear dispersion law was incorrect. The dispersion law only agreed with experiment for small-angled prisms, and the angles would be different for each set of two media. Klingenstierna arranged for a Latin translation of his paper to be delivered to Dollond in London. After a number of experiments, Dollond found a way of producing achromatic lenses for refracting telescopes on a commercial scale, and in April 1758 obtained a patent for the invention, four years after making his first achromatic lens.[83]

In *The System of the World*, §17, Newton proposed a practical method of measuring planetary diameters, by shining a lamp, placed at a great distance, through a circular hole, and decreasing both hole and the light from the lamp so that the image appeared through the telescope just like the planet. The ratio of the diameter of the hole to the distance from the objective glass would be the same as the true diameter of the planet to its distance from the observer. He also suggested diminishing the light of the lamp by pieces of cloth or smoked glass. William Herschel would later describe an arrangement similar to this for measuring the relative brightness of stars as a 'lamp micrometer'.[84] In terms of the most pressing practical astronomical problem of the day, Newton suggested the design of a reflecting octant for measuring angular distances at sea, and followed this with a significantly modified version. Newton's designs were put into practice subsequently by John Hadley, and are still in use in the modified form of Colin Campbell's sextant.[85]

As a public servant, in effect the first Government scientific adviser, he also reported on the various methods suggested for finding longitude at sea and helped to set up the Board of Longitude to oversee this problem. He was instrumental in recommending a prize to be offered by Parliament for finding a solution, with a value to be determined by the quality of the result. His view, as expressed in a letter to John French, of 22 March 1715, was that two methods could be used, accurate astronomical observations and accurate

---

[83] Euler (1747), Dollond (1753, 287), Dollond (1758), Klingenstierna (1754), Dollond (1789).

[84] Herschel (1782).

[85] 'A Description of an Instrument', 1699; Cohen (1958b, 236–238), second edition, 1978, 259–268; Hadley (1731); May (1973). The earlier version of the instrument is described in *Astronomia, Unp.*, 386–388, translation, 388–391.

time-keeping, but, of these, he preferred the former, and his desire to discover a theory of the Moon's motion was partly aimed at solving this problem:

> And I have told you oftner than once that it is not to be found by Clockwork alone. Clock work may be subservient to Astronomy but without Astronomy the longitude is not to be found. Exact instruments for keeping of time can be usefull only for keeping the Longitude while you have it. If it be once lost it cannot be found again by such Instruments. But if you are unwilling to meddle with Astronomy (the only right method & the method pointed at by the Act of Parliament) I am unwilling to meddle with any other methods then the right one.[86]

In principle, Newton seems to be arguing (correctly) that longitude is established in the first instance by an astronomical method, and that using a clock requires that it is kept operational for the whole time from the beginning of a voyage. A clock thus requires a continual operation, while the astronomical method can be used whenever it is required without relying on a prior measurement.

Eventually, both methods were developed in the eighteenth century, though Greenwich Observatory's emphasis on the superiority of the astronomical method delayed John Harrison's award for solving the problem by accurate time-keeping in his series of chronometers. In relation to much that has been said in the past about the 'failure' of Newton's lunar theory in this practical matter, it is interesting that it now seems that Mayer's lunar tables of 1752 that were the foundation of the astronomical method for longitude, and were long thought to have been based on the work of Euler, made very considerable use of the practical lunar theory that Newton had published in 1702. According to S. A. Wepster, 'Newton's theory exercised, through Mayer's tables, a much more profound impact on eighteenth century positional astronomy than has hitherto been thought'.[87] At the same time, Harrison's chronometric method was perfected using Newton's theory of resisted harmonic motion, and, specifically, the anharmonic ratio $2/\pi$, found for the first time in *Principia*, Book II, Proposition 30.[88] Though Harrison

---

[86]*Corr.* VI, 211–212.

[87]Wepster (2010, 91). Kollerstrom (2000) had earlier used a computer analysis to show that Newton's practical method, as used by Halley, gave results accurate to one degree and would have been good enough to be awarded his own suggested £10,000 prize!

[88]Harrison's own contributions to nonlinear dynamics are significant in themselves and allowed him to use the concepts of displacement limiting and velocity limiting, in addition to creating a mechanical system with the characteristics of the Van der Pohl oscillator. In principle, he observed the kind of behaviour in oscillating systems which is now called chaos (Hobden, 1981–1982, 2011; Harrison, 1988).

had already established proof of principle by 1735, his later sea clocks, in particular H4 which was completed in 1759 and tested at sea in 1761 and 1764 at the request of the Board of Longitude, together with Larxum Kendall's copy K1, which accompanied Cook on his second and third voyages, were so reliable in regard to long term usage that Newton's concerns about the need for maintaining continuous operation could be completely overcome.[89]

Related to observational astronomy from the earliest times has been the establishment of a calendar relating different measures of astronomical time, in particular the times for the rotation of the Earth and its orbit round the Sun. In 1582, the Gregorian calendar had been adopted in the Catholic countries of Europe to replace the old Julian calendar established by Julius Caesar in 46 BC, but this reform would not be adopted in England until 1752. Newton's *Considerations upon Rectifying the Julian Kalendar*, written in response to a request sent by Leibniz to the Royal Society in February 1700, proposed a solar calendar which was marginally more accurate than the Gregorian.[90] In principle, the tropical year is 365.2422 days long, while the Julian year was fixed at an average of 365.25 days, and the Gregorian year at an average of 365.2425 days, an advance of about 26 seconds a year. The average year in the Newtonian calendar would be fixed at 365.242 days, which is slightly closer to the 'true' tropical year than the Gregorian year.

Whereas the Gregorian calendar had fixed the vernal equinox at March 21 and removed the leap day from centurial years not divisible by 400, Newton proposed in the first version of his draft

> To divide it [the year] by the four cardinal periods of the equinoxes and solstices, so that the quarters of the year may begin at the equinox and solstice as they ought to do, and then to divide every quarter into three equal months, which will be done by making the six winter months to consist of thirty days each, and the six summer months of thirty one days each, excepting one of them, suppose the last, which in leap years shall have thirty one days, in the others only thirty days. At the end of every hundred years, omit the intercalary day in the leap year, excepting at the end of every five hundred years. For the rule is exacter than the Gregorian,

---

[89]According to Cook's report to the Lords of the Admiralty, made after his return, 'Mr Kendall's Watch exceeded the expectations of its most zealous advocate and by being now and again corrected by lunar observations has been our faithful guide through all vicissitudes of climates'. The lunar observations were made on the occasions when the voyage made landfall, as accurate ones were difficult to obtain at sea (Hobden and Hobden, 2009, May, 1976).

[90]*Considerations upon Rectifying the Julian Kalendar*, 1699; summary Brewster (1855, II, 311–312).

of omitting it at the end of every hundred years, excepting the end of every four hundred years, and thus reckoning by five hundreds and thousands of years is rounder than the other by four hundred, eight hundred and twelve hundred. And I take this to be the simplest, and in all respects the best form of the civil year that can be thought of.

The basic reform (concerning the leap years) made the year approximately 365 days 5 hours 48 minutes 46.2 seconds by comparison with the tropical year of that period of 365 days 5 hours 48 minutes 48.24 seconds. After 2000 tropical years of 730,484.4 days, the Newtonian calendar would have measured 730,484 days, the Gregorian calendar 730,485, and the Julian 730,500.[91]

Further refinements, in later versions, have been studied by Belenkiy and Vichagüe.[92] Newton was apparently the first to apply the statistical technique known as linear regression analysis in investigating Hipparchus's observations of the equinox, and he made the 'remarkable guess', later confirmed by the twentieth-century astronomer Robert Newton, that there were systematic errors in Hipparchus's observations that tended to be 'equal and opposite' at the two equinoxes. Though the solar part of Newton's calendar used a 'different algorithm' to the Gregorian one it was identical in effect until the year 2400 but 'was more precise in the long run within a period of 5,000 years'. His 'lunar algorithm' was also 'more elegant than the Gregorian one'. By 25 April 1700 Newton had provided an answer to Leibniz's request which had been received on 21 February. It was sent to Leibniz on 4 July. Belenkiy and Vichagüe believe that Newton's calendar could have provided 'a viable alternative' for England and other Protestant countries to the Gregorian calendar adopted by the Catholic countries. It would also have countered Leibniz's desire to bring the Protestant nations in line with the reforms already made. However, it appears that Newton abandoned the plan because he had a value of the tropical year from the authorities he quoted which seemingly implied a less good correction than his suggestions would actually have made.[93]

It is interesting that Newton, as so often in other contexts, was able to come up with a more ingenious solution for calendrical reform, which could, in theory, have been adopted by England in preference to the Gregorian

---

[91] Belenkiy and Vichagüe (2005, 226), Gjertsen (1986, 132–133).

[92] Belenkiy and Vichagüe (2005). The version of *Considerations* in Yahuda MS 24 is transcribed on pp. 236–241.

[93] Belenkiy and Vichagüe (2005), especially 223, 226, 227–229, 235, *Corr.* IV, 328–330; R. R. Newton (1977, 83).

one, in the same way as the French Revolutionaries later imposed their own calendar for a short period, but calendars are decided for cultural reasons as much as scientific exactitude, and the already widespread use of the Gregorian calendar would have made any solution with such a marginal advantage impractical to implement. Newton's proposal, then, is not very significant in the larger context of the development of world science, but it does provide yet another indication of his endless versatility and ingenuity.

7.6. Newtonian Gravity and Cosmology

Newton was in fact, a very good astrophysicist, using all the data at his disposal, and some very ingenious arguments to make a wide variety of correct conclusions about stars, planets and comets. Cosmology, however, raised a very different set of problems, and significant data was not readily available. Then, as now, speculation was often the first and only resort. In the seventeenth century, it was also often irretrievably mixed up with a belief in the literal truth of the Scriptures. Newton's cosmological arguments were more often than not extricated in response to questions by correspondents and, in later life, from questions raised in verbal discussions. He at no time set out to write out a system that could be described as cosmology.

It was Richard Bentley, setting out to give the first series of lectures in commemoration of Robert Boyle, only a few years after the publication of the *Principia*,[94] who elicited Newton's most significant comments on cosmology, and they are based purely on a consideration of the physical effects of gravity, without reference to any Scriptural passages. Replying to Bentley's first inquiry, on 10 December 1692, Newton wrote:

> As to your first Query, it seems to me, that if the matter of our Sun & Planets & ye matter of the Universe, was eavenly scattered throughout all the heavens, & every particle had an innate Gravity towards all the rest & the whole Space throughout wch this matter was scattered was but finite: the matter on ye outside of this Space would by its Gravity tend towards all ye matter on the inside & by consequence fall down into ye middle of the whole space & there compose one great spherical mass.

The universe could not be finite, for even ancient arguments by such as Lucretius suggested that a finite universe would necessarily collapse into a central heap of undifferentiated matter. It had, therefore, to be infinite — as

---

[94]Robert Boyle had died almost a year earlier in the very last hours of 31 December 1691.

is now widely believed to be the case, in direct contradiction to much of the cosmology pursued in the twentieth century.[95]

> But if the matter was eavenly disposed throughout an infinite space, it would never convene into one mass but some of it convene into one mass & some into another so as to make an infinite number of great masses, scattered at great distances from one another throughout all yt infinite space. And thus might ye Sun and Fixt stars be formed, supposing the matter were of a lucid nature.[96]

On 17 January 1693 he added an argument against the distribution of stars in an infinite universe being perfectly uniform:

> but that there should be a Central particle so accurately placed in ye middle as to be always equally attracted on all sides & thereby continue without motion seems to me a supposition fully as hard as to make ye sharpest needle stand upright on its point upon a looking glass. For if ye very mathematical center of ye central particle be not accurately in ye very mathematical center of ye attractive power of ye whole mass, ye particle will not be attracted equally on all sides.
>
> And much harder it is to suppose that all ye particles in an infinite space should be so accurately poised one among another as to stand still in a perfect equilibrium. For I reccon this as hard as to make not one needle only but an infinite number of them (so many as there are particles in an infinite space) stand accurately poised upon their points.

The idea is reminiscent of modern calculations of the limits of exactness required of the fundamental constants for the universe we know of to exist at all. A universe in which the stars behaved like needles standing on their points would be unstable to the slightest variation. Newton continues: 'Yet I grant it possible, at least by a divine power; & if they were once so placed I agree with you that they would continue in that posture without motion for ever, unless put into new motion by the same power'. So he says to Bentley: 'When therefore I said that matter eavenly spread through all spaces would convene by its gravity into one or more great masses, I understand it of matter not resting in an angular poise'.[97]

This would explain the massive clumps we now observe, but not the orbits or 'transverse motion' of the planets. Newton knew of no 'power in nature wch would cause this ... motion without ye divine arm',

---

[95] An infinite universe also accords with the views on space extending 'infinitely in all directions' recorded in *De gravitatione*, *Unp.*, 133–134.

[96] *Corr.* III, 234.

[97] *Corr.* III, 238.

writing: 'So then gravity may put ye planets into motion but without ye divine power it could never put them into such a Circulating motion as thay have about ye Sun, & therefore for this as well as other reasons I am compelled to ascribe ye frame of this Systeme to an intelligent agent'.[98] This was ideal for Bentley's purpose in showing God's providence in the design of the universe and is characteristic of this stage in Newton's development, but, in later years, he was inclined to believe that God's providence was manifested through discoverable laws of nature and not in contradiction to them, even if the particular natural cause could not be explained.

Bentley replied on 18 February:

> Sir, In a finite world where there are *outward* fixt starrs, this seems plainly necessary. But in ye supposition of an infinite space, let me ask your opinion. I acquiesce in your authority, yt in matter diffused in an infinite space, 'tis as hard to keep those infinite particles fixt at an equilibrium, as poise infinite needles on their points on an infinite speculum. Instead of particles, let me assume Fixt starrs or great Fixt Masses of opake matter; is it not as hard, yt infinite such Masses in an infinite space should maintain an equilibrium, and not convene together? so yt though our System was infinite, it could not be preserved but by ye power of God.[99]

In the published version of his sermon, he says:

> yet, we say, the continuance of this Frame and Order for so long a duration as the known ages of the World must necessarily infer the Existence of God. For though the Universe was Infinite, the Fixt Starrs could not be fixed, but would naturally convene together, and confound System with System: for, all mutually attracting, every one would move whither it was most powerfully drawn. This, they may say, is indubitable in the case of a Finite World, where some Systems must needs be Outmost, and therefore be drawn toward the Middle: but when Infinite Systems succeed one another through an Infinite Space, and none is either inward or outward; may not all the Systems be situated in an accurate Poise; and, because equally attracted on all sides, remain fixed and unmoved? But to this we reply; That unless the very mathematical Center of Gravity of every System be placed and fixed in the very mathematical Center of the Attractive Power of all the rest; they cannot be evenly attracted on all sides, but must preponderate some way or other. Now he that considers, what a mathematical Center is, and that Quantity is infinitly divisible; will never be persuaded, that such an Universal Equilibrium arising from the coincidence of Infinite Centers can naturally be acquired or maintain'd.[100]

---

[98] *Corr.* III, 238.
[99] *Corr.* III, 238.
[100] Bentley (1693), Cohen (1958b, 351).

The letters to Bentley present Newton's own cosmological model, based on the 'symmetry principle' that there is no preferred direction for the gravitational field in an infinite uniform universe, and the model can be made to fit either expanding, contracting or static universes, in exactly the same way as in general relativity, even without using the equivalence principle.[101] The fate of the universe depends only on its overall density. Though the three conditions were originally derived by Alexander Friedmann, in 1922, as possible solutions of Einstein's field equations, they have nothing to do with the model used, in particular with whether the gravitational field is linear or nonlinear.[102] This was realised as early as 1934, in the 'Newtonian' cosmology of Edward A. Milne and William McCrea.[103]

Milne and McCrea were able to show that the equations for the expanding universe, worked out with great labour from general relativity, could be derived without difficulty from a Newtonian theory. In fact the equation for the expansion

$$\frac{1}{r^2} \left(\frac{dr}{dt} \right)^2 = \frac{8\pi}{3} G\rho$$

is nothing more advanced than a rearrangement of the classical kinetic energy equation

$$\frac{1}{2} mv^2 = \frac{GMm}{r}$$

already available as Proposition 41 of *Principia*, Book I. It is clear that this is not a result of amazing coincidence, but rather because the key factors involved in the equations derive (as do all the gravitational, rather than kinematic, consequences of general relativity) from the Newtonian gravitational potential, and not from the relativistic corrections. The only thing that could change this in any way would be the *direct* observation of effects that could emerge only from a gravitational singularity, not the presumption that certain indirect indicators suggest that one must exist.

Milne and McCrea used the equivalence principle, but only the Newtonian 'symmetry principle' is required. Don S. Lemons states that 'This method for defining gravitational forces in an infinite universe is sometimes discounted as 'wrong' or 'inadequate'. If so, I claim it is only because the method leads to a cosmological model different from that of general

---

[101] Lemons (1988).
[102] Friedmann (1922).
[103] McCrea and Milne (1934), Milne (1934, 1935).

relativity and not because of an internal inconsistency'.[104] In addition, more recent measurements have shown that certain inherent assumptions in Newton's 'symmetry principle' — that the geometry of the universe is Euclidean, with zero curvature, and probably infinite in extent, and that its density is fixed at the critical value which prevents either runaway inflation or future collapse — were justified, despite contrary indications from relativistic cosmology.[105] The assumption that modern cosmology is a consequence of general relativity, and a further demonstration of the theory's validity, though very prevalent, is totally unjustified. In addition the general relativistic model has been found to be extremely inconvenient in practice.

Reviewing a textbook on mathematical cosmology, John D. Barrow has written:

> There have been times in the past, for example during the 1960s and 1970s, when there was very great emphasis upon mathematical cosmology to determine the conditions under which a singularity might have existed in the past and to discover the range of non-uniform universes that might evolve to look like our own after billions of years of expansion. But since the early 1980s the focus has shifted to the examination of particle physics processes within the simplest possible model of the expanding universe — a model which is provided for quite adequately by Newton's theory of gravitation — and this means that one can teach most of modern cosmology without developing the general theory of relativity at all. Despite this trend there is still a tendency for authors to treat cosmology simply as an elaborate worked example in a course on general relativity.[106]

7.7. Cosmological Speculations

Other cosmological ideas put forward by Newton introduce us to a remark-able peculiarity of his thought process, the rationalisation of extra-scientific ideas to the point where they begin to look rational. By assuming, by some unconscious process, that earlier sources could be treated on the same rational basis as scientific work of his own time, an idea certainly fuelled by his belief in a *prisca sapientia*, or ancient tradition of wisdom needing to be recovered, he had a tendency to look for meanings in ancient texts that were more a reflection of his own type of reasoning, and seemingly gradually

---

[104] Lemons (1988).

[105] Riess *et al.* (1998). The significance of a Euclidean universe with respect to earlier philosophical claims that general relativity shows space-time to be 'really' curved has yet to be investigated.

[106] Barrow (1993).

morphed into it, often using his own scientific advances. He did this in his use of alchemical authors and also in theology, where he pursued a rationalising agenda that led him to a very singular view of religion and church history. Many of his cosmological speculations seem to have had their origins in Scriptural readings, though they are made to conform with more naturalistic explanations. A process of 'accommodation' of biblical texts to more naturalistic explanations had been going on since at least the time of St. Augustine in the fifth century, but Newton took the process to much greater lengths to accommodate to the more advanced scientific knowledge being acquired in the seventeenth century. The theory outlined in the letter to Burnet, in which Newton supposed that the Solar System could have originated from the gradual action of gravity on the primary Mosaical Chaos, may perhaps be taken as a distant ancestor of the eighteenth century nebular hypothesis: 'suppose that all the Planets about the Sun were created together, that they and the Sun had one common Chaos'. He attempted to show that the seven 'days' of creation involved the earth being 'accelerated from rest by divine force until it reached its present rate of rotation'.[107] In fact, conservation of angular momentum would mean that the rotation rates of the objects in the Solar System and those of associated neighbouring stars would certainly have increased as they condensed out of the original nebular material.

The various statements on cosmology in such works as the *Hypothesis on Light*, the letters to Burnet and Bentley, *Principia*, Book III and the conversations recorded by John Conduitt seem to suggest that cosmological processes, for Newton, were natural (in our terms) to the extent that they could be explained by mathematical laws, but not to the extent that they were due to the persistence of mechanism. Thus it was correct to say that 'There exists an infinite and omnipresent spirit by which matter is moved according to mathematical laws',[108] but not that, given a conserved quantity of motion, 'matter . . . would in infinite time run through all variety of forms one of wch is that of a man'.[109] The former allowed the proper action of God in the world, the latter was an atheistic notion. The passage shows clear knowledge of the hypothesis of an ergodic universe, but refutes it.

Concerning the long-term future of the universe, and in particular the solar system, Newton claimed that comets produced instability, an anticipation of chaos theory. The same orbit was never produced twice by any

[107]Newton to Burnet, January 1681, *Corr.* II, 329–334.

[108]CUL, Add MS 3956.6, f. 266^{v}; Westfall (1971, 399).

[109]draft Q 23, 1706; quoted by Westfall (1971, 419–420).

planet because of perturbations. There could be no perfection in an infinitely many body system. Certainly, there was a semi-literalist interpretation of the Bible involved in Newton's thinking. Already, in 'Of Earth' from his early notebook, he was interpreting the Bible, somewhat idiosyncratically, in terms of the inevitable decay and destruction of the world, and its probable renewal: 'Its conflagration testified 2 peter 3^d vers 6, 7, 10, 11, 12, ... The succession of worlds, probable from Pet. $3^c.13^v$. ...' He also saw Revelation 20:10 as prophesying 'a succession of worlds'.[110] Newton's later cosmological work, however, appears to show an increasing 'rationalism', becoming considerably more 'naturalistic' after the *Principia* showed the success of doing it that way.

Unfortunately, Newton's views on the instability of the Solar System have suffered serious distortion from the better-known 'interpretations' of Leibniz and Laplace. These seem to have had their origin in the statement in Query 31 of an argument from design, in which this claim was made:

> Now by the help of these Principles, all material Things seem to have been composed of the hard and solid Particles above-mention'd, variously associated in the first Creation by the Counsel of an intelligent Agent. For it became him who created them to set them in order. And if he did so, it's unphilosophical to look for any other Origin of the World, or to pretend that it might arise out of Chaos by the mere Laws of Nature; though being once form'd, it may continue by these Laws for many Ages. For while Comets move in very excentrick Orbs in all manner of Positions, blind Fate could never make all the Planets move one and the same way in Orbs concentrick, some inconsiderable Irregularities excepted, which may arise from the mutual actions of Comets and Planets upon one another, and which will be apt to increase, till this System wants a Reformation. Such a wonderful Uniformity in the Planetary System must be allowed the Effect of Choice.

The interesting thing about this statement — apart from its early description of a deterministic mechanical system becoming chaotic and of relatively small perturbations ultimately leading to dynamical complexity — is the impression that it gives of Newton's belief in the long-term stability of the system, even though he had sufficient intellectual honesty to realise that the many-body nature of the system would eventually make it unstable. Similarly, in the *Principia*, he went out of his way to minimise the possibility of an aethereal medium in space in order to conclude Book III, Proposition 10 as follows: 'And therefore, the celestial regions being perfectly void of air

---

[110]QQP, 'Of Earth', f. 101^r, NP.

and exhalations, the planets and comets meeting no sensible resistance in those spaces will continue their motions through them for an immense tract of time'. And yet the impression has somehow developed that he believed in the exact opposite, that the universe and the Solar System, in particular, were subject to major instabilities, and that stability could only be restored in them by the continual interference of a *deus ex machina*.

It seems that a remark made by Leibniz at the height of their bitter antagonism has become more familiar than the single phrase by Newton on which it is based. Leibniz wrote: 'Sir Isaac Newton, and his followers, have also a very odd opinion concerning the work of God. According to their doctrine God Almighty wants to wind up his watch from time to time: otherwise it would cease to move. He had not, it seems, sufficient foresight to make it a perpetual motion'.[111] But when Newton wrote 'till this System wants a Reformation',[112] he in no way intended to imply a regular interference by 'a clockmaker mend(ing) his work'; he meant to imply that the Solar System would eventually fall into such disorder that it would require a complete reformation, probably along the lines of the (natural) cyclic processes he had described in the section on comets in book III of the *Principia*, and subsequently intimated in conversation with Conduitt. He also seemed to be suggesting, here and elsewhere, that the 'action of God' (whether 'natural' or 'supernatural' in effect, he did not say) would maintain the universe in motion, even if it allowed the Solar System to fall into disorder — a possibility which would require an infinite open universe. But many points of Newton's cosmology are still obscure; it is an area of his work which has been little studied.

Samuel Clarke replied to Leibniz that, 'The word *correction*, or *amendment*, is to be understood, not with regard to God, but to us only. The present frame of the solar system (for instance,) according to the present laws of motion, will in time fall into confusion; and perhaps, after that, will be amended or put into a new form. But this amendment is only relative, with regard to our conceptions'.[113] Also, as we have seen in 7.4, Newton's remarks on the comet of 1680, in Proposition 41, show that he had glimpsed the possibility of an even longer time span for the Solar System than he had at first anticipated from the strict application of Biblical chronology (if not entirely comfortably).

---

[111] Alexander (1956, 11).

[112] Q 31.

[113] Alexander (1956, 22).

God's action had to be continual, and providential, not simply that of the unmoved mover, but this governing role was built into the system directly from the beginning. '...if there be an universal life and all space be the sensorium of a thinking being...then laws of motion arising from life or will may be of universal extent'.[114] Because of God's providence, as manifested by the active principles, the solar system remained stable for long periods, though it would eventually collapse. The same Providence answered the question 'whence is it that Planets move all one and the same way in Orbs concentrick, while Comets move all manner of ways in Orbs very excentrick, and what hinders the fix'd Stars from falling upon one another?'[115] Providence had to restore order to prevent the instability of a system of stars displaying a less than perfect symmetry in their distribution.

Newton's universe, in effect, was dynamic and in a state of permanent change; the alternative that Leibniz seemed to be promoting, was, for all its dependence on the preservation of motion, static. For Newton, the fixed stars lost mass, though very slowly, through the emission of light and vapours, but gained mass from the absorption of material from comets. And the fixed stars were not 'fixed', for many changes could be observed in them; some had disappeared altogether, others had flared up anew, others were consistently variable in brightness. Vapours from the Sun, fixed stars and comets were absorbed into the atmospheres of planets and gradually condensed into liquids and solids. Following up some arguments by Boyle, Newton wrote in Book III, Proposition 42:

> The vapours which arise from the sun, the fixed stars, and the tails of the comets, may meet at last with, and fall into, the atmospheres of the planets by their gravity, and there be condensed and turned into water and humid spirits; and from thence, by a slow heat, pass gradually into the form of salts, and sulphurs, and tinctures, and mud, and clay, and sand, and stones, and coral, and other terrestial substances.

The very bodies of the Sun, fixed stars and comets were formed by the gravitational condensation of light and vapours in various places throughout the heavens. The Solar System itself must have been formed in this way from an original chaotic state, though it could not easily be explained why the Sun had become a lucid body while the planets had not; the lucidity of the Sun must be a function of its great mass.

---

[114]draft Q 23, for *Opticks*, 1705; McGuire (1968, 196), Westfall (1971, 397).
[115]Q 31.

Newton's explanations of cosmological processes would seem to have become increasingly naturalistic as he extended his speculations into new areas. In the letters to Bentley, for instance, the lucidity of the Sun is a fact which cannot be explained by naturalistic means, but in the *Opticks* twelve years later it becomes attributable to the process whereby massive bodies spontaneously heat themselves by absorbing their own light. The later astronomical manuscripts such as *Atmosphaera Solis, Cosmography,* and *Phaenomena* contain no suggestion of divine intervention or divine providence.[116] Even the General Scholium does not say much more than that the motions of the planets and comets could not have been derived from the laws of gravity alone. In a draft Newton specifies that 'these motions cannot at their beginning arise from mechanical causes'.[117] The emphasis is on the inability of mechanical causes or the random associations produced by matter in motion to explain the harmony and order in the Solar System, because, as Newton was aware, mechanism alone cannot lead to anything other than a progressive loss of motion and consequent disorder.

The increasingly naturalistic tone of such writings, which really represents a change of emphasis rather than a change in the fundamental nature of Newton's beliefs, appears to coincide with his renewed interest in physical models of the aether. By the time he came to draft out his final series of Queries he was already asking 'Is there not something diffused through all space in & through wch they act upon one another at a distance in harmonical proportions of their distances'.[118] The 'active principles' so harmoniously at work in nature undoubtedly revealed the direct action of God in maintaining order in the universe but now they could also be accommodated within a naturalistic explanation.

But 'naturalistic' did not mean 'mechanistic'. The active principles had been integrated into a universal scheme of nature governed by mathematical laws, but the causes of these principles remained abstract and fundamental (just as they are in twenty-first century physics). It was in the assimilation of such abstract and nonmechanistic entities into a naturalistic mode of physical explanation that Newton made his major contribution to science. If few contemporary scientists share his desire to show the direct action of God

---

[116] *Atmosphaera Solis, Unp.,* 318–319; *Cosmography. Chap. 1. Of the Sun & fixt Stars, Unp.,* 374–377; *Phaenomena,* early 1690s, *Unp.,* 378–385.

[117] General Scholium, draft A, Add. 3965, fols. 357–358; *Unp.,* 349–352, transl. 352–355, 353.

[118] draft Query 17, UL Cambridge, MS Add. 3970.3, 234^r, NP.

in the many processes of nature, they nevertheless make use of the abstract and universal laws of nature which this desire generated.

Laplace, at the end of the eighteenth century, showed that, to a large extent, the perturbations described by Newton due to planets on each other would cancel each other out over a long enough time period and that the stability of the Solar System was much greater than Newton had imagined, but he did not thereby negate the principle that all physical systems are inherently unstable.[119] It would seem that he also deliberately overstated the instabilities assumed by Newton in order to emphasise the stabilising factors which he himself had found. We still don't know if the Solar System is stable on the level of the lifetime of the Sun, or whether it is really heading towards the chaotic.

At the end of his life, Newton produced some final speculations on cosmology, imagined as a cyclic process, with some kind of 'revolution in the heavenly bodies'. He discussed them with John Conduitt in a long conversation held on 7 March 1725.

> He *conjectured* [Conduitt's emphasis], refusing to affirm anything, that 'vapours and light emitted by the sun...had gathered themselves by degrees in to a body and attracted more matter from the planets and at last made a secondary planett [a moon]...and then by gathering to them and attracting more matter became a primary planet, and then by increasing still became a comet, which after certain revolutions by coming nearer and nearer the sun had all its volatile parts condensed and became a matter set to recruit and replenish the Sun...and that would probably be the effect of the comet in 1680 sooner or later...[120]

This might be after perhaps 5 or 6 revolutions; the increased heat of the Sun would then destroy life on Earth.

He speculated that the 'new stars' seen by Tycho Brahe in 1572 and Kepler in 1604 might be due to stars becoming brighter by recruiting the matter from comets.[121] No one at the time had any idea how new stars could have appeared in the heavens, though we now know that they are produced by the gravitational collapse and subsequent implosion of supermassive stars. Interestingly, both Tycho's and Kepler's stars were supernovae of type Ia, in which the collapse is brought on by the 'recruiting' of matter by an invisible white dwarf star from an orbiting companion object, though in this case it is a red giant star rather than a comet. The principle put forward by Newton

---

[119]Laplace (1799–1825).

[120]Conversation with John Conduitt, 7 March 1725, KCC, Keynes MS 130.11, NP.

[121]*Principia*, III, Proposition 42.

is correct and it is the beginning of the idea that gravity not only causes the motions of the heavenly bodies but is responsible for constitutional change within them; but the reality, of course, is on an unimaginably larger and more cataclysmic scale than he could ever have contemplated.

Newton was also correct in separating out these spectacular arrivals from the kind of variable stars 'that appear and disappear by turns, and slowly and by degrees, and scarcely ever exceed the stars of the third magnitude'. He thought these were 'of another kind' entirely, and that, in revolving about their axes, they appeared to alternate between showing 'a light and a dark side'. There are, of course, many reasons why stars may appear to be variable, but pulsars are neutron stars that operate exactly on the principle that Newton describes.[122]

Newton thought that humans on Earth were of relatively recent date; all human invention ('arts as letters') had been made 'within the memory of History which could not have happened if the world had been eternal'; signs of ruin on the Earth's surface indicated previous cataclysms of the same kind, and not just because of the biblical Flood. After such a 'mass extinction', repeopling the Earth would require a creator.[123] He had previously told David Gregory in 1694 that the satellites of Jupiter or Saturn could replace Earth, Venus and Mars if they were destroyed, 'and be held in reserve for a new Creation'.[124] Gregory very probably reported Newton's thinking in his *Elements of Physical and Geometrical Astronomy*, when he said that a comet passing near a planet could change its orbit and its period, and so disturb a planetary satellite that it would 'leave its Primary Planet and itself become a Primary Planet about the Sun'.[125] It may be of interest in this context to note that in a manuscript draft preserved by Brewster Newton wrote of the possibility of inhabitants existing in other parts of the universe: 'For if all places to which we have access are filled with living creatures, why should all these immense spaces of the heavens above the clouds be incapable of inhabitants?'[126] In his conversation with Conduitt: 'He seemed to doubt [whether?] there were not intelligent beings superior to us who superintended these revolutions of the heavenly bodies by the direction of the supreme being...'[127]

---

[122] *Principia*, III, Proposition 42.

[123] Conversation with John Conduitt, 7 March 1725, KCC, Keynes MS 130.11, NP.

[124] Memorandum by David Gregory, 5–7 May 1694, *Corr.* III, 336.

[125] David Gregory (1702, 1715).

[126] MS draft in Brewster (1855, II, 354).

[127] Conversation with John Conduitt, 7 March 1725, KCC, Keynes MS 130.11, NP.

Such passages as the one recorded by Conduitt show that many of Newton's cosmological speculations had their ultimate origins in the Scriptures and Biblical revelation, though most had been drastically modified and interpreted by a rationalising process that he could not avoid adopting. Conduitt asked him why, unlike Kepler, he had chosen not to publish his conjectures. 'I do not deal in conjecture', he replied. He showed Conduitt the passage in the second edition of the *Principia* where he had discussed how the fixed stars, including the Sun, 'were recruited and replenished by Comets'. Conduitt asked why he wasn't more forthright in his views. He laughed, adding that 'he had said enough for people to know his meaning'.[128] He was certainly not going to risk the charge of heresy that would inevitably go with the implication that there had been Earth-like planets before the present one, that there had been men before Adam, that the Creation described in *Genesis* was only one in a continual cycle of creation events, and that the world might conceivably be eternal, even if it was assumed that its maintenance also required divine intervention.[129]

Cosmology, even today, is a much less certain science than astrophysics. It relies on interpretations of an experiment which cannot be repeated. Though it has established an account of the universe which has gained a large degree of acceptance, it has left us with many unanswered questions and a number of anomalies which are yet to be resolved. It is likely that future observations will require significant changes to the models currently favoured, as has happened on a regular basis over the last few decades. The well-known joke that 'cosmologists are often error, but never in doubt' points to the quasi-theological nature of a subject which cannot be put to the test of rigorous experimental verification under controlled conditions. Newton was fully aware that his cosmological conjectures, based ultimately on a quasi-theology of his own, were a significantly different kind of exercise to the more certainly established results based on mathematical laws and experimental demonstration.

---

[128] KCC, Keynes MS 130. 11, NP.

[129] Kubrin (1967), reprinted in Cohen and Westfall (1995, 294).

Chapter 8

Gravity and Inertia

8.1. Gravitomagnetism

As we have seen earlier (6.6), the introduction of general relativity as the new theory of gravity in 1915 has been widely considered as a 'revolution', a total 'replacement' of the Newtonian theory by one conceived in entirely different terms. This has left us with a philosophical problem. The contextual connection between the two theories has never been established. While quantum theory and special relativity can be accommodated with Newtonian theory if we choose the right form of the quantum and relativistic quantities to create a formal similarity with Newtonian equations, general relativity has been thought to be different. Eddington, as part of his promotional strategy, very strongly proclaimed a discontinuity in 1919. Unfortunately, from the beginning philosophers decided that the theory was too mathematically complex for them to disentangle and diverted their efforts into analyses of how well-established 'paradigms' could be overthrown by minute pieces of physical evidence. It led in the twentieth century to the privileging once again of hypothesis-led research since no principles now remained which indicated how a theory should be constructed — theories became, simply, best fits to data with no obvious process of evolutionary development from earlier ones which had proved successful in all previous cases. It also put general relativity into an anomalous position with respect to the rest of physics, and, in particular quantum mechanics, as being based on pure geometry rather than dynamical principles.

The dislocation between general relativity and the rest of physics need never have happened. What was needed but never sought, as the supporters of the new theory deemed it irrelevant, was an *organic* connection with Newtonian theory. General relativity required Newtonian theory for all its translations of geometry into physical meaning but the reason for this was left unexplained and the 'Newtonian limit' made purely arbitrary.

To restore the connection, we need to see how Newtonian theory was actually modified by Newton's successors in the direction of what would become general relativity. Pursuing this line could make possible a more meaningful transition between the theories. In fact, the geometrical aspect was not as revolutionary as it must have seemed when first presented. It was not a choice between pure real geometry and a non-existent 'physical' force of attraction between massive objects. As we have seen, Newton had already used very similar geometrical methods in perturbation theory, and Cartan later showed that they could be used in any field theory, whatever the origin of the field.[1] In addition, Newtonian theory did not specify a physical origin for what emerged as gravitational effects. Both Newtonian theory and general relativity were abstract generic theories not based on 'physical' assumptions. Those originally assumed by general relativity were abandoned when the final mathematical structure was created.

The most significant modification of Newtonian theory was not, in fact, the specification of the field in terms of geometry but the introduction of the quantity generally known as the velocity of light (c) and the fundamental 4-vector relation between space and time now known as special relativity. Ultimately this reflected the time-delayed interactions which emerged between localised point sources or charges, in particular from electric charges. The time-delayed component introduced the magnetic, as well as electric field, and extended the single equation describing Coulomb's inverse square law of electrostatic force, to a system of four Maxwell equations, which could be shown to require the generation of electromagnetic waves travelling at c. The equations can be taken as equivalent to a combination of Coulomb's law and special relativity, with its extensive apparatus of Lorentz transformations, length contraction, time dilation and mass increase. If a quantum picture is added, then the force transmission also involves the emission and reabsorption of a virtual massless spin 1 boson, while the generation of waves becomes equivalent to the emission of real bosons.

What would become special relativity was first incorporated into gravitational theory via an equivalent set of Maxwell-type equations generated by a combination of the standard gravitational field and a gravitomagnetic quantity equivalent to the magnetic field of electromagnetic theory and represented by a rotational term ω. Such gravitomagnetic equations, leading to gravitational waves and gravitomagnetic precession effects on planetary

---

[1]E. Cartan, *Ann. Ecole Norm.*, 40, 325, 1923; 41, 1, 1924.

orbits, were worked out by Oliver Heaviside considerably before either form of relativity.[2] At the present time they are used to generate effects such as rotational frame dragging, geodesic deviation and precession of gyroscopes from general relativity although these are also derivable more simply from the special theory.

Despite the appearance of c in the Maxwell-type equations, however, it has become increasingly accepted in recent years that both the static gravitational and electrostatic fields surrounding any given discrete source, featuring in the first of the Maxwellian equations, extend to infinity and move instantaneously with the source. Experimental testing has confirmed this in the electrical case.[3] Anything that can be defined as a static 'influence' or a 'Coulomb interaction' is always transmitted instantaneously. It is not 'propagated'. This is why it is possible to experience the mass and charge of a black hole when the finite-velocity radiation it generates is suppressed. The naïve view that the Newtonian or 'Coulombic' static gravitational force is transmitted at the velocity of light, which was supported by Eddington[4] and was once widely held as a consequence of general relativity, seems now no longer tenable. Instantaneous transmission of the force between static objects or objects in uniform relative motion is actually demanded by *relativity*, to maintain Lorentz invariance. The time-delayed transmission refers only to gravitational or electromagnetic waves and their real carrier boson equivalents, and these are only produced when systems are disturbed.

8.2. Early Calculations of Spectral Shift and Time Delay

Though the creation of gravitational equations involving the velocity of light in a relativistic sense occurred only at the end of the nineteenth century, the effects of gravity *on* light were explored at a much earlier date, and this could have led to a more coherent relation between general relativity and Newtonian theory. Newton knew that a finite velocity of light would necessarily affect astronomical measurements and he knew that light rays must be deflected by gravity in a manner analogous to optical refraction. A number of significant calculations made in the eighteenth century extended the Newtonian account of the interactions between gravity and light corpuscles. The explicit calculations relating to all these processes were carried out by Newton's successors, Michell, Cavendish, Blair, Laplace

---

[2]'A Gravitational and Electric Analogy', Parts I–II, *The Electrician*, 31, 281–282, 359, 1893; reprinted *Electromagnetic Theory*, volume II, Appendix, 1894.

[3]Calcaterra *et al.* (2012).

[4]Eddington (1920).

and Soldner, though elements of them can be found in Newton's own work. In addition, the denial by some modern authors that the gravito-optical effects are in any sense Newtonian because modern relativistic mass is not mass in the Newtonian sense is falsified by the precedent of Newton's own work. John Michell, in the first such calculation in 1772, showed that a light particle emitted by the Sun would be slowed down by the attraction of the Sun's gravitational field.[5] Michell made specific use of Newton's Proposition 39, which equates the work done in a gravitational field to the change in kinetic energy for a particle of unit mass, and is equivalent to the equation:

$$\int_{r_0}^{r} g dr = \int_{r_0}^{r} \frac{Gm}{r^2} dr = \frac{1}{2} \left(v^2 \left(r_0 \right) - v^2 \left(r \right) \right).$$

He calculated that the speed acquired by a particle, initially at speed 0, within a star of density ρ and radius r_0, escaping to infinity, would be

$$v \left(r_0 \right) = \sqrt{\frac{2GM}{r_0}} = \sqrt{\frac{8 \pi G \rho r^2}{3}}.$$

Using the connection between velocity and refractive index assumed by Newton, in which the refractive index in a medium was assumed to be inversely proportional to the speed of a light corpuscle, Michell proposed, in 1783,[6] to measure a difference in the refractive indices in the two components in a double star system, and so calculate their respective velocities and masses, and use this to determine the distance of the system from Earth. Michell expected, in effect, to see that a more massive object would slow down the speed of the light corpuscles more than a less massive one, and so produce a greater angle of refraction, and this would be true for single stars as well as double star systems. It would have been equivalent to a 'Doppler-Fizeau effect', or a gravitational spectral shift.[7]

Michell's argument led Cavendish to a calculation of gravitational light bending. It also led to a paper by Robert Blair, professor of practical astronomy at the University of Edinburgh, which was read to the Royal Society of London in April 1786, and which applied the Newtonian principle of relativity to calculate the velocity of light, as received from a distant source, in relation to the velocities of source and observer, though measurements of aberration had already suggested that the velocity of light was not dependent

---

[5] Priestley (1772, 1, 786–791).
[6] Michell (1784).
[7] Eisenstaedt (2005).

on that of the source.[8] According to the Newtonian argument[9] the refractive index depended on the incident velocity according to the formula

$$n = \frac{\sin i}{\sin r} = \sqrt{1 + \frac{k}{v_i^2}}$$

which, with v_i expressed, for convenience, as c, we can rearrange as

$$\sin r = \left(1 + k/c^2\right)^{-1/2} \sin i.$$

Then a change in refraction Δr produced by a change in velocity Δc, could be calculated from

$$\Delta r \cos r = -k/c^2 \left(1 + k/c^2\right)^{-3/2} \sin i \Delta c/c,$$

from which

$$\frac{\Delta r}{\tan r} = -\left(1 - 1/n^2\right) \Delta c/c.$$

A similar change in refractive index occurs in the wave theory where $\Delta r/\tan r$ is again proportional to $-\Delta c/c$ as a result of the Doppler effect, the shift in wavelength $\Delta \lambda$ due to the relative motion between the source and observer, with $\Delta \lambda/\lambda = \Delta c/c$.

Blair considered measuring the spectral shift between astronomical and terrestrial sources comparing the light from the Sun with that from a candle. This would allow an observer to measure the radial velocity of the Sun, and to determine whether it was approaching or receding. It would also detect any shift due to the Sun's gravitation. He also proposed to use it to measure the rotations of Jupiter, Saturn, Mercury and the Sun and to determine whether the objects rotated from left to right or from right to left. A memorandum by William Herschel from this period on 'Hints. Desiderata. Experiments to be made' considers related experiments: 'Would not a star having a strong proper motion from the eye, have the intervals of its colours separated by a prism larger than one whose proper motion is towards it?' 'Are the prismatic intervals of the light of a star in the ecliptic different when the velocity of the Earth's motion is from or towards the star?'[10] Herschel and Nevil Maskelyne tried to detect the effect in the summer of 1783, using a prism, but were unable to obtain a positive result.[11]

---

[8] Blair (1784).

[9] *Principia*, Book I, Proposition 94.

[10] Eisenstaedt (2005, 372).

[11] Eisenstaedt (2005, 351); Jungnickel and McCormmach (1999), quoting Michell to Cavendish, 20 April 1784.

A rotational spectral shift of light is, in principle, an effect of the same kind as the rotation of the Earth producing a bulging at the equator and flattening at the poles. In both cases, it can be treated as due to the action of rotation in producing a noninertial frame of reference for the observations, just as the gravitational redshift of light produces a noninertial frame for the photon. In the Newtonian calculation of the effect of rotation on the figure of the Earth, the rotation acts to increase the negative energy of gravity at the equator where the rotational velocity is highest, while leaving that at the poles unaltered, since the rotational velocity here reduces to zero. The increase in negative energy at each point on the Earth's surface is determined by the ratio of the squared tangential velocities of the Earth's rotation, say $v = \omega r \sin\theta$, and orbital motion, say u, at each latitude, such that the total potential energy at the equator becomes $-GM(1+v^2/2u^2)/r$. The $1/2$ factor in the centrifugal potential component $v^2/2r$ is due to the kinetic origin of the rotational term.

In the case of light, the positive energy of a light photon emerging from the Sun $h\nu$ will be reduced by a gravitational potential energy $-GM/r$, equivalent to a negative v^2 term produced by an acceleration $-v^2/r$. For the rotational spectral shift the effect is determined by whether the actual rotational acceleration is towards or away from the observer, that is, whether it is given by v^2/r or $-v^2/r$. In each case, we can consider the effect as equivalent to a fictitious centrifugal acceleration produced by a rotating or linearly accelerating reference frame.

Though Blair seems never to have put his theory to the test beyond some preliminary work with Jupiter, and never published his paper, François Arago, who had some knowledge of Blair's work through an article published later by John Robison, tried a similar experiment around 1806–1810, once again with a negative result.[12] To observe a shift in refraction due to the relative motion of the observed body requires a shift in frequency of the light by comparison with that from a terrestrial source. The shift in refraction would have been tiny, and precision measurements would have been extremely difficult, but the observers, in both this and the Herschel-Maskelyne experiment, seem to have made it virtually impossible to detect at all by using achromatic prisms. Ironically, Newton (who had once observed the spectrum of Venus) would have had a better chance with his chromatic prisms, for a dispersed spectrum means that the wavelengths are spread out and individual colours can be identified and

---

[12]Arago (1810).

used for the comparison — Herschel notably saw the importance of using 'intervals' of the 'colours'. The effect would be enhanced if the distant source produced a line spectrum, where individual frequencies could be easily identified, but line spectra from astronomical sources were not identified until well into the nineteenth century, and the gravitational spectral shift proposed by Michell was not observed until the twentieth century. It is, in principle, the same as that required by general relativity and equivalent to a gravitational time dilation. To derive it, we need only apply the Newtonian gravitational potential $-GM/r$ to a light photon with relativistic mass $h\nu/c^2$, to obtain a redshift $-GMh\nu/rc^2$.

One measurement proposed by Blair would have been possible with achromatic prisms. He supposed here that, if the velocity of light did not depend upon that of the source: 'The same difference of refrangibility ought to take place in the light of a candle in consequence of the earth's motion as in that of a fixed star; and it would be possible in this case to determine the velocity and direction of the Earth's motion, and even that of the Solar System, as far as the instrument can be depended on, in a close room'.[13] The idea is similar to that of the Michelson–Morley experiment of 1887. The velocity of the rays of light from the candle would vary only with the Earth's motion relative to the aether, and the effect would be identified by the direction of the ray rather than its frequency. Arago and Mascart both carried out this experiment during the nineteenth century, with negative results.[14] As Jean Eisenstaedt points out, Blair saw that the effect would only be observed in the case of a (nonrelativistic) wave theory, because, for a Newtonian emission theory in which the velocity of light varied with that of the source, Newtonian relativity would hold, so 'The motion of the prism, and the difference in the absolute velocity of light arising from the motion of the candle, counterbalance each other, and destroy the effect, which would arise from either of these causes acting apart'.[15]

To make the result negative in a wave theory of light, we require Einsteinian relativity, as would become clear from the Michelson–Morley experiment and its subsequent explanations. It is interesting that both relativistic particle and wave theories produce a negative result, though nonrelativistic ones would not, and that the form of relativity actually used has to be the one appropriate to the particular theory being tested. In principle, there is no *physical* reason for choosing Einsteinian relativity with

---

[13]Blair, quoted Eisenstaedt (2005, 373).
[14]Mascart (1874).
[15]Eisenstaedt (2005).

the Lorentz transformations over Newtonian relativity using the Galilean transformations. It is purely a convention, though the convenience of retaining the invariant nature of the velocity of light has led us to adopt the Einsteinian approach as the standard. In addition, the variation of c with the source in Newtonian emission theory is needed only when the measurement is taken with respect to the aether. In any other measurement, the variation of c can only be considered with respect to the *relative* velocity of observer and source, and the nonvariation with respect to the source observed in aberration (and accepted by many emission theorists) does not affect the assumed variation with respect to the observer.

The refractive index equation, as proposed by Michell, has also led to another modern test of general relativity. Writing the equation in the form

$$c' = c(1 - 2GM/rc^2)^{1/2},$$

together with the one for gravitational time dilation,

$$dt' = dt(1 - GM/rc^2),$$

we can use

$$dt' = ds/c$$

and

$$ds^2 = dr^2 + r^2 d\phi^2$$

and the standard approximation

$$r \approx \frac{d}{\cos \phi},$$

where d is the distance of closest approach, to obtain

$$dt = \left(1 + \frac{GM}{rc^2}\right) \left(\frac{dr^2 + r^2 d\phi^2}{c'}\right)^{1/2}$$

and integrate to obtain the total delay. The modern test, first carried out by Ross and Schiff in 1966,[16] calculated and successfully measured the delay on the basis of the equation

$$c^2 dt^2 \left(1 - \frac{2GM}{Rc^2}\right) = dr^2 + r^2 d\phi^2.$$

---

[16]Ross and Schiff (1966).

8.3. The Schwarzschild Singularity and the Gravitational Deflection of Light

Another eighteenth-century calculation relating to the effect of gravity on light was the definition of what we now call the Schwarzschild singularity, or the radius, determined at $2GM/c^2$, at which a gravitating system becomes a black hole, unable to emit light. In the special case of a body with the same density as the Sun but a radius about 500 times, and, more accurately, 497 times greater, the escape velocity as calculated by Michell would be greater than that of light. 'Hence', he wrote, 'all light emitted from such a body would be made to return toward it, by its own proper gravity', meaning that it would become invisible, while even a body of the same density but smaller radius would emit light with diminished velocity.[17] With the ratio R_{star}/R_{sun} written as α, and using

$$\frac{2GM}{R_{Sub}c^2} = \frac{8\pi G r^2}{3c^2} = \frac{1}{497^2},$$

we find that

$$c' = c\sqrt{1 - \frac{2GM}{R_{Star}c^2}} = \sqrt{1 - \frac{\alpha^2}{497^2}}.$$

Michell recognised that the invisible bodies he proposed could be identified from the motions of a double star system.[18] He had already argued, on statistical grounds in 1767, that gravitationally-bound double star systems must exist, although this was only finally demonstrated by Herschel in 1802.[19]

Apart from Michell's discussion of such invisible bodies, the other eighteenth-century calculation of the Schwarzschild singularity emerged from Laplace's discussions of what he called 'corps obscurs' or 'corps invisibles' in the first edition of *L'Exposition du Systeme du Monde* of 1796, and in the *Allgemeine geographische Ephemeriden* of 1799.[20] Laplace certainly knew of Michell's statistical argument for double stars, which he cited in the second edition of *La Théorie Analytique des Probabilités* in 1814,[21] and he very likely knew also of Michell's paper in the *Philosophical Transactions* of 1784. Laplace, however, structured his calculation on the Earth rather than the

---

[17] Michell (1784, 37).
[18] Michell (1784, 50).
[19] Michell (1767), Herschel (1803–1804).
[20] Laplace (1796).
[21] Laplace (1814).

Sun. He argued that a star of the same density as the earth and 20 times the radius would have an escape velocity at its surface greater than the velocity of light, and so would be a 'corps obscur'.[22]

Laplace's calculation was significant in inspiring a published derivation of the gravitational light deflection. Its author, Johann von Soldner denied Laplace's assumption that the velocity of light was constant, and considered the possibility of a 'corps obscur' at the centre of the Milky Way, though he thought that the absence of any rapid motion of the stars around the centre of the galaxy went against this hypothesis.[23] For Soldner, as for Cavendish, the gravitational deflection of light followed naturally from its having a variable speed.

According to a modern interpretation of the calculations by Clifford Will,[24] both Cavendish and Soldner assumed that the orbit of a fast-moving light corpuscle in the gravitational field of the Sun would be a hyperbola. For a distance of closest approach R and eccentricity e, the equation of motion of the light particle could be written, in parametric form, as

$$r = R \frac{(1+e)}{(1 + e \cos \phi)}$$

and

$$r^2 \frac{d\phi}{dt} = (GMR(1+e))^{1/2}.$$

Then

$$v^2 = \frac{GM}{R(1+e)} \left(1 + 2e \cos \phi + e^2\right)$$

and the half-deflection angle δ will be given by

$$\sin \delta = \frac{1}{e}.$$

To find e, we can assume, as Soldner did, that the light particle has the velocity c at the distance of closest approach, when ϕ, the angle made with

---

[22]Laplace (1796).
[23]Soldner (1801).
[24]Will (1988).

the horizontal axis, is 0. Then

$$c^2 = \frac{GM\,(1+e)}{R},$$

$$e = \frac{Rc^2}{GM} - 1,$$

and

$$\sin\delta = \frac{GM}{Rc^2 - GM}.$$

Alternatively, we can assume, with Cavendish,[25] that the velocity of the particle is c at infinity, when $\cos\phi = -1/e$. Then

$$c^2 = \frac{GM\,(e-1)}{R},$$

$$e = \frac{Rc^2}{GM} + 1,$$

$$\sin\delta = \frac{GM}{Rc^2 + GM}.$$

In either case, with $Rc^2/GM \gg 1$,

$$\delta \approx \frac{1}{e} \approx \frac{GM}{Rc^2}$$

leading to a full deflection of

$$2\delta \approx \frac{2}{e} \approx \frac{2GM}{Rc^2}.$$

However, for a hyperbolic orbit *in the process of formation*, the kinetic, rather than potential, energy equation could be used, leading to

$$\frac{1}{2}c^2 = \frac{GM\,(1+e)}{R}$$

or

$$\frac{1}{2}c^2 = \frac{GM\,(e-1)}{R}$$

and a full deflection of

$$2\delta \approx \frac{2}{e} \approx \frac{4GM}{Rc^2}.$$

---

[25]Cavendish (c 1784/1921).

Significantly, Soldner *did* use the kinetic equation or its equivalent, as he based his calculation on Laplace's argument about the escape velocity from a black hole, which involves kinetic, not potential, energy. Stanley Jaki,[26] who republished the paper in the twentieth century and had presumably been influenced by Eddington's original presentation about the difference between relativistic and Newtonian predictions, thought that this was an error, but it would seem that it was not. Cavendish also used a black hole calculation but, as we lack the details and particularly a numerical prediction, we don't know whether his u^2 was equivalent to rc^2/GM or $rc^2/2GM$. Soldner, it has been said, obtained only half the correct value, but it would seem that the source of *this* supposed 'error' was that he took his result over δ, rather than 2δ, which may mean merely that he *specified* his result in terms of the half-deflection.

In a sense, the calculation was genuinely Newtonian, for Cavendish and Soldner's use of hyperbolic orbits went back to equivalent calculations by Newton on the force of optical refraction.[27] Newton's assumption of a constant refracting field f at height h above the Earth's surface is virtually identical in action to the gravitational field $g(= GM/r^2)$, and Newton uses it with the conservation of energy equation, Proposition 41, in much the same way as Michell did to calculate hyperbolic orbits for light rays in the Earth's atmosphere, similar to those which would apply to certain comets in the Solar System. In these calculations, mc^2 becomes a potential energy term $mfr(1 + \cos\phi)$, analogous to the gravitational $GMm(1 + \cos\phi)/r$. The equivalent modification of c^2 by the factor $(1 - 2fh/c^2)$ can be taken as comparable to the principle of equivalence modifying c^2 by $(1 - 2gr/c^2)$ or γ^{-2} in gravitational bending, where $c^2(dt)^2$ becomes $\gamma^{-2}c^2(dt)^2$. But relativity makes c a constant, and so requires a dilation in measured values of time.

It would seem that, all the direct interactions between gravity and light can be derived using Newtonian theory adapted to the appropriate circumstances. In the case of light deflection, we assume that the orbit is created by the interaction with the gravitating body and not through

---

[26] Jaki (1978). The same objection had been made by Robert Trumpler (*Science*, 31 August 1923, 161–162) and H. von Kluber (*Vistas in Astronomy*, Pergamon Press, London, 1960, 3, 47–77). It is notable that the factor 2 in Soldner's calculation was obtained using the *Schwarzschild singularity*.

[27] Newton to Flamsteed, 17 November 1694, January, 26 January, 16 February 1695, 15 March 1695, *Corr.* IV, 46–49, 67–69, 73–75, 86–88, 93–95. 32; draft Q 22 for *Opticks*, 1717, Add. MS 3970, f. 621ʳ; Bechler (1974).

the emission of the light from the source. So the initial velocity of the light corpuscle *in the orbit* is 0, not c. This is also compatible with special relativity, where c is not a mechanical velocity, and the effect can also be derived from special relativity by a number of methods. There is certainly no need to use the full apparatus of general relativity, as is also the case for the time delay, the gravitational redshift and the derivation of the Schwarzschild radius.

8.4. The Dynamics of Material Bodies

The previous sections show that certain immediately post-Newtonian physicists were already attempting to incorporate the effects that we now attribute to localization to Newton's nonlocal theory. Despite all its elaborate geometrical and tensor apparatus, general relativity tends to operate in much the same way. Though the theory is described by relatively simple mathematical equations, they are extraordinarily difficult to solve for real situations. Solutions have been obtained only for cases where extreme symmetry removes most of the complications. In these cases the solutions reduce largely to the theories which the theory assumes as its limiting cases, namely special relativity and Newtonian gravity. Virtually the only analytic solution of the field equations so far obtained with testable results different from standard Newtonian gravity is the simplest possible, the Schwarzschild solution for the spherical gravitational field round a point source,[28] which leads to a line element of the form

$$ds^2 = -\gamma^2 dr^2 - r^2 d\theta^2 - r^2\sin^2\theta d\phi^2 + \gamma^{-2}c^2 dt^2,$$

where $\gamma^{-2} = (1 - 2GM/c^2)$ by comparison with

$$ds^2 = -dr^2 - r^2 d\theta^2 - r^2\sin^2\theta d\phi^2 + c^2 dt^2$$

in the field's absence. As has been pointed out on a number of occasions, if the factor γ^{-2}, applied to $c^2 dt^2$, is responsible for gravitational time dilation, which we know only requires the application of the Newtonian potential to the photon, then a factor γ^2 will need to be applied simultaneously to the term dr^2 to maintain the Lorentz invariance and create the corresponding length contraction — which becomes 'curvature' in the Einstein language. Although this is only a heuristic argument, all the systems so far considered

---

[28] *Sitzungsberichte, Preussische Akademie der Wissenschaften*, 1916, 189–196 and 424–434.

can be shown by more rigorous argument to be explicable in terms of the special theory of relativity applied to the optical component of the system, with no direct effect upon the gravity.

The Schwarzschild solution also leads to predicted dynamical effects on purely material bodies, not previously obtained by the immediately post-Newtonian physicists, the most easily observed of which are an additional perihelion precession produced in planetary orbits and a mathematically and physically equivalent periastron precession in binary pulsars. We can see immediately from the Schwarzschild line element that the basic effect is only special relativistic, and again this heuristic insight can be backed up by more rigorous calculations, but, this time the special relativistic effect is now applied for the first time to the gravitational interactions between material bodies, and not confined to the gravitational effect on an optical system, which we already know is special relativistic. That is, special relativistic aspects are directly associated with a system not previously defined as special relativistic.

Now, in general relativity we start with the field equations as a complete generic description of all possible sources of curvature of the flat 4-dimensional space-time that exists without the field and then impose a series of restrictions to isolate the particular conditions being investigated — which ultimately leads to the introduction of Newtonian potential terms as the physical manifestation. We could, however, imagine working from the opposite direction beginning with the Newtonian terms and see what modifications are successively required to obtain the same result. Neither route requires specific assumptions about whether the equations involve c because the space-time is intrinsically Lorentzian for gravity or whether the instantaneous transmission of the static force also implies that the speed of light and gravitomagnetism are not intrinsic to the gravitational force, although the first position has always been associated with relativity and the second has often been assumed to be the natural one associated with Newtonian gravity. In the case of the second position, the gravitational interaction will not *itself* be affected by the finite speed of light, but our observations of gravitating systems will have to be corrected for the gravitationally-affected observational delay in exactly the same way as stellar positions were corrected for aberration after Bradley's discovery of the effect in 1728.

The modification of the Newtonian theory to obtain the equivalent of the Schwarzschild solution is relatively straightforward. Here, with no gravitational or other deflecting field, or source of dynamic energy, a ray of light or a particle of unit mass will travel in a straight line according to the equation:

$$\left(\frac{dr}{dt}\right)^2 + r^2 \left(\frac{d\phi}{dt}\right)^2 = 0.$$

If we now apply a Newtonian potential $-GM/r$ to a material particle in a system of total mechanical energy E, we will expect a deviation from straight line motion, governed by a dynamic equation of the form:

$$\left(\frac{dr}{dt}\right)^2 + r^2 \left(\frac{d\phi}{dt}\right)^2 = \frac{GM}{r} - E.$$

Now, in a gravitational field, with Newtonian potential $-GM/r$, both the Newtonian calculation and the Schwarzschild solution lead to an equation for the light ray *geodesic* (or path in the absence of dynamical energy) of the form:

$$\left(\frac{dr}{dt}\right)^2 + \left(1 - \frac{2GM}{rc^2}\right) r^2 \left(\frac{d\phi}{dt}\right)^2 = 0.$$

If we now suppose that this describes what happens in a gravitational field to the straight line of a measurement made optically, or by using any process transmitted at the gravitationally affected speed of light, we will expect the same effect in a system with a material body responding to a Newtonian potential $-GM/r$ and a total mechanical energy E, and so generate a dynamical equation incorporating this modification:

$$\left(\frac{dr}{dt}\right)^2 + \left(1 - \frac{2GM}{rc^2}\right) r^2 \left(\frac{d\phi}{dt}\right)^2 = \frac{GM}{r} - E.$$

This is recognisable as the equation which leads to the relativistic perihelion precession. We also recognise it as being equivalent to a rotation by the gravitational potential of the local coordinate system based on Lorentzian space-time, by an amount which adds the term $-(2GM/c^2)r^2(d\phi/dt)^2$ to the equation of motion. The term added to the 'pure' Newtonian equation for this system is essentially the same as would be produced by a fictitious inertial force or 'inertial energy', suggesting that, if the assumption that an instantaneous transmission of gravity implies an extrinsic c is correct, then a process of measurement which depends on c will generate a noninertial frame for a gravitational system. We can consider it as a kind of 'aberration', in this case of space rather than light, and it will produce an effective curvature of space and time of the same form as the curvature which is assumed to be responsible for gravity in general relativity. Of course, if we were able to use an 'absolute' system of coordinates entirely separate from the process of measurement, then no such rotation would occur.

Here, we are saying that, should the presumed Newtonian default position of instantaneous transmission of gravity with no intrinsic c be our starting point, we would still obtain the additional perihelion precession when we made a measurement using any technique dependent on the velocity of light, firstly, because of the finite speed involved, and, secondly, because of the gravitational effect on this speed. For an orbiting planet, there would be a deflection of the orbit resulting from a rotation or twist of the coordinate system, which would have the same effect as adding an extra perturbing force or forces, centrifugal and/or Coriolis, exactly as happens to the Earth's gravity when the Earth rotates. Newton, of course, was very familiar with these kinds of effects.

In this case, adding the deflection of the coordinate system to the orbit would give the correct solution for the observed motion of the planet, as would special relativity with the velocity of the frame equal to that generated by the Newtonian potential. The gravitomagnetic component would also be similar to the magnetic correction to the Coulomb's law analogue to Newton's law, although we might expect slight differences because the velocity of light is itself gravitationally affected. So, *even in purely Newtonian terms*, if we accept the Newtonian calculation of the light deflection, and realise that inertial forces are part of the Newtonian system, we can account for the extra perihelion of Mercury that general relativity predicts. We can also describe the effects on measurement at the speed of light as curvature and express it in tensor notation via covariant derivatives, as we can any force, real or fictitious, including Newtonian gravity.[29]

8.5. Gravity and Inertia

Like Newton's theory of gravity, Einstein's theory is purely generic. It is a set of linked mathematical equations, the general relativistic field equations, with no other meaning than the geometric structure it describes. Of course, when we use the equations in a particular physical situation, we *interpret* them to have physical meaning, and the meaning that we supply is that

---

[29]E. Cartan, *Ann. Ecole Norm.*, 40, 325, 1923; 41, 1, 1924. The contributions to the perihelion precession produced by ordinary perturbation terms are, of course, of the same form as the relativistic ones, with the γ^{-2} replaced by an equivalent numerical coefficient dependent on kinetic and potential energies, and so could be equally represented in Newtonian theory by distortions in the metric. Heaviside could have predicted the precession as a gravitomagnetic effect.

defined by the Newtonian potential and the mass-energy which is the source of the field. As we have seen, the Newtonian theory of gravity is just as much a theory of curvature and of pure geometry as is Einstein's theory. Newtonian perturbation theory is often about the changes to the coordinate system necessary to find the perturbations. The rewriting of Newton's mathematical structures in the form of algebraic calculus has obscured the fact that his geometrical methods often necessitated the replacement of physical force with geometrical structures equivalent to curved spaces. In effect, Newton rewrites physics as changes in local spaces, just as navigation requires a *de facto* use of spherical geometry to define the world in which it operates. In fact, from the experimental results so far obtained, it seems virtually impossible to know the difference between effects that are due to a finite speed of propagation in fact (ontological) and a finite speed of measurement (epistemological) and whether the distinction means anything, especially as the relevant equations are insensitive to the speed.[30]

Now, general relativity was constructed by abandoning the physical principles on which it was founded. Only the mathematical structure has any meaning. Historically, at the point when this was discovered, we could have accepted that the physical principles had been falsified, and that the mathematical structure did not support them. However, this was not what happened. *The mathematical structure continued to be interpreted in terms of physical principles which it had already proved to be false.* The result was the prediction of a series of physical consequences contradictory to some of the most important fundamental principles, such as the conservation laws. In addition, the ultimate consequence was proclaimed to be the self-destruction, through nonlinear terms, of the very theory which had supposedly been their origin.

Now, modifications to Newtonian equations can only be accomplished by introducing a gravitomagnetic component. If we are correct in seeing this as an aspect of the local *inertial* side of the gravitational/inertial duality rather than of the nonlocal side, then the origin of the time delay will lie in the localisation produced by the measurement process and the presumed nonlinear terms will no longer be a consequence. Since all phenomena ultimately have both local and nonlocal descriptions, it will become a

---

[30]It would probably require the discovery of a spin 2 rather than spin 1 intermediate boson (see 6.6) or a significant deviation from general relativity in strong fields. Perhaps, it could be argued that the fact that the universe is not geocentric means that universal inertial forces must be fictitious and that the kind of nonlinearity leading to a deviation from general relativity will never be discovered (see 5.8). Perhaps the dark energy as predicted from inertial forces is already the expression of this.

question of which aspects of any given phenomenon should be described by the nonlocal (Newtonian, absolute or ontological) space-time and which should be described by the local (Lorentzian, relative or epistemological) space-time of measurement. Ultimately, also, as in quantum mechanics, equations will have to be developed showing the connection between the two descriptions.

The general relativistic field equations say nothing about the physical nature of gravity, or even gravity at all. They are not about the nonlocal aspects of the gravitational field. They merely say that the *local* space-time coordinate structure can be described as curved in the presence of mass-energy. They don't actually justify saying that the curvature *is* mass-energy, a consequence of the strong principle of equivalence, which the theory actually *rejected* in arriving at the equations.

As far as we know, the general relativistic field equations hold up perfectly, even in relatively strong fields involving binary pulsars[31] and gravitational waves from colliding black holes.[32] There is no evidence of self-destruction or nonlinearity. Rather than being alternative paradigms, Newtonian physics and relativity (special and general) can be seen as parts of a single fundamental theory. With the universe at the critical density, even the cosmology is identical. This fact has been obscured by the 'revolutionary' rhetoric introduced immediately after the First World War by Eddington and his followers.

Ultimately, we see that a fault-line appeared in modern physics at the point where Eddington and his followers proclaimed that general relativity was a new physical paradigm that replaced the Newtonian theory of gravity. In order to maintain this position, we need to interpret the theory in a way that is not inherent in the equations, and that is directly contradicted by them. To do this, we have to assume that Einstein's theory is actually only an approximation and is fundamentally flawed in a way that will be accessible by experiment.

In fact, I believe both statements are incorrect because both theories are fundamentally true and not in contradiction with each other. Both are generic mathematical structures derived by elimination of hypothetical model-dependent elements almost by attrition. The general relativistic field equations could be used to link gravity and inertia by describing the inertial effects due to the nonlocal nature of the gravitational effect being measured using local interactions. They are not designed to explain gravity and they

[31] J. Antoniadis *et al.*, *Science*, 340, no. 6131, 499, 2013.
[32] B. P. Abbott *et al.*, *PRL* 116, 061102, 2016.

do not, in themselves, provide an 'alternative' to the Newtonian theory. However, they are a mathematically perfect description of such connections and are unlikely to need significant modification any time in the near future. By reconciling them to Newtonian theory through showing how the latter can be extended within the Newtonian paradigm, we are also reconciling general relativity and quantum mechanics, and removing the main barrier to a more profound unification.

Bibliography

Abbreviations

Corr. H. W. Turnbull, J. F. Scott, A. R. Hall and L. Tilling, *The Correspondence of Isaac Newton*, 7 vols., Cambridge University Press, 1959–1977

CUL Cambridge University Library

Hyp. *An Hypothesis Explaining the Properties of Light*, 1675, *Correspondence*, I, 361–386

KCC King's College, Cambridge

MP D. T. Whiteside, *Mathematical Papers*, 8 volumes, Cambridge University Press, 1967–1981

NP Rob Iliffe and Scott Mandelbrote, The Newton Project, http://www.newtonproject.sussex.ac.uk/

NPA William R. Newman, The Chymistry of Isaac Newton Project, http://webapp1.dlib.indiana.edu/newton/

NPC Stephen D. Snobelen, The Newton Project Canada, http://www.isaacnewton.ca

OP Alan E. Shapiro, *The Optical Papers of Isaiac Newton*, vol. 1, Cambridge University Press, 1984

QQP *Quaestiones quaedam philosophicae*, NP, *Certain Philosophical Questions: Newton's Trinity Notebook*, c 1664, ed. J. E. McGuire and M. Tamny, Cambridge, 1983

Q1-31 *Opticks*, Queries 1–31

Unp. A. Rupert Hall and Marie Boas Hall, *The Unpublished Scientific Papers of Isaac Newton*, Cambridge University Press, 1962

Works Cited

Adams, J. C. *Monthly Notices of the Royal Astronomical Society*, 63, 43, 1882.

Aiton, E. J., 'The inverse problem of central forces', *Annals of Science*, 20, 81–99, 1965.

Aiton, E. J., 'Newton's aether-stream hypothesis and the inverse square law of gravitation', *Annals of Science*, 25, 255–260, 1969.

Alexander, H. G. *The Leibniz-Clarke Correspondence*, 1956.

Arago, D. F. J., 'Mémoire sur la vitesse de la lumière, lu à la première Classe de l'Institut le 10 décembre 1810', *Académie des Sciences (Paris). Comptes Rendus*, 36, 38–49, 1853.

Arbuthnot, John, 'An Argument for Divine Providence Taken from the Constant Regularity Observ'd in the Birth of Both Sexes', *Philosophical Transactions*, 27, 186–190, 1710–1712.

Arnol'd, V. I., 'Topological proof of the transcendence of the abelian integrals in *Newton's Principia*', *Istoriko-Matematicheskie Issledovaniya*, 31, 7–17, 1989.

Arnol'd, V. I., *Huygens and Barrow, Newton and Hooke*, Birkhäuser Verlag, Basel, 1990.

Arnol'd, V. I., *Real Algebraic Geometry*, Springer-Verlag, Berlin Heidelberg, 2013.

Arnol'd, V. I. and Vasil'ev, V. A., '*Newton's Principia* read 300 years later', *Notices of American Mathematical Society*, 36 (9), 1148–1154, 1989.

Baily, Francis, *An Account of the Revd John Flamsteed, the first Astronomer Royal, compiled from his own Manuscripts, and other authentic documents, never before published*, London, 1835, photo-reprint, London: Dawsons, 1966.

Barrow, Isaac, *Lectiones XVIII, Cantabrigiae in scholis publicis habitae; in quibus opticorum phenomenon genuinae rationes investigantur, ac exponuntur (Lectiones opticae)* (*Optical Lectures*), London, 1669.

Barrow, John D., 'Inflationary Universe', *Times Higher Educational Supplement*, 14 May 1993, 21, a review of *An Introduction to Mathematical Cosmology* by J. N. Islam (Cambridge, 1992).

Bechler, Zev, 'Newton's Search for a Mechanistic Model of Colour Dispersion: A Suggested Interpretation', Archive for History of Exact Sciences, 11, 1–37, 1973.

Bechler, Zev, 'Newton's law of forces which are inversely as the mass: a suggested interpretation of his later efforts to normalise a mechanistic model of optical dispersion', *Centaurus*, 18, 184–222, 1974.

Bechler, Zev, "A less agreeable matter': The Disagreeable Case of Newton and Achromatic Refraction', *British Journal for the History of Science*, 8, 101–126, 1975.

Belenkiy, Ari and Vichagüe, Eduardo Vila, 'History of one defeat: reform of the Julian calendar as envisaged by Isaac Newton', *Notes and Records of the Royal Society*, 59, 223–254, 2005.

Bennett, J. A., *The Mathematical Science of Christopher Wren*, Cambridge University Press, 1982.

Bentley, Richard, *A Confutation of Atheism from the Origin and Frame of the World*, London, 1693 (the Boyle lectures for 1692–1693).

Bernoulli, Johann, *Acta Eruditorum*, May 1697, 206–211.

Bertrand, J., 'Theoreme relatif au movement d'un point attire vers un center fixe', *Comptes Rendus*, 77, 849–853, 1875.

Birch, T., *The History of the Royal Society of London*, 4 vols., London, 1756–1757.

Blair, Robert, Royal Society, Manuscript L & P, VIII, 182, read 6 April 1786., unpublished, discussed in Jean Eisenstaedt, 'Light and Relativity, a Previously Unknown Eighteenth-Century Manuscript by Robert Blair' (1748–1828), *Annals of Science*, Vol. 62, No. 3, July 2005, 347–376.

Bloye, Nicole and Huggett, Stephen, 'Newton, the Geometer', *European Maths Society Newsletter*, 82, 19–27, December 2011.

Bochner, S., *The Role of Mathematics in the Rise of Science*, Princeton University Press, 1966.

Bohlin, K., 'Note sur le problème des deux corps et sur une integration nouvelle dans le problème des trois corps, *Bulletin Astronomique*, 28, 113–119, 1911.

Boulliau, Ismail, *Astronomia Philolaica*, Paris, 1645, 1657.

Brackenridge, J. Bruce, 'The Critical Role of Curvature in Newton's Developing Dynamics', in P. M. Harman and Alan E. Shapiro (eds.), *An Investigation of Difficult Things: Essays on Newton and the History of the Exact Sciences*, Cambridge University Press, 1992, 231–260.

Brackenridge, J. Bruce, *The Key to Newton's Dynamics: The Kepler Problem and the Principia*, Berkeley: University of California Press, 1995.

Brackenridge, J. Bruce and Nauenberg, Michael, 'Curvature in Newton's dynamics', in I. Bernard Cohen and George E. Smith (eds.), *The Cambridge Companion to Newton*, Cambridge University Press, 2002, 85–137.

Bradley, James, 'Nutation', *Philosophical Transactions*, 45, no. 485, 1, 1748 (dated Greenwich, 31 December 1747).

Breger, H., 'Über den von Samuel König veröffentlichten Brief zum Prinzip der kleinstenn Wirkung', in Hartmut Hecht (ed.), *Pierre Louis Moreau de Maupertuis. Eine Bilanz nach 300 Jahren*. Schriftenreihe des Frankreich-Zentrums der Technischen Universität Berlin. Verlag, Arno Spitz GmbH/Nomos Verlagsgesellschaft, Berlin, 1999, 363–381.

Brewster, Sir David, *Memoirs of the Life, Writings, and Discoveries of Isaac Newton*, 2 vols., Edinburgh Thomas Constable & Co., 1855, New York: Johnson Reprint Corporation, 1865.

Brouwer, D. and Clemence, G. M., *Methods of Celestial Mechanics*, Academic Press, New York, 1961.

Brown, E. W., *An Introductory Treatise on the Lunar Theory*, Cambridge University Press, 1896, republished by Dover, 1960.

Brown, E. W., *Tables of the Motion of the Moon* Yale University Press, New Haven, CT, 1919.

Buchwald, Jed Z. and Feingold, Mordechai, *Newton and the Origin of Civilization*, Princeton University Press, 2012.

Buttazzo, Giuseppi and Fredianao, Aldo (eds.), *Variational Analysis and Aerospace Engineering*, Springer, 2009.

Calcaterra, A., de Sangro, R., Finocchiaro, G., Patteri, P., Piccolo, M. and Pizzella, G., 'Measuring Propagation Speed of Coulomb Fields', arXiv:1211. 2913, 2012.

Calder, Lucy and Lahav, Ofer, 'Dark energy: Back to Newton?', *A & G*, February 2008, vol. 49, 1.13–1.18 and arXiv:1712.2196v2.

Cartwright, D. A., *Tides: A Scientific History*, Cambridge University Press, 1999.

Cauchy, A. L., *Cours d'Analyse*, 1821.

Cavendish, Henry, c 1784, in E. Thorpe (ed.), *The Scientific Papers of the Honourable Henry Cavendish, F.R.S., Volume II: Chemical and Dynamical*, London, 1921, 433–437.

Cavendish, Henry, *Philosophical Transactions*, 88, 1798, 469–526.

Cavendish, Henry, *The Electrical Researches of the Honourable Henry Cavendish*, ed. J. C. Maxwell, Cambridge University Press, 1879.

Challis, James, *Philosophical Magazine* (4), 18, 442–451, 1859, 450.

Chandrasekhar, S., *Newton's Principia for the Common Reader*, 1995, 201.

Chasles, Michel, *Les Trois Livres de Porismes d'Euclide*, Paris, 1860.

Chéseaux, Jean-Phillippe Loys de, *Traité de la Comète qui Paru en Décembre 1743 & en Janvier, Février & Mars 1744*, Lausanne, 1744, 223–229, reprinted in Harrison, 1987, 221–222.

Clairaut, A.-C., *Histoire et Mémoires de l'Académie Royale des Sciences*, 58, 329–364, 1745 (read 15 November 1747).

Clairaut, A,.-C., *Théorie de la Lune*, 1752.

Clairaut, A.-C., Théorie du movement des Comètes, Paris, 1760.

Clark, R. W., *Einstein: The Life and Times*, London, 1973, first edition, 1971, 227.

Cohen, I. Bernard, *Archives Internationales d'Historie des Sciences*, 11, 357–375, 1958.

Cohen, I. Bernard, *Isaac Newton's Papers and Letters of Natural Philosophy*, Cambridge, Massachussetts, Harvard University Press, 1958.

Cohen, I. Bernard, *Introduction to Newton's 'Principia'*, Cambridge, Massachussetts, 1971.

Cohen, I. Bernard, 'Newton, Isaac', *Dictionary of Scientific Biography*, Charles Scribner's Sons, 1975–1980, 12, 42–101.

Cohen, I. Bernard, 'A Guide to Newton's *Principia*', in I. Bernard Cohen and Anne Whitman, *Isaac Newton The Principia A New Translation*, Berkeley and Los Angeles and London: University of California Press, 1999, 1–370.

Cohen, I. Bernard and Westfall, Richard S. (eds.), *Newton A Norton Critical Reader*, W. W. Norton & Company, 1995, General Introduction, xi–xii.

Comte, M. and Diaz, J. I., 'On the Newton partially flat minimal resistance body type problems', 20 December 2004.

Cotes, Roger, 'Logometria', *Philosophical Transactions*, 29, no. 338, 5–45, January-March 1714, *Harmonia Mensurarum, sive Analysis & Synthesis per Rationum & Angulorum Mensuras*, Cambridge, 1722.

de Morgan, A., *Essays on the Life and Work of Newton*, ed. Philip E. B. Jourdain, Chicago, 1914.

Densmore, Dana, *Newton's Principia The Central Argument*, Green Lion Press, Santa Fe, New Mexico, third edition, 2003.

Descartes, René, *La Géométrie*, in *Discours de la Méthode*, Leiden, 1637, 297–413.

Descartes, René, *Principia Philosophiae* (1644), translated by Valentine Rodger Miller and Reese P. Miller, D. Reidel, Dordrecht, 1983.

Descartes, René, *Geometria, à Renato Des Cartes Anno 1637 Gallicè Edita*, Leiden, 1649, first Latin edition of *La Géométrie*.

Descartes, René, *Geometria, à Renato Des Cartes Anno 1637 Gallicè Edita*, Amsterdam, 1659–1661, second Latin edition of *La Géométrie*.

Digges, Thomas, 'Perfect Description of the Celestial Spheres', appendix to L. Digges, L. Digges, *A Prognostication everlasting*, 1576 (first published in 1555); reprinted in Harrison, 1987, 211–217.

DiSalle, Robert, 'Newton's philosophical analysis of space and time', in I. Bernard Cohen and George E. Smith (eds.), *The Cambridge Companion to Newton*, Cambridge University Press, 2002, 33–56.

DiSalle, Robert, 'Newton's Views on Space, Time, and Motion', *Stanford Encyclopedia of Philosophy*, 12 August 2004, revised 22 August 2011.

Di Steno, Simonetta and Galuzzi, Massimo, 'La Qunita Sezione del Primo Libro dei *Principia*: Newton e il Problema di Pappo', *Archives Internationales d'Histoire des Sciences*, 39, 51–68, 1989.

Dobbs, Betty Jo Teeter, *The Janus Faces of Genius*, Cambridge University Press, 1991.

Dobson, Geoffrey J., 'Newton's Problems with Rigid Body Dynamics in the Light of his Treatment of the Precession of the Equinoxes', *Archive for History of the Exact Sciences*, 53, 125–145, 1998.

Dobson, Geoffrey J., 'Against Chandrasekhar's Interpretation of Newton's Treatment of the Precession of the Equinoxes', *Archive for History of the Exact Sciences*, 53, 577–597, 1999.

Dobson, Geoffrey J., 'On Lemmas 1 and 2 to Proposition 39 of Book 3 of Newton's *Principia*', *Archive for History of the Exact Sciences*, 55, 345–363, 2001.

Dollond, John, 'A mistake in Euler's theorem for correcting aberration in object-glasses for refracting telescopes', *Philosophical Transactions*, 1753, 287.

Dollond, John, 'An account of some experiments concerning the different refrangibility of light', *Philosophical Transactions*, 50, 733–743, 1758.

Dollond, Peter, *Some Account of the Discovery, Made by the Late Mr. John Dollond, F.R.S. which Led to the Grand Improvement of Refracting Telescopes, in Order to Correct Some Misrepresentations, in Foreign Publications, of that Discovery: With an Attempt to Account for the Mistake in an Experiment Made by Sir Isaac Newton; on which Experiment, the Improvement of the Refracting Telescope Entirely Depended*, London: J. Johnson, 1789.

Drexler, M., Sobey, I. J. and Bracher, C., 'Fractal Characteristics of Newton's method on Polynomials', Report No. 96/14, 1996.

Eberty, Felix, *Die Gestitne und die Weltgeschichte, Gedanken über Raum, Zeit und Ewigkeit*, translated anonymously as *The Stars and the Earth; or Thoughts upon Space, Time, and Eternity*, 1846.

Edwards, C. Henry, 'Newton's nose-cone problem', *The Mathematica Journal*, 7, 64–71, 1997.

Ehrlichson, H., 'Comment on "Longburied dismantling of a centuriesold myth: Newton's bPrincipia and inversesquare orbits," ' by Robert Weinstock [American Journal of Physics 57, 846–849 (1989)]', *American Journal of Physics*, 58, 882–884, 1990.

Einstein, Albert, 'On the electrodynamics of moving bodies', *Annalen der Physik*, 17, 891–921, 26 Sep 1905 (received 30 June); translated in A. Sommerfeld (ed.), *The Principle of Relativity*, New York, 35–65.

Eisenstaedt, Jean, 'Light and Relativity, a Previously Unknown Eighteenth-Century Manuscript by Robert Blair (1748–1828)', *Annals of Science*, Vol. 62, No. 3, 347–376, July 2005.

Engelsman, S. B., *Families of curves and the origins of partial differentiation*, North Holland, 1984.

Euler, Leonhard, *Sur la perfection des verres objectifs des lunettes*, Mem. de l'acad. de Berlin, 1747 (1753 and 1754).

Fatio de Duillier, Nicolas, *Lineae brevissimi Descensus Investigatio Geometrica Duplex (Line of Quickest Descent)*, London, 1699.

Fellmann, E. A., 'The *Principia* and Continental Mathematicians', in D. C. King-Hele and A. R. Hall (eds.), *Newton's Principia and its Legacy, Notes and Records of the Royal Society*, 42, no. 1, 13–34, 1988.

Ferguson, James, 'A brief survey of the history of the calculus of variations and its applications', arxiv.org/pdf/math/0402357.

Flamsteed, John, *Historia Coelestis Britannica*, 1712.

Flamsteed, John, *Historia Coelestis Britannica*, 1725.

Flamsteed, John, *History of his own Life and Labors*, in F. Baily, *An Account of the Revd John Flamsteed, the first Astronomer Royal, compiled from his own Manuscripts, and other authentic documents, never before published*, London, 1835, photo-reprint, London: Dawsons, 1966, 7–105.

Frank, Philipp, 'Das Relativitätsprinzip der Mechanik und die Gleichungen für die elektromagnetischen Vorgänge in bewegten Körpern', *Annalen der Physik*, 27, 897–902, 1908, 898.

Friedmann, Alexander, 'Über die Krümmung des Raumes', *Zeitschrift für Physik*, 10 (1), 377–386, 1922; 'On the curvature of space', *General Relativity and Gravitation*, 31 (12), 1991–2000, 1999.

Gjertsen, Derek, *The Newton Handbook*, Routledge & Kegan Paul, 1986.

Goldstine, H., *A History of the Calculus of Variations from the 17th through the 19th Century*, New York: Springer-Verlag, 1980.

Grant, Aaron K. and Rosner, Jonathan L., 'Classical orbits in power law potentials', *EFI 92-69-Rev*, 1–12, 1993.

Greene, Brian, *The Fabric of the Cosmos*, Penguin, 2004.

Gregory, David, *Catoptricae et dioptricae sphaericae elementa*, 1695.

Gregory, David, *Astronomiae physicae et geometricae elementa*, Oxford, 1702.

Gregory, David, *Elements of Physical and Geometrical Astronomy*, 1715 (translation of Latin of 1702), second edition 1726, reprinted Johnson Reprint Corporation, New York, 1972.

Gregory, James, *Optica promota*, 1663.

Gregory, James, *Geometriae pars universalis*, 1668.

Guicciardini, Niccolò, *Reading the Principia*, Cambridge University Press, 1999.

Guicciardini, Niccolò, 'Introduction Conceptualism and contextualism in the recent historiography of Newton's Principia', *Historia Mathematica*, 30, 407–431, November 2003.

Guicciardini, Niccolò, *Isaac Newton on Mathematical Certainty and Method*, The MIT Press, Cambridge, Mass. and London, 2009.

Gutzwiller, Martin, '*Quantum chaos*', *Scientific American*, 266, no. 1, 78–84, January 1992, reprinted 27 October 2008.

Hadley, John, 'Description of a new instrument for taking angles', *Philosophical Transactions*, 37, 147–57, 1731–1732 (1731).

Hall, A. Rupert, 'Newton on the calculation of central forces', *Annals of Science*, 13, 62–71, 1957.

Hall, A. Rupert, *Philosophers at War: The Quarrel between Newton and Leibniz*, Cambridge, 1980.

Hall, A. Rupert, *History of Technology*, 1985, 23 ff.

Hall, A. Rupert, *Isaac Newton Adventurer in Thought*, Cambridge University Press, 1992.

Hall, Rachel W. and Josić, Krešimir, 'Planetary Motion and the Duality of Force Laws', *SIAM Review*, 42, No. 1, 115–124, 2000.

Halley, Edmond, 'Emendationes ac notae in vetustas Albatenii observationes astronomicas', *Philosophical Transactions*, 17, 913–21, 1693.

Halley, Edmond, *Astronomiae cometicae synopsis*, Oxford, 1705; *A Synopsis of the Astronomy of Comets*, London, 1705; *Philosophical Transactions*, 24, 1882–1889, 1704–1705.

Halley, Edmond, 'Considerations of the Change of Latitudes of Some of the Principal Fixt Stars', *Philosophical Transactions*, 30, 736–738, 1717–1719, January–April 1718.

Halley, Edmond, 'On the infinity of the sphere of fix'd stars', *Philosophical Transactions*, 31, 22–24, 1720–1721, January–April 1720.

Halley, Edmond, 'Of the Number, Order and Light of the Fx'd stars', *Philosophical Transactions*, 31, 24–26, 1720–1721, January–April 1720.

Halley, Edmond, 'The Number, Order, and Light of the Fix'd Stars', *Philosophical Transactions*, 31, 1720–1721, 24–26, read 16 March 1721.

Halley, Edmond, *Tabulae astronomicae*, 1749, translated 1752.

Harman, P. M. , 'Newton to Maxwell: the *Principia* and British physics', in D. C. King-Hele and A. R. Hall (eds.), *Newton's Principia and its Legacy*, *Newton's Principia and its Legacy*, *Notes and Records of the Royal Society*, 42, no. 1, 75–96, 1988.

Harrison, David Mark, *The Influence of Non-linearities on Oscillator Noise Performance*, PhD Thesis, September 1988, Leeds University.

Harrison, Edward, *Darkness at Night: A Riddle of the Universe*, Harvard University Press, 1987, 218–219 and 219–220.

Harvey, A., *American Journal of Physics*, 46, 928–929, 1978.

Hawking, Stephen, *Black Holes and Baby Universes and Other Essays*, 1993.

Hayes, Charles, *A Treatise of Fluxions: or, an Introduction to Mathematical Philosophy*, 1704.

Herivel, John, *The Background to Newton's Principia*, Oxford: Clarendon Press, 1965, 257–289.

Herivel, John, 'Newton's Achievement in Dynamics', *The Texas Quarterly*, 10, no. 3, autumn 1967, and in Robert Palter (ed.), *The Annus Mirabilis of Sir Isaac Newton 1666–1966*, M.I.T. Press, Cambridge, Massachussetts and London, 1970, 120–135.

Hermann, Jacob, *Giornale dei Letterati d'Italia*, 7, 173–229, 1711.

Hermann, Jacob, *Phoronomia*, 1716.

Herschel, William, *Philosophical Transactions*, 1782, 163 ff.

Herschel, William, *Philosophical Transactions*, 1803, 339.

Herschel, William, *Philosophical Transactions*, 1804, 353.

Hill, G. W., *The collected mathematical works of George William Hill*, 4 vols., Carnegie Institution of Washington, 1905–1907.

Hobden, Heather and Mervyn, *The Cosmic Elk*, eighth edition 2009.

Hobden, M. K. 'John Harrison, Balthazar van der Pol and the Non-Linear Oscillator', *Horological Journal*, July 1981, 16–18, November 1981, 16–19, February 1982, 26–28, April 1982, 15–18, June 1982, 14–15.

Hobden M. K. 'As 3 is to 2', *Horological Science News*, November 2011, NAWCC Chapter. 161.

Hooke, Robert, *An Attempt to Prove the Motion of the Earth by Observations*, 1674, reprinted in *Lectiones Cutlerianae*, 1679; reproduced in R. T. Gunther, *Early Science in Oxford*, Oxford University, Oxford, 1930, VII, 1–28.

Hooke, Robert, *Lectures and Collections*, London, 1678.

Hooke, Robert, *Discourse of the Nature of Comets* (1680), *Posthumous Works*, 1705, reprinted London: Frank Cass & Co.Ltd., second edition, 1974.

Hoskin, M. A., 'Newton, Providence and the Universe of Stars', *Journal for the History of Astronomy*, 8, 77–101, 1977.

Hoskin, M. A., 'Stukeley's Cosmology and the Newtonian Origins of Olbers's Paradox', *Journal for the History of Astronomy*, 8, 77–112., 1985.

Hoskin, M. A., *Mem. S. A. It.*, 60, 687–694, 1989.

Hoyle, F. and Wickramasinghe, N. C., *Lifecloud*, J. M. Dent, London, 1978.

Hoyle, F. and Wickramasinghe, N. C., *Diseases from Space*, J. M. Dent, London, 1979.

Hughes, D. W., 'The *Principia* and Comets', in D. C. King-Hele and A. R. Hall (eds.), *Newton's Principia and its Legacy, Newton's Principia and its Legacy*, 42, no. 1, 53–74, 1988.

Hutton, James, *Abstract of a dissertation read in the Royal Society of Edinburgh, upon the seventh of March, and fourth of April, MDCCLXXXV, Concerning the System of the Earth, Its Duration, and Stability*, Edinburgh, 1785.

Hutton, James, *Theory of the Earth; or an investigation of the laws observable in the composition, dissolution, and restoration of land upon the Globe, Transactions of the Royal Society of Edinburgh*, vol. 1, Part 2, 209–304, 1788.

Hutton, James, *Theory of the Earth; with proofs and illustrations*, Edinburgh: Creech, 2 vols, 1795.

Huygens, Christiaan, *De vi centrifuga*, 1673, in *Oeuvres Complètes de Christiaan Huygens*, The Hague, 1929, XVI, 253–301.

Iliffe, Rob, 'The religion of Isaac Newton', in Rob Iliffe and George E. Smith (eds.), *The Cambridge Companion to Newton*, second edition, Cambridge University Press, 2016, 485–523.

Jaki, S. L., *Found. Phys.*, 8, 927–950, 1978.

Janiak, Andrew, *Newton as Philosopher*, Cambridge University Press, 2008.

Jungnickel, Christa and McCormmach, Russell, *Cavendish The Experimental Life*, Cranbury, 1999.

Kant, Immanuel, *Gedanken von der Wahren Schatzung der lebendigen Krafte*, 1747.

Kassner, Klaus, 'Classroom reconstruction of the Schwarzschild metric', arXiv: 1502.00149v2.

Klingenstierna, S., 'Anmerkung über das Gesetz der Brechung bey Lichtstrahlen von verschiedener Art, wenn sie durch ein durchsichtiges Mittel in verschiedene andere gehen', *Abhandlungen aus der Naturlehre, Haushaltungskunst und Mechanik*, 16, 300–309, 1754.

Knobloch, Eberhard, 'Leibniz et Son Manuscrit Incdité sur la Quadrature des Sections Coniques', in *The Leibniz Renaissance*, Olschki, Florence, 1989.

Kollerstrom, Nicholas, *Newton's Forgotten Lunar Theory: His Contribution to the Quest for Longitude*, Green Lion Press, Santa Fé, 2000.

Koyré, A., 'Newton's Regulae Philosophandi', *Newtonian Studies*, 1965.

Koyré, A. and Cohen, I. B., 'Newton and the Leibniz-Clarke correspondence, with notes on Newton, Conti and des Maizeaux', *Archives internationals d'histoire des sciences*, 15, 63–126, 1962.

Kriloff, A. N., *Monthly Notices of the Royal Astronomical Society*, 84, 392–395, 1926.

Kubrin, David, 'Newton and the Cyclical Cosmos: Providence and the Mechanical Philosophy', *Journal of the History of Ideas*, 28, 325–346, 1967, reprinted in I. Bernard Cohen and Richard S. Westfall, *Newton A Norton Critical Edition*, New York and London: W. W. Norton & Company, 281–296.

Kulviecas, Liubomiras, 'Two unnoticed moments in I. Newton's scientific works', delivered at the conference of the Lithuanian Association of history and philosophy of science, dedicated to the 350th birth anniversary of I. Newton, Vilnius, 28 January 1993, *Evolution of Science and Technology / Mokslo ir technikos raida*, 4, no. 1, 39–46, 2012 (in Lithuanian).

Lacroix, S. F., *Traité elémentaire de calcul différentiel et de calcul intégral*, Paris, 1802; translated as *An Elementary Treatise on the Differential and Integral Calculus*, with an Appendix and Notes, by C. Babbage, J. F. W. Herschel, and G. Peacock, Cambridge: J. Deighton and Sons, 1816.

Laplace, P. S., *Exposition du Système du Monde*, 1796, 2: 304–6, and in F. X. von Zach, *Allgemeine geographische Ephemeriden*, Weimar, 1799, 4, 1–6 (dated May 1798), translated in S. W. Hawking and G. F. R. Ellis, *The Large-Scale Structure of Space-Time*, London, 1973, 365–8.

Laplace, P. S., *Traité de Mécanique Céleste*, 5 volumes, 1799–1825.

Laplace, P. S., *La Théorie Analytique des Probabilités*, second edition, 1814; *A Philosophical Essay on Probabilities*, 1814, translated by F. W. Truscott and F. L. Emory, 1902, Dover, 1951.

Laplace, P. S., *Traité de Mécanique Céleste*, vol. 5, Paris: Bachelier Libraire, 1825, 409. 'Parmi les inégalités da mouvement de la Lune en longitude, Newton n'a développé que la *variation*. La méthode qu' il suivie me parait être une des choses le plus remarquables de l'Ouvrage des Principes.'.

Lehn, Waldemar H., 'Isaac Newton and the astronomical refraction', *Applied Optics*, 47, No. 34, H95–105, 1 December 2008.

Leibniz, G. W., *Tentamen de motuum coelestium causis*, Acta Eruditorum, 2, 82–86, February 1689.

Leibniz, G. W., *De quadrature arithmetica circuli ellipseos et hyperbolae cujus corollarium est trigonometria sine tabulis*, ed. Eberhard Knobloch, Vandenhoeck and Ruprecht, Göttingen, 1993.

Lemons, D. S., *American Journal of Physics*, 56, 502–504, 1988.

Le Verrier, U. J. J., *Comptes Rendus Acad. Sci.*, 49, 379, 1859.

Levitin, Dmitri, 'Scholarly and Scholastic contexts for Newton's General Scholium', presented at 'Isaac Newton's General Scholium to the Principia: science, religion and metaphysics: a tercentenary symposium', King's College, Halifax Nova Scotia, 24–26 October 2013.

Lictenstein, L., *Sitzungsber. Pruess. Akad. Wiss. Phys. Math. Kl.*, p. 1120, 1918.

Littlewood, J. E., 'Newton and the Attraction of a Sphere', *Mathematical Gazette*, 32, 1948, reprinted in B. Bollobas (ed.), *Littlewood's Misecellany*, Cambridge University Press, 1982, 179–181.

Lynden-Bell, Donald, 'The Wonderful Geometry of Dynamics Newton's Principia for the Common Reader by S. Chandrasekhar', *Notes and Records of the Royal Society*, 50, No. 2, 253–255, July 1996, 253–254.

Lynden-Bell, Donald, 'The Newton wonder in mechanics', *The Observatory*, 120, 131–136, 2000.

Lyons, Sir Henry, *The Royal Society 1660–1940*, Cambridge University Press, 1944.

Mach, Ernst, 'History and Root of the Principle of the Conservation of Energy', *The Science of Mechanics; a Critical and Historical Account of its Development, by Ernst Mach: supplement to the 3rd English edition containing the author's additions to the 7th German edition*, translated by Philip Edward Bertrand Jourdain, The Open Court Publishing Co., Chicago, 1911.

Marsden, B. G., Sekanina, Z. and Yeomans, D. K., 'Comets and Nongravitational Forces', *Astronomical Journal*, 78, 211–223, 1973.

Maskelyne, Nevil, *Philosophical Transactions*, 65, 495–499, 1775.

Maupertuis, P. L. de, *Sur la figure de la terre*, 1738.

Maupertuis, P. L. de, *Mémoires de l'Académie des Sciences*, 417, 1744, and 'Recherches des lois du mouvement' (read to the Berlin Academy, 1746); both in *Oeuvres de Maupertuis*, Alyon, 1768, vol. IV.

May, William Edward, *A History of Marine Navigation*, G. T. Foulis & Co. Ltd., Henley-on-Thames, Oxfordshire, 1973.

May, William Edward, 'How the Chronometer went to Sea', *Antiquarian Horology*, March 1976, 638–662.

McCrea, W. H. and Milne, E. A., *Quart. J. Math.*, 5, 73, 1934.

McGuire, J. E., 'Force, Active Principles, and Newton's Invisible Realm', *Ambix*, 15, 1968, 154–208.

McGuire, J. E., 'Newton on Place, Time, and God: An Unpublished Source', *The British Journal for the History of Science*, 11, No. 2, 114–129, July 1978.

Mercator, Nicholas, *Institutiones astronomicae*, in two books, 1676.

Michell, John, *An Inquiry into the Probable Parallax and Magnitude of the Fixed Stars*, *Phil. Trans.*, 57, 234–264, 1767.

Michell, John, 'On the Means of Discovering the Distance, Magnitude, & C. of the Fixed Stars, in Consequence of the Diminution of the Velocity of Their Light, in Case Such a Diminution Should Be Found to Take Place in Any of Them, and Such Other Data Should Be Procured from Observations, As Would Be Farther Necessary for That Purpose', *Royal Society of London Philosophical Transactions*, 74, 35–57, 1784, read 27 November 1783.

Milne, E. A., *Quarterly Journal of Mathematics*, 5, 64, 1934.

Milne, E. A., *Relativity, Gravitation and World Structure*, Oxford University Press, 1935.

Minkowski, Hermann, *Physikalische Zeitschrift*, 10, 104–111, 1 February 1909 (read at Cologne, 21 September, received 23 December 1908); reprinted in *Das Relativitätsprinzip* (Lorentz, Einstein and Minkowski, 1913); translated into English in *The Principle of Relativity* (Lorentz, Einstein, Minkowski and Weyl, 1923), 104.

Motte, Andrew, *The Mathematical Principles of Natural Philosophy. By Sir Isaac Newton. Translated into English*, 2 volumes, London, 1729.

Nauenberg, Michael, 'Newton's Early Computational Method for Dynamics', *Archives for History of Exact Science*, 46, 221–251, 1994.

Nauenberg, Michael, 'Newton's Curvature Measure of Force: New Findings', in I. Bernard Cohen and Anne Whitman, *Isaac Newton The Principia A New Translation*, Berkeley and Los Angeles and London: University of California Press, 1999, 76–82.

Nauenberg, Michael, 'Newton's Portsmouth Perturbation Theory and its application to Lunar Motion', in R. H. Dalitz and M. Nauenberg (eds.), *The Foundations of Newtonian Scholarship*, World Scientific, 2000, 167–194.

Nauenberg, Michael, 'Newton's Perturbation Methods for the three-Body Problem and their application to Lunar motion', in J. Z. Buchwald and I. B. Cohen (eds.), *Isaac Newton's Natural Philosophy*, Cambridge, MA and London: The M.I.T. Press, 2001, 189–224.

Nauenberg, Michael, 'Newton's Lunar Theory', a review of Nicholas Kollerstrom, *Newton's Forgotten Lunar Theory: His Contribution to the Quest for Longitude*, *Journal for History of Astronomy*, 32, 162–168, 2001.

Nauenberg, Michael, 'Kepler's Area Law in the *Principia*: Filling in Some Details in Newton's Proof of Proposition 1', *Historia Mathemaica*, 30, 441–456, 2003.

Nauenberg, Michael, 'Robert Hooke's Seminal Contribution to Orbital Dynamics', *Physics in Perspective*, 7, 1–31, 2005.

Nauenberg, Michael, 'The early application of calculus to the inverse square problem', *Archive for History of Exact Sciences*, 64, 269–300, 2010.

Nauenberg, Michael, 'The Reception of Newton's *Principia*', on line, 2011.

Nauenberg, Michael, 'Comment on 'Is Newton's Second Law Really Newton's?' by Pruce Pourciau [Am. J. Phys. 79(10), 1015–1022, 2011]', *American Jounal of Physics*, 80, 931–933, 2012.

Néményi, P. F., 'The Main Concepts and Ideas of Fluid Dynamics in their Historical Development', *Archive for History of Exact Sciences*, 2, 52–86, 1963.

Newcomb, Simon, *Astronomical Papers of the American Ephemeris*, 1, 472, 1882.

Newton, Isaac, *Opticks*, first edition, 1704, Latin edition, 1706, second English edition, 1717, NP; fourth edition, 1730 (edition cited); *Opticks* (preface by I. B. Cohen, foreword by A. Einstein, introduction by E. T. Whittaker), Dover Publications, 1952 (based on the fourth edition, 1730).

Newton, Isaac, *Principia*, first edition, 1687, second edition, 1713, third edition, 1726, NP; English translation by Andrew Motte, 1729 (edition cited); revised by Florian Cajori, 1934; I. Bernard Cohen and Anne Whitman, *Isaac Newton*

The Principia A New Translation, Berkeley and Los Angeles and London: University of California Press, 1999.

Newton, R. R., *Crime of Claudius Ptolemy*, Johns Hopkins University Press, 1977.

Nicholas of Cusa On Learned Ignorance A Translation and Appraisal of De Docta Ignorantia by Jasper Hopkins, The Arthur J. Banning Press, Minneapolis, 1981, second edition, 1985.

Olbers, H. M. W., *Astronomisches Jahrbuch für das Jahr 1826*, Berlin, 1823, 110–121 reprinted in Harrison, 1987, 223–226.

Palmieri, Paolo, Review *of Isaac Newton's Natural Philosophy*, edited by J. Z. Buchwald and I. B. Cohen, Dibner Institute Studies in the History of Scoence and Technology, Cambridge, MA, 2004.

Pannekoek, A., *A History of Astronomy*, London, 1961.

Pask, Colin, *Magnificent Principia Exploring Isaac Newton's Masterpiece*, Prometheus Books, 2013.

Pearson, Karl, *Nature*, 117, 551–552, 1926.

Pesic, Peter, 'The validity of Newton's lemma 28', *Hist. Math.*, 28, No. 3, 215–219, 2001.

Poincaré, Henri, *Comptes Rendus de l'Académie des Sciences*, 140, 1504–1508, 1905 (communicated 5 Jun); *Oeuvres de Henri Poincaré*, 11 vols. (Paris, 1936–1956), 9: 489.

Pourciau, Bruce, 'On Newton's proof that inverse-square orbits must be conics', *Annals of Science*, 48, 159–172, 1991.

Pourciau, Bruce, 'The Preliminary Mathematical Lemmas of Newton's *Principia*', *Archive for the History of Exact Sciences*, 52, 279–295, 1998.

Pourciau, Bruce, 'The Integrability of Ovals: Newton's Lemma 28 and Its Counterexamples', *Archive for History of the Exact Sciences*, 55, 479–499, 2001.

Pourciau, Bruce, 'Newton and the notion of limit', *Historia Mathematica*, 28, 18–30, 2001.

Pourciau, Bruce, 'Newton's Argument for Proposition I of the *Principia*', *Archive for History of the Exact Sciences*, 57, 267–311, 2003.

Pourciau, Bruce, 'Instantaneous and continuous force: the foundations of Newton's *Principia*', in Rob Iliffe and George E. Smith (eds.), The Cambridge Companion to Newton, second edition, Cambridge University Press, 2016, 93–186.

Poynting, J. H., *Mason College Magazine*, 1, 107–117, 1883; *Collected Scientific Papers*, Cambridge, 1920, 552–556.

Priestley, Joseph, *A History of Electricity*, 1767.

Priestley, Joseph, *The History and Present State of Discoveries Relating to Vision, Light and Colours*, 1772.

Pugliese, Patri J., 'Hooke and the dynamics of motion in a curved path', in M. Hunter and S. Schaffer (eds.), *Robert Hooke: New Studies*, Woodbridge, The Boydell Press, 1989, 181–205.

Pugliese, Patri P., *Hooke, Robert, Oxford Dictionary of National Biography*, 2004.

Ramm, Ekkehard, 'Principles of Least Action and of Least Constraint', GAMM-Mitt. 34, No. 2, 164–182 (2011)/DOI 10.1002/gamm.201110026.

Rayleigh, Lord, 'Optical topics in part connected with Charles Parsons', *Nature*, 152, 676–682, 1948.

Riess, A. G. *et al.*, *Astronomical Journal*, 116, 1009, 1998.

Rindler, W., 'Counterexample to the Lenz-Schiff Argument', *Am. J. Phys.*, 36, 540–544, 1968.

Robinson, Abraham, *Non-stamdard Analysis*, American Elsevier Pub. Co., New York, 1974, second edition, Princeton University Press, 1996.

Roche, John, 'Newton's *Principia*', in John Fauvel, Raymond Flood, Michael Shortland and Robin Wilson, *Let Newton Be! A New Perspective on his Life and Works*, Oxford University Press, 1988, 43–61.

Rogers, A. J., 'The System of Locke and Newton', in Zev Bechler (ed.), *Contemporary Newtonian Research*, London, 1982, 223.

Roinila, Markku, *Leibniz on Rational Decision–Making*, Department of Philosophy, University of Helsinki, 2007.

Rømer, O., 1676, 'Démonstration touchant le movement de la lumière trouvé par M. Römer de l'Academie Royale des Sciences', *Journal des Sçavans*, 7 December 1676, 223–36.

Rømer, O., A Demonstration concerning the Motion of Light', *Philosophical Transactions*, 12, No. 136, 25 June 1677, 893–894.

Rosenfeld, L., 'Newton and the Law of Gravitation', *Archive for History of Exact Sciences*, 2, 365–86, 1962–1965.

Ross, D. K. and Schiff, L. I., *Phys. Rev.*, 141, 1215–1218, 1966.

Rouse Ball, W. W., 'Sir Isaac Newton', from *A Short Account of the History of Mathematics*, 4th Edition, 1908, chapter XVI, 263–287.

Rouse Ball, W. W., *An Essay on Newton's Principia*, 1893, reprinted Johnson Reprint Corporation, New York, 1972.

Rowlands, Peter, *A Revolution Too Far: The Establishment of General Relativity*, PD Publications, Liverpool, 1994.

Rowlands, Peter, *Zero to Infinity The Foundations of Physics*, World Scientific, 2007.

Rowlands, Peter, 'Physical Interpretations of Nilpotent Quantum Mechanics', arXiv:1004.1523, 2010.

Rynasiewicz, Robert, 'Newton's Views on Space, Time, and Motion', in Edward N. Zalta (ed.), *The Stanford Encyclopedia of Philosophy, Fall 2011 edition*, http://plato.stanford.edu/archives/fall2011/entries/newton-stm/.

Saari, Donald G., 'A visit to the Newtonian n-body problem via elementary complex variables', *Amer. Math. Monthly*, 97, 105–119, 1990.

Sacks, W. M. and Ball, J. A., 'Simple derivations of the Schwarzschild metric', *Am. J. Phys.*, 36, 240–245, 1968.

Schooten, Frans van, *Exercitationum Mathematicarum Libri Quinque*, Leiden 1657.

Scriba, C. J. and Torres, D. F. M., 'Two-dimensional Newton's problem of minimum resistance', PAM Preprint, 2006.

Seeliger, H., 'Über das Newton'sche Gravitationsgesetz', *Astronomische Nachrichten*, 137, 129–136, 1895.

Seeliger, H., 'Über das Newton'sche Gravitationsgesetz', *Sitzungsberichte der math.-phys. Classe der köngl. Bayrischen Akademie der Wissenschaften zu München*, 26, 373–400, 1896.

Shapiro, Alan E., 'The optical lectures and the foundations of the theory of optical imagery', chapter 2 of Mordecai Feingold (ed.), *Before Newton: The Life and Times of Isaac Barrow*, Cambridge University Press, 1990, 105–178.

Shapiro, Alan E., 'Skating on the edge: Newton's investigation of chromatic dispersion and achromatic prisms and lenses', in J. Z. Buchwald and A. Franklin (eds.), *Wrong for the Right Reasons*, Springer, 2005, 99–125.

Sherburne, Sir Edward, *The Sphere of Marcus Manilius*, 1675.

Sheynin, O. B., 'Newton and Classical Theory of Probability', *Archive for History of the Exact Sciences*, 7, 217–243, 1972.

Shkolenok, Galina, 'Geometrical Constructions Equivalent to Non-Linear Algebraic Transformations of the Plane in Newton's Early Papers', *Archive for History of Exact Sciences*, 9(1), 22–44, 1972.

Silverman, Mark P., Review of *David H. Clark and Stephen P. H. Clark, Newton's Tyranny: The Suppressed Scientific Discoveries of Stephen Gray and John Flamsteed*, American Journal of Physics., 71, no. 5, 508, May 2003.

Simms, D. L., 'Newton's Contribution to the Science of Heat', *Annals of Science*, 61, 33–77, 2004.

Simson, Robert, *Opera Quaedam Reliqua ... Nunc Primum Post Auctoris Mortem in Lucem Edita*, Glagow, 1776.

Smeenk, Chris and Schliesser, Eric, 'Newton's *Principia*', in Jed Buchwald and Robert Fox (eds.), *The Oxford Handbook of the History of Physics*, Clarendon Press, Oxford, 2013, 109–165.

Smith, George E., 'Essay Review: *Newton's Principia for the Common Reader*, by S. Chandrasekhar', *Journal for the History of Astronomy*, 17, 353–361, 1996.

Smith, George E., 'Newton and the Problem of the Moon's Motion', in I. Bernard Cohen, 'A Guide to Newton's *Principia*', in I. Bernard Cohen and Anne Whitman, *Isaac Newton The Principia A New Translation*, Berkeley and Los Angeles and London: University of California Press, 1999, 252–257.

Smith, George E., 'The methodology of the *Principia*', in I. Bernard Cohen and George E. Smith (eds.), *The Cambridge Companion to Newton*, Cambridge University Press, 2002, 138–173.

Smith, George E., 'How Newton's *Principia* Changed Physics', in Andrew Janiak and Eric Schliesser (eds.), *Interpreting Newton: Critical Essays*, Cambridge University press, 2012, 360–395.

Smyth, Charles Piazzi, *Teneriffe, An Astronomer's Experiment: or, Specialities of a Residence Above the Clouds*, London: L. Reeve, 1858.

Soldner, Johann von, *Astronomisches Jahrbuch für das Jahr 1804*, Späthen, Berlin, 161, 1801, dated March 1801, reprinted by P. Lenard in *Analen der Physik*, 65, 1921, 593, translated in S. L. Jaki, *Found. Phys.*, 8, 927–950, 1978.

Sommerfeld, Arnold, *Mechanics: Lectures on Theoretical Physics*, vol. 1, 1942, reprinted Academic press, New York, 1964.

Spencer, John H., *The Eternal Law: Ancient Greek Philosophy, Modern Physics, and Ultimate Reality*, Param Media, 2012.

Stein, Howard, 'On the Notion of Field in Newton, Maxwell, and Beyond', in Roger H. Stuewer (ed.), *Historical and Philosophical Perspectives of Science*, Minneapolis: University of Minnesota Press, 1970, 274.

Stein, Howard, 'Newton's Metaphysics', in I. Bernard Cohen and George E. Smith (eds.), *The Cambridge Companion to Newton*, Cambridge University Press, 2002, 256–307.

Stigler, Stephen M., 'Isaac Newton as a Probabilist', arXiv:math/0701089v1, 3 January 2007.

Strandberg, M. W. P., *Am. J. Phys.*, 54, 321–331, 1986, 321–322.

Streete, Thomas, *Astronomia Carolina: a New Theorie of Coelestial Motions*, 1661.

Stukeley, William, *Memoirs of Sir Isaac Newton's Life* (1752), ed. A. Hastings White, London, Taylor & Francis, 1936.

Sundman, K. F., 'Recherches sur la problème des trois corps', *Acta Soc. Sci. Fennicae*, 34, 6, 1–43, 1907.

Tait, P. G., *Historical Sketch of the Science of Energy*, Edinburgh, 1877.

Taylor, Brook, *Methodus Incrementorum Directa & Inversa*, London, 1715.

Thomson, William (Lord Kelvin), 'On Ether and Gravitational Matter through Infinite Space', *Philosophical Magazine*, Series 6, 2, 161–177, 1901, in Baltimore Lectures, London: Oxford University Press, 1904, Lecture 16, 260–278; Harrison, 1987, 227–228.

Tisserand, Félix, *Traité de Mécanique Céleste*, Paris 1889.

Torres, D. M. F. and Plakhov, A. Yu., 'Optimal control of Newton-type problems of minimum resistance', *Rend. Sem. Mat. Univ. Pol. Torino*, 64, 1, 79–95, 2006.

Turnbull, H. W., *James Gregory Tercentenary Memorial Volume*, Royal Society of Edinburgh, 1939, 306–307.

Turnbull, H. W., *The Mathematical Discoveries of Newton*, London-Glasgow, 1945.

Turnbull, H. W., 'Newton: the Algebraist and Geometer', *The Royal Society Newton Tercentenary Celebrations*, Cambridge University Press, 1947, 62–72.

Voltaire, *Essay on the Civil War in France*, 1727.

Voltaire, *Lettres philosophiques*, 1733; *Letters Concerning the English Nation*, London, 1733; *Letters on England*, Harmondsworth, Penguin, 1980, 75.

Waller, Richard, *The Posthumous Works of Robert Hooke*, 1705.

Wallis, John, *Treatise of Algebra, Both Historical and Practical*, Oxford, 1685.

Ward, Seth, *In Ismaelis Bullialdi astronomiae philolaicae fundamenta inquisitio brevis*, 1653.

Waterston, J. J., *Monthly Notices of the Royal Astronomical Society*, 22, 60, 1862.

Weinberg, Steven, *The First Three Minutes: A Modern View of the Origin of the Universe*, Fontana / Collins, Glasgow, 1981.

Weinberg, Steven, 'Physics & history', *Daedalus*, vol. 234, issue 4, 1998.

Wepster, S. A., *Between Theory and Observations*, Springer, 2010.

Westfall, Richard S., *Force in Newton's Physics*, Macdonald, London and New York, 1971.

Westfall, Richard S., *Never at Rest*, Cambridge University Press, 1980.

Whiston, William, *Memoirs of the life and writings of Mr. William Whiston ... Written by himself*, London, 1749; second edition, 2 vols., London, 1753.

Whiteside, D. T., *The Mathematical Works of Isaac Newton*, New York and London: Johnson Reprint Corporation, 1964.

Whiteside, D. T., 'Newton's early thoughts on planetary motion: a fresh look', *BJHS*, 2, 117–137, 1964.

Whiteside, D. T., 'Sources and strengths of Newton's early mathematical thought', *The Texas Quarterly*, 10, no. 3, autumn 1967, and in Robert Palter (ed.), *The Annus Mirabilis of Sir Isaac Newton 1666–1966*, M.I.T. Press, Cambridge, Massachussetts and London, 1970, 69–85.

Whiteside, D. T., 'The mathematical principles underlying Newton's *Principia Muthematica*', *Journal for the History of Astronomy*, 1, 116–138, 1970.

Whiteside, D. T., 'Newton's Lunar Theory: From High Hope to Disenchantment', *Vistas in Astronomy*, 19, 317–328, 1976.

Whiteside, D. T., *Centaurus*, 24, 288–315, 1980.

Whiteside, D. T., 'Newton the Mathematician', in Z. Bechler (ed.), *Contemporary Newtonian Research, Studies in the History of Modern Science volume 9*, D. Reidel Publishing Co, Dordrecht, Boston, London, 1982, 109–127.

Whiteside, D. T., *The Preliminary Manuscripts for Newton's 1687 Principia*, 1684–1685, Cambridge University Press, 1989.

Whiteside, D. T., The prehistory of the 'Principia' from 1664 to 1686', *Notes and Records of the Royal Society*, 45, No. 1, 11–61, January 1991.

Wigner, Eugene, *Symmetries and Reflections: Scientific Essays of Eugene P. Wigner*, Indiana University Press, Bloomington, 1967.

Will, Clifford M., *American Journal of Physics*, 56, 413–415, 1988.

Wilson, Curtis, 'Predictive Astronomy in the Century after Kepler', in *Planetary Astronomy from the Renaissance to the Rise of Astrophysics, Part A: Tycho Brahe to Newton, The General History of Astronomy*, vol. 2A, eds. R. Taton and C. Wilson, Cambridge: Cambridge University Press, 1989, 161–206.

Wilson, Curtis, 'From Kepler to Newton: Telling the Tale', in R. H. Dalitz and Michael Nauenberg (eds.), *The Foundations of Newtonian Scholarship*, Singapore: World Scientific, 2000, 223–242.

Wilson, Curtis, 'Newton on the Moon's variation and apsidal motion: The need for a newer 'new analysis', J. Z. Buchwald and I. B. Cohen (eds.), *Isaac Newton's Natural Philosophy*, Cambridge, MA and London: The M.I.T. Press, 2001, 139–188.

Wilson, Curtis, 'Newton and celestial mechanics', in I. Bernard Cohen and George E. Smith (eds.), *The Cambridge Companion to Newton*, Cambridge University Press, 2002, 202–226.

Wing, Vincent, *Harmonicon Coeleste, or The Celestial Harmony of the Visible World*, London, 1651.

Wing, Vincent, *Astronomia Britannica*, London, 1669.

Wolfers, J. Ph., *Mathematische Principien der Naturlehre*, German edition of the *Principia*, Berlin, 1872, facsimile, Darmstadt, 1963.

Young, Thomas, *A Course of Lectures on Natural Philosophy and the Mechanical Arts*, 2 vols., London, 1807.

Index

A

A New Theory about Light and Colours, 3
aberration of light, 272, 278–279
aberration of space, 198, 222, 279
aberration, chromatic, 34, 36, 245–246
aberration, spherical, 34–35, 245–246
absolute acceleration, 69
absolute motion, 59–60, 69, 71, 74
absolute place, 59
absolute quantities, 57, 62
absolute rotation, 71–73
absolute space, 57, 59, 61–62, 64–65,
  68–69, 71–73
absolute time, 57–58, 61–62, 65–69, 71–72,
  81, 102
absolute velocity of light, 271
abstraction, 266
acceleration, 12, 25, 63, 67, 69, 74, 82–83,
  91, 93, 109–110, 128, 150, 187, 193, 242,
  270
Account of the Commercium Epistolicum,
  9
achromatic lens, 246–248
achromatic prism, 246–247, 270–271
achromatic telescope, 247–248
acids, 236
action, 38, 74, 81, 88, 93–94, 102–103,
  113, 139, 144–145
action and reaction, 85, 90, 95, 109, 120
action at a distance, 69, 81, 84, 113, 116,
  123, 126, 150, 213
active principles, 15, 89–90, 103, 116–119,
  153, 214, 235, 260–261
Adams, John Couch, 165
aether, 7, 10, 18, 64, 72, 74, 88, 113, 127,
  129–133, 150, 214, 239, 242–243, 261,
  271–272
aethereal density gradient, 131–133, 214

affine translation, 50
age of the Earth, 244–245
air, 18, 59, 214, 228, 236, 242
air resistance, 113, 131, 151, 171, 239
Airy, George Biddell, 205
Aiton, E. J., 107
alchemy, 21, 126, 131, 135, 141, 229, 257
algebra, 24, 26, 30, 40, 44, 52, 80, 82, 107,
  162, 165, 206, 281
algebraic curves, 32, 42–43, 50, 206
algebraic epsilon-argument, 28
algebraic nonintegrability of curves, 42
anaclastic problem, 34
analysis, 15–16, 26, 44, 52–53, 93, 175
analytic geometry, 24, 47
angular momentum, 73–74, 81, 96, 98,
  100, 102, 108–109, 144–145, 166–167,
  181, 192–193, 195
angular velocity, 92, 100
*An Hypothesis Explaining the Properties
  of Light*, 10, 129, 131–135, 242, 257
annus mirabilis, 126
antidifferentiation, 26
Apollonius, 47, 49–52
Aquinas, Thomas, 12–13
Arago, D. F. J., 270–271
Arbuthnot, John, 55
Archimedes, 185–186
argument from design, 258
Arianism, 21
Aristotle, 3, 11, 14–15, 61, 83
Arius, 5
Arnol'd, V. I., 164
artificial satellites, 168, 170–171
asteroids, 240, 242
astrophysics, 264
Atmosphaera Solis, 261
atmosphere, scattering of light by, 231

Printed in the United States
By Bookmasters